T0252237

Arduino: A Technical Reference

A Handbook for Technicians, Engineers, and Makers

J. M. Hughes

Beijing · Boston · Farnham · Sebastopol · Tokyo

Arduino: A Technical Reference

by J. M. Hughes

Printed in the United States of America.

Published by O'Reilly Media, Inc., 1005 Gravenstein Highway North, Sebastopol, CA 95472.

O'Reilly books may be purchased for educational, business, or sales promotional use. Online editions are also available for most titles (*http://safaribooksonline.com*). For more information, contact our corporate/institutional sales department: 800-998-9938 or *corporate@oreilly.com*.

Editor: Dawn Schanafelt
Production Editor: Colleen Lobner
Copyeditor: Rachel Head
Proofreader: Kim Cofer
Indexer: Ellen Troutman-Zaig

Interior Designer: David Futato
Cover Designer: Randy Comer
Illustrator: John M. Hughes and Rebecca Demarest

May 2016:   First Edition

Revision History for the First Edition
2016-05-05:   First Release

See *http://oreilly.com/catalog/errata.csp?isbn=9781491921760* for release details.

978-1-491-92176-0

[LSI]

Table of Contents

Preface

Since its introduction in 2005 the Arduino has become one of the most successful (some might argue the most successful) open source hardware projects in the world. Boards based on the open designs released by the Arduino team have been fabricated in countries around the world, including Italy, Brazil, China, the Netherlands, India, and the United States. One can purchase a fully functional Arduino-compatible board for around $15, and the Arduino development environment is readily available for download and is completely free. Originally based on the 8-bit AVR family of microcontrollers (the AVR is itself an interesting device with an equally interesting history), the Arduino has moved into the realm of 32-bit processing with the addition of the Due model with an ARM processor, the Yún with an on-board module running the OpenWrt version of Linux, and the upcoming Zero model. Arduinos have been used for everything from interactive art to robotics, and from environmental sensors to the smarts in small "CubeSat" satellites built by small teams and launched for a fraction of what a full-size satellite would cost.

I bought my first Arduino (a Duemilanove) many years ago, more out of curiosity than anything else. I had worked with microprocessor and microcontroller development systems since the early 1980s, starting with the 6502, 6800, and 8051, and then moving on to the 8086, the Z80, the 80186, and the 68000 family. Early on I usually programmed these devices in assembly language or PL/M, since these were really the only rational choices at the time for embedded systems. Later it became feasible to use C or Ada, as the capabilities of the microprocessors improved and the software tools matured. In each case, however, I came to expect having loads of reference material available: datasheets, hefty manuals, handbooks, and reference design documentation that would arrive along with the development circuit board and its accessories. It usually showed up in a large, heavy box.

When my new Arduino arrived and I opened the very small box I found that there was only a circuit board, a plug-in power pack, a few LEDs and resistors, some jumper wires, and a solderless breadboard block. No manuals, no handbooks, and no datasheets. Not even a CD with documents and software on it. Nothing more than a few sheets of paper with a manifest of what was in the box and a URL to a web page where I could read some "how to get started" material and find links to the software I needed. I was, to say the least, surprised.

I was also ignorant. When I bought the Arduino I didn't know its full backstory, nor was I aware of its intended audience. It turns out that it was originally meant primarily for people with little or no hardcore technical background who just wanted to experiment (in a playful sense) with something cool and make things go. In other words, artists and tinkerers, not engineers who have a penchant for technical details and an addiction to project plans, specifications, and, of course, manuals.

Once I understood this, it all made a lot more sense. Reading Massimo Banzi's book *Getting Started with Arduino* (O'Reilly) gave me a better understanding of the Arduino philosophy, and it was a good starting place in my quest for additional details. Also, unlike semiconductor manufacturers with their development kits, the folks on the Arduino team aren't in the business of selling chips—they're working to inspire creativity. The AVR microcontroller was chosen because it was inexpensive and it could be easily integrated into their vision for a device that could be readily applied to a creative endeavor. The AVR has enough computational "horsepower" and sufficient built-in memory to do complex and interesting things, unlike earlier generations of microcontrollers that usually needed expensive development tools and provided only scant amounts of internal on-chip memory.

Simplicity and low cost aside, the real secret to the success of the Arduino is the firmware bootloader on the AVR chip, coupled with a simple and easy-to-use integrated development environment (IDE) and the code libraries supplied with it—all provided free under open source and Creative Commons licensing. The Arduino philosophy is to make it as easy as possible to use an Arduino. By clearing away the bulk of the technical details and simplifying the development process, the Arduino invites the user to experiment, try new things, and yes, play. For the first time in a long time I found myself actually having a lot of fun just connecting things in different combinations to see what I could make it do. I wish that the Arduino had been around when I was teaching introductory embedded systems design—it would have helped reduce a lot of frustration for the people in the class trying to wade through assembly language listings, memory maps, and flowcharts.

Since my first Arduino arrived I've found many sources for useful and interesting add-on components for the Arduino family, some of them quite amazing in terms of both price and capabilities.[1] In fact, I became something of an Arduino pack rat, buying inexpensive shields and modules and building up a sizable collection of various bits. But, sadly, I have to say that many times I've opened a package with a nifty new gizmo in it, only to discover that there is no documentation of any kind. Not even a simple wiring diagram.

As an engineer, it is particularly frustrating to me to purchase something interesting, only to have it show up with no documentation. I then have to embark on a quest to determine if any documentation actually does exist, and in a form that I can read (I can't read Chinese). Sometimes that search is fruitless, and I'm forced to resort to component datasheets and reverse engineering the circuit board to figure out the wiring. Other times the information I'm seeking does exist, but it is scattered across multiple websites in various bits and pieces. This situation is slowly improving, but it can still be a real pain. After years of collecting stacks of notes, web page links, and datasheets, I finally decided to get organized and assemble it in one place.

So, what does this book have that can't be found somewhere on the Internet? Not that much, to be perfectly honest, but hopefully what it will do is reduce a lot of the potential frustration and wasted time—and of course there are all the bits that I've discovered on my own. The official technical data comes from manufacturers such as Atmel, the Arduino team, and numerous other sources, both well known and obscure. Some overseas vendors do have at least rudimentary websites, whereas others have very nice web pages with links to other sites for the documentation. This book contains as much of the essential data as I could find or reverse engineer, all in one convenient place, with as many of the holes plugged as I could manage. I wanted to save others the frustration I've experienced trying to run down some mundane little technical detail about the USB interface, or figure out why a shield wasn't working correctly, or why that really neat sensor I bought from someone through eBay didn't seem to work at all.

The result of my frustrations is this book, the one I've been wanting for working with the Arduino boards and shields. I really wanted something physical that I could keep near at hand on my workbench. It isn't always practical to have to refer to a web page to look something up, and to make things even more interesting, sometimes access to the Internet just isn't available (like when you're trying to debug a remote data logging device on a mountaintop with just a little netbook for company and no wireless service for 60 miles in any direction). I wanted something that I could use to quickly

---

1 Entering "Arduino" into the search field on eBay will return a multitude of things like ultrasonic range sensors, temperature and humidity sensors, various Arduino clones, Bluetooth and ZigBee shields, and much more. But, unfortunately, some of the items come with little or no documentation, and even if there is some it may not be particularly up-to-date or accurate. That doesn't mean you shouldn't consider these sellers (the prices are usually excellent and the build quality is generally very good), but as always when buying things online, caveat emptor.

look up the answer to a question when working with the Arduino and associated add-on components, no matter where I was. As far as I know, such a thing didn't exist until now. I hope you find it as useful as I have as I assembled my notes for this book.

Intended Audience

This book is intended for those people who need or want to know the technical details. Perhaps you've gone as far as you can with the introductory material and the "99 amazing projects" type of books, and you now need to know how to do something novel and unique. Or, you might be a working engineer or researcher who would like to incorporate an Arduino into your experimental setup in the lab. You might even be someone who wants to install an Arduino into an RC airplane, use one as part of a DIY weather station,[2] or maybe do something even more ambitious (a CubeSat, perhaps?).

Ideally you should have a basic knowledge of C or C++, some idea of how electrons move around in a circuit, and some experience building electronic devices. If I may be so bold, I would suggest that you have a copy of my book *Practical Electronics: Components and Techniques* (also from O'Reilly) on hand, along with some reference texts on programming and electronics (see Appendix D for some suggestions).

What This Book Is

This book is a reference and a handbook. I have attempted to organize it such that you can quickly and easily find what you are looking for. I have included the sources of my information whenever possible, and those insights that are the result of my own investigations are noted as such.

What This Book Is Not

This book is not a tutorial. That is not its primary purpose. I don't cover basic electronics, nor is there any discussion of the dialect of the C++ language that is used to create the so-called "sketches" for programming an Arduino. There are some excellent tutorial texts available that cover general electronics theory and programming, and I would refer the reader to those as a place to get started with those topics.

This book is also not an Arduino-sanctioned guide to the products produced by the Arduino team. It is based on information gleaned from various sources, some more

---

2 The books *Environmental Monitoring with Arduino* and *Atmospheric Monitoring with Arduino* (O'Reilly), both by Emily Gertz and Patrick Di Justo, offer some good ideas for doing just this with cheap and readily available sensors and an Arduino.

obscure than others, along with my own notes and comments based on my experiences with the Arduino. As such, I am solely responsible for any errors or omissions.

About Terminology

The distinctions between processors, microprocessors, and microcontrollers arose sometime in the early 1980s as manufacturers were trying to distinguish their products based on size and amount of external circuitry required for the devices to do something useful. Full-size mainframe processors and the smaller microprocessors like those found in desktop PCs both typically require some external components (sometimes a lot of them) in order to be useful. A microcontroller, on the other hand, has everything it needs to do its job already built in. Also, a microprocessor will typically support external memory, whereas a microcontroller may have only limited support (if any at all) for adding additional memory to what is already on the chip itself.

Throughout this book I will use the terms "microcontroller" and "processor" interchangeably. Although "microcontroller" might be considered to be technically more correct, in my mind it is still a processor of data, just a smaller version of the huge machines I once worked with in the distant past. They all do essentially the same thing, just at different scales and processing speeds.

What's in This Book

Chapter 1 presents a brief history of the Arduino in its various forms. It also introduces the AVR microcontrollers used in the Arduino boards, and discusses the differences between software-compatible and hardware-compatible products based on the Arduino.

The Atmel AVR microcontroller is the subject of Chapter 2. This is intended as an overview of what is actually a very sophisticated device, and so this chapter provides a quick tour of the highlights. This includes the timer logic, the analog comparator, the analog input, the SPI interface, and other primary subsystems on the chip.

Chapter 3 takes a closer look at the AVR microcontrollers used in Arduino boards, namely the ATmega168/328, the ATmega1280/2560, and the ATmega32U4 devices. It builds on the overview presented in Chapter 2, and provides additional low-level details such as internal architecture, electrical characteristics, and chip pinouts.

Chapter 4 covers the physical characteristics and interface functions of various Arduino boards. This includes the USB interface types, printed circuit board (PCB) dimensions, and board pinout diagrams.

What really makes the Arduino unique is its programming environment, and that is the subject of Chapter 5. This chapter defines the Arduino sketch concept and how it utilizes the C and C++ languages to create sketches. It introduces the Arduino bootloader and the Arduino main() function. This chapter also describes how you can download the Arduino source code and see for yourself what it looks like under the hood.

Chapter 6 describes the AVR-GCC toolchain and presents techniques for programming an Arduino board without using the Arduino IDE. It also covers makefiles and includes a brief overview of assembly language programming. The chapter wraps up with a look at the tools available to upload code into an AVR.

The focus of Chapter 7 is on the standard libraries supplied with the Arduino IDE. The Arduino IDE is supplied with numerous libraries, and more are being added all the time. If you want to know if a library module exists for a particular sensor or for a specific operation, then this is a good starting point.

Chapter 8 presents the various types of shields available for the Arduino. It covers many of the commonly available types, such as flash memory, prototyping, input/output, Ethernet, Bluetooth, ZigBee, servo control, stepper motor control, LED displays, and LCD displays. It also covers using multiple shields, and presents some hints and tips for getting the most from a shield.

Chapter 9 describes some of the various add-on components available that can be used with an Arduino. These include sensors, relay modules, keypads, and other items that aren't specific to the Arduino, but work with it quite nicely. Electrical pin-out information and schematics are provided for many of the items discussed.

Sometimes there just isn't a readily available shield to do what needs to be done. Chapter 10 describes the steps involved in creating a custom shield. It also describes how to use an AVR microcontroller without an Arduino-type circuit board but still take advantage of the Arduino IDE.

Chapters 11, 12, and 13 cover some projects that illustrate the capabilities of the AVR microcontrollers and the Arduino boards. They are intended to demonstrate how an Arduino can be applied in various situations, not as how-to guides for building a board or device. You can, however, build any of the items described yourself, if you feel so inclined, and they might serve as jumping-off points for your own projects. Each example project description includes theory of operation, schematics, a detailed parts list, PCB layouts (if required), and an overview of the software necessary to make it go.

Because the main emphasis of this book is on the Arduino hardware and related modules, sensors, and components, the software shown is intended only to highlight key points; my aim was not to present complete, ready-to-run examples. The full software listings for the examples and projects can be found on GitHub (*https://www.github.com/ardnut*).

Chapter 11 describes a basic programmable signal generator, a handy thing to have around when working with electronic circuits. With this project you can generate pulses at various duty cycles, output a sequence of pulses in response to a trigger input, generate sine waves, and also create programmable patterns of pulses.

Chapter 12 covers the design and construction of a "smart" thermostat suitable for use with a home HVAC (heating, ventilation, and air conditioning) system. Instead of paying for something that is already built and sealed into a plastic case, you can build it yourself and program it to behave exactly the way you want it to. I'll show you how to incorporate more than just a temperature sensor: features include multiple temperature and humidity sensors, and the use of your HVAC system's fan to provide a comfortable environment without the cost of running the compressor or lighting up the heater.

In Chapter 13 we will look at how to build an automatic model rocket launcher with a programmable sequencer and automatic system checks. Even if you don't happen to have a model rocket handy, this project describes techniques that can be applied to many types of sequentially controlled processes, be it on a production line or a robotic material handling device in a laboratory.

Appendix A is an overview of the basic tools and accessories you may need if you want to go beyond prefabricated circuit boards and modules.

Appendix B is a compilation of the control registers for the ATmega168/328, the ATmega1280/2560, and the ATmega32U4 microcontrollers.

Appendix C is a summary listing of the Arduino and compatible products distributors and manufacturers listed in this book. It is by no means exhaustive, but hopefully it will be enough to get you started on your quest to find what you need.

Appendix D lists some recommended books that cover not just the Arduino, but also C and C++ programming and general electronics.

Finally, Appendix E is a summary of some of the readily available Arduino and AVR software development tools that are currently out there.

Endorsements

Other than references to the Arduino team and the folks at Arduino.cc, there aren't any specific endorsements in this book—at least, not intentionally. I've made reference to many different component manufacturers, suppliers, and other authors, but I've tried to be evenhanded about it, and I don't specifically prefer any one over another. My only criteria in selecting those I do mention are that I own one or more of their products and that I've successfully managed to use a shield, module, sensor, or Arduino main PCB (or clone PCB in some cases) from the supplier in something, even if just in a demonstration of some type. Any trademarks mentioned are the property of their respective owners; they appear here solely for reference. As for the photography, I tried to use my own components, tools, circuit boards, modules, and other items as much as possible, and although an image may show a particular brand or model, that doesn't mean it's the only type available—it likely just happens to be the one that I own and use in my own shop. In some cases I've used images with permission from the vendor or creator, works in the public domain, or images with a liberal Creative Commons (CC) license, and these are noted and credited as appropriate. I created all the diagrams, schematics, and other nonphotographic artwork, and I am solely responsible for any errors or omissions in these figures.

Conventions Used in This Book

The following typographical conventions are used in this book:

Italic

> Indicates new terms, URLs, email addresses, filenames, and file extensions.

`Constant width`

> Used for program listings, as well as within paragraphs to refer to program elements such as variable or function names, databases, data types, environment variables, statements, and keywords.

`Constant width italic`

> Shows text that should be replaced with user-supplied values or by values determined by context.

This element signifies a tip or suggestion.

This element signifies a general note.

This element indicates a warning or caution.

Safari® Books Online

 Safari Books Online is an on-demand digital library that delivers expert content in both book and video form from the world's leading authors in technology and business.

Technology professionals, software developers, web designers, and business and creative professionals use Safari Books Online as their primary resource for research, problem solving, learning, and certification training.

Safari Books Online offers a range of product mixes and pricing programs for organizations, government agencies, and individuals. Subscribers have access to thousands of books, training videos, and prepublication manuscripts in one fully searchable database from publishers like O'Reilly Media, Prentice Hall Professional, Addison-Wesley Professional, Microsoft Press, Sams, Que, Peachpit Press, Focal Press, Cisco Press, John Wiley & Sons, Syngress, Morgan Kaufmann, IBM Redbooks, Packt, Adobe Press, FT Press, Apress, Manning, New Riders, McGraw-Hill, Jones & Bartlett, Course Technology, and dozens more. For more information about Safari Books Online, please visit us online.

How to Contact Us

Please address comments and questions concerning this book to the publisher:

O'Reilly Media, Inc.
1005 Gravenstein Highway North
Sebastopol, CA 95472
800-998-9938 (in the United States or Canada)
707-829-0515 (international or local)
707-829-0104 (fax)

We have a web page for this book, where we list errata, examples, and any additional information. You can access this page at *http://bit.ly/arduino-a-technical-reference*.

To comment or ask technical questions about this book, send email to *bookquestions@oreilly.com*.

For more information about our books, courses, conferences, and news, see our website at *http://www.oreilly.com*.

Find us on Facebook: *http://facebook.com/oreilly*

Follow us on Twitter: *http://twitter.com/oreillymedia*

Watch us on YouTube: *http://www.youtube.com/oreillymedia*

Acknowledgments

This book would not have been possible without the enduring patience and support of my family. Writing appears to be habit-forming, but they've been very encouraging and supportive, and will even bring me things to eat and occasionally check to make sure I'm still alive and kicking in my office. It doesn't get much better than that. I would particularly like to thank my daughter Seren for her photographic assistance and help in keeping my collection of various bits and pieces cataloged and organized.

I would also like to thank the editorial staff at O'Reilly for the opportunity to work with them once again. As always, they have been helpful, patient, and willing to put up with me. Special thanks goes to Brian Sawyer and Dawn Schanafelt for their excellent editorial support and guidance, and to Mike Westerfield for his insightful technical review of the material.

CHAPTER 1

The Arduino Family

This chapter provides a brief history of the Arduino, along with a terse genealogy of the various board types created since 2007. It doesn't cover boards produced before 2007, nor does it attempt to be comprehensive in its coverage of the various clones and derivatives that have been produced. The main focus here is on the differences between the various primary types of Arduino boards, with a specific focus on the types of processors used and the physical design of the boards. It also takes a quick look at the range of possible applications for Arduino circuit boards.

Chapter 2 provides general information about the internal functions of the Atmel AVR processors, and Chapter 3 covers the specific processors used in Arduino boards. With the exception of the Yún, Chapter 4 describes the physical characteristics of different official Arduino boards introduced in this chapter.

A Brief History

In 2005, building upon the work of Hernando Barragán (creator of Wiring), Massimo Banzi and David Cuartielles created Arduino, an easy-to-use programmable device for interactive art design projects, at the Interaction Design Institute Ivrea in Ivrea, Italy. David Mellis developed the Arduino software, which was based on Wiring. Before long, Gianluca Martino and Tom Igoe joined the project, and the five are known as the original founders of Arduino. They wanted a device that was simple, easy to connect to various things (such as relays, motors, and sensors), and easy to program. It also needed to be inexpensive, as students and artists aren't known for having lots of spare cash. They selected the AVR family of 8-bit microcontroller (MCU or μC) devices from Atmel and designed a self-contained circuit board with easy-to-use connections, wrote bootloader firmware for the microcontroller, and packaged it all into a simple integrated development environment (IDE) that used programs called "sketches." The result was the Arduino.

Since then the Arduino has grown in several different directions, with some versions getting smaller than the original, and some getting larger. Each has a specific intended niche to fill. The common element among all of them is the Arduino run-time AVR-GCC library that is supplied with the Arduino development environment, and the on-board bootloader firmware that comes preloaded on the microcontroller of every Arduino board.

The Arduino family of boards use processors developed by the Atmel Corporation of San Jose, California. Most of the Arduino designs utilize the 8-bit AVR series of microcontrollers, with the Due being the primary exception with its ARM Cortex-M3 32-bit processor. We don't cover the Due in this book, since it is radically different from the AVR devices in many fundamental ways and really deserves a separate discussion devoted to it and similar microcontrollers based on the ARM Cortex-M3 design.

Although an Arduino board is, as the Arduino team states, just a basic Atmel AVR development board, it is the Arduino software environment that sets it apart. This is the common experience for all Arduino users, and the cornerstone of the Arduino concept. Chapter 5 covers the Arduino IDE, the libraries supplied with the IDE, and the bootloader. Chapter 6 describes the process of creating software for an AVR MCU without using the Arduino IDE.

Types of Arduino Devices

Over the years the designers at Arduino.cc have developed a number of board designs. The first widely distributed Arduino board, the Diecimila, was released in 2007, and since its initial release the Arduino family has evolved to take advantage of the various types of Atmel AVR MCU devices. The Due, released in 2012, is the first Arduino to utilize a 32-bit ARM Cortex-M3 processor, and it breaks from the rest of the family in terms of both processing power and board pinout configuration. Other boards, like the LilyPad and the Nano, also do not have the same pinout as the other members of the family, and are intended for a different range of applications—weara-bles in the case of the LilyPad; handheld devices for the Esplora; and compact size in the case of the Mini, Micro, and Nano.

With each passing year new types of Arduino boards appear, so what is listed here may be out of date by the time you're reading it. The newer versions have more advanced processors with more memory and enhanced input/output (I/O) features, but for the most part they use the same pinout arrangements and will work with existing add-on boards, called *shields*, and various add-on components such as sen-sors, relays, and actuators. Table 1-1 lists the Arduino types that have appeared since 2005. The newer versions of the Arduino will also run most of the sketches created for older models, perhaps with a few minor tweaks and newer libraries, but sketches written for the latest versions may or may not work with older models.

Table 1-1 is not a buyer's guide. It is provided to give a sense of historical context to the Arduino. As you can see, the years 2007 and 2008 saw the introduction of the LilyPad; the small form-factor boards like the Nano, Mini, and Mini Pro; and the introduction of the Duemilanove as a natural evolutionary step based on the Diecimila. While there are no significant physical differences between the Diecimila and the Duemilanove, the Duemilanove incorporates some refinements in the power supply, most notably in its automatic switchover between USB power and an external DC (direct current) power supply. Later versions of the Duemilanove also utilize the ATmega328 MCU, which provides more memory for programs.

Table 1-1 doesn't include the Arduino Robot, which is a PCB with motors and wheels attached. One of the newest boards in the Arduino lineup is the Yún, an interesting beast that has both an ATmega32U4 microcontroller and a Linino module with an Atheros AR9331 MIPS-based processor capable of running a version of the Linux-based OpenWrt operating system. I won't get into the OpenWrt end of the Yún, but the Arduino side is basically just a standard Arduino (a Leonardo, to be specific). If you want to learn more about the Yún, I would suggest checking it out on the Arduino website (*https://www.arduino.cc/en/Main/ArduinoBoardYun*).

Table 1-1. Timeline of Arduino products

Board name	Year	Microcontroller	Board name	Year	Microcontroller
Diecimila	2007	ATmega168V	Mega 2560	2010	ATmega2560
LilyPad	2007	ATmega168V/ATmega328V	Uno	2010	ATmega328P
Nano	2008	ATmega328/ATmega168	Ethernet	2011	ATmega328
Mini	2008	ATmega168	Mega ADK	2011	ATmega2560
Mini Pro	2008	ATmega328	Leonardo	2012	ATmega32U4
Duemilanove	2008	ATmega168/ATmega328	Esplora	2012	ATmega32U4
Mega	2009	ATmega1280	Micro	2012	ATmega32U4
Fio	2010	ATmega328P	Yún	2013	ATmega32U4 + Linino

When more than one microcontroller type is shown in Table 1-1, it indicates that a particular version of an Arduino board was made initially with one microcontroller, and later with the other (usually more capable) device. For example, an older version of the Duemilanove will have an ATmega168, whereas newer models have the ATmega328. Functionally the ATmega168 and the ATmega328 are identical, but the ATmega328 has more internal memory.

The latest additions to the Arduino family, the Leonardo, Esplora, Micro, and Yún, all use the ATmega32U4. While this part is similar to an ATmega328 it also incorporates an integrated USB-to-serial interface component, which eliminates one of the integrated circuit (IC) parts found on boards like the Uno and Duemilanove.

The programming interface also behaves slightly differently with the boards that use the ATmega32U4, but for most people this should be largely transparent. Chapter 2 describes the general functionality of AVR microcontrollers, Chapter 3 contains descriptions of the specific AVR MCU types found in Arduino devices, and Chapter 4 provides descriptions of the primary Arduino circuit boards and their pinout definitions.

Arduino Galleries

Tables 1-1 through 1-5 show some of the various types of Arduino boards, both past and present. It is not completely inclusive, since new types and updates to existing types occur periodically. The following images show the wide diversity in physical shapes and intended applications of the Arduino.

Physically, an Arduino is not a large circuit board. The baseline boards, which have the physical pin arrangement commonly used for add-on boards (called shields, described in Chapter 8), are about 2.1 by 2.7 inches (53.3 by 68.6 mm) in size. Figure 1-1 shows a selection of Arduino boards with a ruler for scale, and Figure 1-2 shows a Nano mounted on a solderless breadboard.

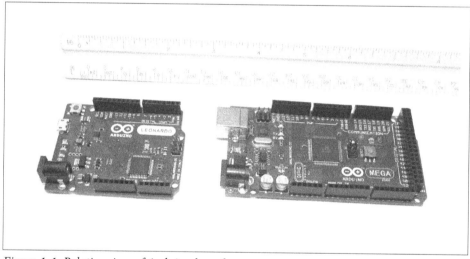

Figure 1-1. Relative sizes of Arduino boards

Chapter 4 contains reference drawings with dimensions and pin definitions for most of the common Arduino boards. Note that while it is small, a board like the Nano has all of the same capabilities as a Duemilanove, except for the convenient pin sockets and regular (type B) USB connector. It is ideal for applications where it will not be disturbed once it is installed, and where small size is a requirement. Some applications that come to mind are autonomous environmental data collection devices (automated solar-powered weather data stations or ocean data collection buoys, for example), timing and data collection for model rockets, security systems, and perhaps even a "smart" coffee maker.

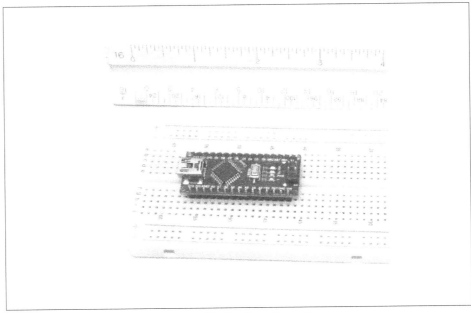

Figure 1-2. An Arduino Nano on a solderless breadboard

Table 1-2. Baseline layout of Arduino boards

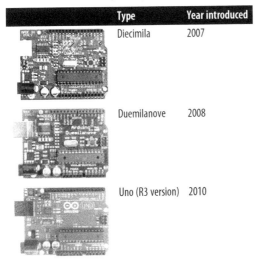

	Type	Year introduced
	Diecimila	2007
	Duemilanove	2008
	Uno (R3 version)	2010

	Type	Year introduced
	Ethernet	2011
	Leonardo	2012

Table 1-3. Mega layout of Arduino boards

	Type	Year introduced
	Mega	2009
	Mega 2560	2009
	Mega ADK	2011

Table 1-4. Small form factor Arduino boards

	Type	Year introduced
	Nano	2008
	Mini	2008
	Fio	2010
	Micro	2012

Table 1-5. Special form factor Arduino boards

	Type	Year introduced
	LilyPad	2007
	Esplora	2012`

Arduino-Compatible Devices

In addition to the various board types designed or sanctioned by Arduino.cc, there are many devices that are either hardware compatible or software compatible. What makes these devices Arduino compatible is that they incorporate the Arduino bootloader (or something that works like it), and they can be programmed with the Arduino IDE by selecting the appropriate compatible Arduino board type from the IDE's drop-down list.

Hardware-Compatible Devices

An Arduino hardware–compatible device is one where the various I/O pins on the board have been arranged to match one of the existing Arduino form factors. A hardware-compatible board can (usually) accept any of the shields and plug-in modules created for an official Arduino board. The reasons behind this are covered in "The Arduino Naming Convention" on page 9.

In most cases hardware-compatible boards look like any other Arduino board, except that the official Arduino logo and silkscreen graphics are missing. Other hardware-compatible products might not look anything like a typical Arduino board, but do provide the pin sockets in the correct arrangement for using a standard Arduino-type shield board. Some hardware-compatible products include additional connections, like the SainSmart version of the Uno with additional connectors for I/O functions. Table 1-6 lists a few Arduino clones and compatible boards that are available. There are many more than what is shown here, but this should give some idea of what is available.

Table 1-6. Arduino hardware–compatible devices

	Name	Type	Origin
	SainSmart UNO	Uno clone	China
	SainSmart Mega2560	Mega 2560 clone	China
	Brasuino	Similar to the Uno, with minor changes	Brazil
	Diavolino	An Arduino layout–compatible clone kit	USA

Note that Diavolino is a kit and requires assembly.

Software-Compatible Devices

There are many Arduino software–compatible boards available. These utilize the Arduino bootloader and development environment, but do not have a completely Arduino-compatible physical form factor. Software-compatible devices can be programmed with the Arduino development tools, but may use a different arrangement of I/O pins, or perhaps use some other types of connectors in place of the pin sockets found on stock Arduino boards. Custom circuits based on an AVR microcontroller and built into some larger device or system would fall into the software-compatible category if the Arduino bootloader is installed in the microcontroller.

The core of the Arduino is the processor and the preinstalled bootloader. Using that definition, one could have just a bare ATmega AVR IC with the Arduino firmware loaded into it. It could then be used with a solderless breadboard and the Arduino development environment. AVR MCU ICs with preloaded bootloader code are available for purchase from multiple sources, or you could do it yourself. Chapter 5 describes the steps necessary to load an AVR MCU with the Arduino bootloader firmware.

It is interesting to note that some of the boards from Arduino, such as the Mini, Micro, Nano, LilyPad, and Esplora, are not hardware compatible in terms of using the "standard" I/O connector layout. They can't be used directly with a conventional shield, but they are still Arduino boards, and they are supported by the Arduino IDE.

The Boarduino from Adafruit Industries is one example of an Arduino software–compatible device. This board is designed to mount on a standard solderless bread-board much like a full-size 40-pin IC. It is available in two styles: DC and USB. The DC version does not have an on-board USB chip, so an external USB adapter is needed to program it. Another example of a software-compatible board is the Drag-onfly from Circuit Monkey, which uses standard Molex-type connectors instead of the pins and sockets used on a conventional Arduino. It is intended for high-vibration environments, such as unmanned aerial vehicles (UAVs) and robotics.

The Raspduino is designed to mount onto a Raspberry Pi board, and it is functionally equivalent to an Arduino Leonardo. This results in a combination that is roughly equivalent to the Yún, but not exactly the same. Each setup has its own strengths and weaknesses. Table 1-7 lists a few Arduino software–compatible boards.

Table 1-7. Arduino software–compatible devices

	Name	Description	Origin
	Boarduino DC	Designed to fit on a solderless breadboard	USA
	Boarduino USB	Designed to fit on a solderless breadboard	USA
	Dragonfly	Uses Molex-type connectors for I/O	USA
	Raspduino	Designed to fit on a Raspberry Pi board	Netherlands

This is just a small selection of the various boards that are available. Because the AVR microcontroller is easy to integrate into a design, it has found its way into numerous applications. With the Arduino bootloader firmware, programming a device is greatly simplified and the design possibilities are vast.

The Arduino Naming Convention

While the circuit design and software for the Arduino are open source, the Arduino team has reserved the use of the term "Arduino" for its own designs, and the Arduino logo is trademarked. For this reason you will sometimes find things that behave and look like official Arduino devices, but which are not branded Arduino and have not been produced by the Arduino team. Some of them use "-duino" or "-ino" as part of the product name, such as Freeduino, Funduino, Diavolino, Youduino, and so on. Some, like the boards made by SainSmart, use just the model name (Uno and Mega2560, for example).

 At the time of this writing, there was an ongoing dispute between the company created by the original founders (Arduino LLC) and a different company started by one of the original founders (Arduino SRL). As a result, Arduino LLC uses the trademark Arduino within the United States and Genuino elsewhere.

Occasionally someone will produce a board that claims to be an Arduino, but is in fact just a copy that uses the Arduino trademark without permission. The silkscreen mask used to put the logo and other information on an official Arduino is also copyrighted, and the Arduino folks don't release the silkscreen with the PCB layout files. Massimo Banzi has a section of his blog (*http://www.massimobanzi.com*) devoted specifically to these unauthorized boards, and his examination of blatant and shameless copies is interesting, to say the least. Just search for the "hall of shame" tag.

The bottom line here is that you are welcome to copy the schematics, the bootloader code, and the Arduino IDE, and use these to create your own version of an Arduino. It is, after all, open source. Just don't call it an Arduino or use the artwork from Arduino.cc without permission.

What Can You Do with an Arduino?

In addition to the ease of programming made possible by the Arduino IDE, the other big feature of the Arduino is the power and capability of the AVR microcontroller it is based on. With a handful of readily available add-on shields (described in Chapter 8) and a wide selection of low-cost sensor and actuator modules (these are described in detail in Chapter 9), there really isn't a whole lot you can't do with an Arduino provided that you keep a few basic constraints in mind.

The first constraint is memory. The AVR MCU just doesn't have a whole lot of memory available for program storage and variables, and many of the AVR parts don't have any way to add more. That being said, the ATmega32 and ATmega128 types can use external memory, but then the I/O functions for those pins are no longer readily available. Arduino boards were not designed to accommodate external memory, since one of the basic design assumptions was that the AVR chip itself would have the necessary I/O and that the user would be running a relatively short program. The Arduino was not intended to be a replacement for a full-on computer system with gigabytes of RAM and a hard disk drive (HDD). There are inexpensive Intel-based single-board computers that fit that description, but they won't fit into an old mint tin, a section of PVC tubing strapped to a pole or a tree, a small robot, or the payload section of a model rocket. An Arduino will.

The second constraint is speed. The Arduino CPU clock rate is typically between 8 and 20 MHz (see Chapter 4 for a detailed comparison of Arduino AVR device types). While this may sound slow, you should bear in mind two key facts: first, the AVR is a very efficient RISC (reduced instruction set computer) design, and second, things in the real world generally don't happen very quickly from a microcontroller's perspective. For example, how often does a so-called smart thermostat need to sample the temperature in a home or office? Once a second is probably overkill, and once every 5 or even 10 seconds will work just fine. How often does a robot need to emit an ultrasonic pulse to determine if there is an obstacle ahead? A pulse every 100 ms is probably more than enough (unless the robot is moving very, very fast). So, for an Arduino running at 16 MHz (like the Leonardo, for example), there will be on the order of 1,000,000 or more CPU clock ticks between sensor pulses, depending on whatever else the CPU is doing with the pulses. Given that an AVR can execute many instructions in one or two clock cycles, that's a lot of available CPU activity in between each pulse of the ultrasonic sensor.

The third main constraint is electrical power. Since the Arduino hardware is actually nothing more than a PCB for an AVR IC to sit on, there is no buffering between the microcontroller and the external world. You can perform a fast "charcoal conversion" of an AVR (in other words, overheat the IC and destroy it) if some care isn't taken to make sure that you aren't sourcing or sinking more current than the device can handle. Voltage is also something to consider, since some of the AVR types have 3.3V I/O, whereas others are 5V tolerant. Connecting 5V transistor-transistor logic (TTL) to a 3.3V device usually results in unhappy hardware, and the potential for some smoke.

With the preceding constraints in mind, here are just a few possible applications for an Arduino:

- Real-world monitoring
  - Automated weather station
  - Lightning detector
  - Sun tracking for solar panels
  - Background radiation monitor
  - Automatic wildlife detector
  - Home or business security system
- Small-scale control
  - Small robots

- Model rockets
- Model aircraft
- Quadrotor UAVs
- Simple CNC for small machine tools
- Small-scale automation
  - Automated greenhouse
  - Automated aquarium
  - Laboratory sample shuttle robot
  - Precision thermal chamber
  - Automated electronic test system
- Performance art
  - Dynamic lighting control
  - Dynamic sound control
  - Kinematic structures
  - Audience-responsive artwork

In Chapters 11, 12, and 13 we will look at applications such as a smart thermostat, a programmable signal generator, and an automated rocket launch control system to help fulfill your suborbital yearnings. These are just the tip of the iceberg. The possibilities are vast, and are limited only by your imagination. So long as you don't try to make an Arduino do the job of a full-on computer system, you can integrate one into all sorts of interesting applications—which is exactly what the folks at Arduino.cc want you to do with it.

For More Information

The boards listed in this chapter are just a small selection of what is available, and there is much more to the story of the Arduino. Entering "Arduino" into Google's search bar will produce thousands of references to explore.

The official Arduino website can be found at *http://www.arduino.cc*.

Massimo Banzi's blog is located at *http://www.massimobanzi.com*.

Also, check the appendixes for more website links and book recommendations.

The AVR Microcontroller

Because an AVR-based Arduino is really just a physical platform for an AVR micro-controller (i.e., a breakout board), the electrical characteristics of an Arduino are essentially those of the AVR device on the PCB. Understanding the low-level details of an Arduino is really a matter of understanding the AVR device that powers it. To that end, this chapter presents broadly applicable material consisting of high-level descriptions of the main functions utilized in the AVR family. This includes the AVR CPU and the so-called peripheral functions such as timers, counters, serial interface logic, analog-to-digital (A/D) converters, analog comparators, and discrete digital I/O ports.

AVR microcontrollers are available in a wide variety of configurations and package types, which makes writing a chapter like this something of a challenge. Fortunately, the various types of 8-bit AVR devices use a common central processing unit (CPU) and a modular internal architecture built around an internal data bus. This allows for each variant to incorporate different combinations and quantities of functional mod-ules into the AVR's internal circuitry to meet specific design requirements and sup-port different intended applications.

Due to practical limitations of space, the descriptions in this chapter are necessarily terse and focused on the essential characteristics, and don't provide may of the low-level details that can be found in the reference documentation available from Atmel (*http://www.atmel.com*). If you need or want to know the logic circuit and register-level details of what's inside a particular AVR microcontroller, datasheets, user's guides, and application notes are available from Atmel free of charge.

Background

The AVR microcontroller began life in the early 1990s as a student project at the Norwegian Institute of Technology. Two students, Alf-Egil Bogen and Vegard Wollan, devised an 8-bit device with a RISC-type internal architecture while working at a local semiconductor facility in Trondheim, Norway. The design was later sold to Atmel, where Bogen and Wollan continued to work on it and refine it.

The AVR microcontrollers are highly configurable and very versatile, and they embody several unique features that set them apart from other 8-bit microcontrollers like the 8051 or 68HC05 components. The AVR is a modified Harvard architecture 8-bit RISC microcontroller. In a Harvard architecture read-only program, code and modifiable data (variables) are stored in separate memory spaces. By way of comparison, a microprocessor like the 68040 uses the Von Neumann architecture, in which programs and data share the same memory space.

The AVR family of devices was one of the first to incorporate on-board flash memory for program storage, instead of the one-time programmable ROM (read-only memory), EPROM (erasable programmable read-only memory), or EEPROM (electrically erasable programmable read-only memory) found on other microcontrollers. This makes reprogramming an AVR microcontroller simply a matter of loading new program code into the device's internal flash memory. Most AVR parts do have a small amount of EEPROM for storing things like operating parameters that must persist between changes in the flash memory.

Internal Architecture

Internally, an AVR ATmega microcontroller consists of an AVR CPU and various input/output, timing, analog-to-digital conversion, counter/timer, and serial interface functions, along with other functions depending on the part number. These are referred to by Atmel as peripheral functions. Besides the I/O functions, the main differences between the AVR microcontroller types lie in the amount of on-board flash memory and available I/O functions. The 8-bit parts all use essentially the same AVR CPU core. The following list shows just some of the basic features of AVR microcontrollers:

- RISC architecture
  - 131 instructions
  - 32 8-bit general-purpose registers
  - Up to 20 MHz clock rate (20 MIPS operation)
- On-board memory
  - Flash program memory (up to 256K)

— On-board EEPROM (up to 4K)

— Internal SRAM (up to 32K)

- Operating voltage

— VCC = 1.8 to 5.5V DC

Figure 2-1 shows a simplified block diagram of the AVR CPU core found on 8-bit AVR devices. Figure 2-2 shows a generic high-level block diagram of an AVR device. This is not intended to represent any specific AVR device, just an AVR in general.

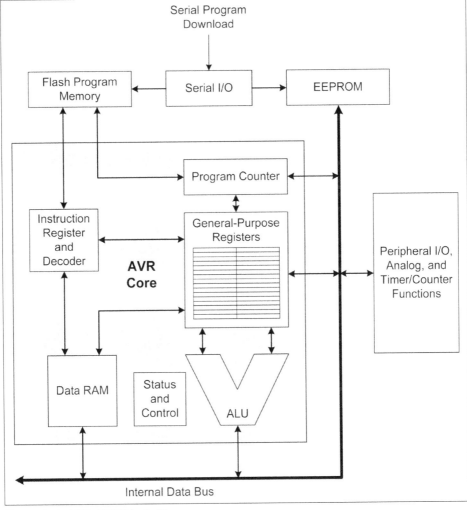

Figure 2-1. AVR CPU block diagram

The peripheral functions are controlled by the CPU via an internal high-speed data bus. Control registers (separate from the CPU registers) are used to configure the operation of the peripherals. All peripheral functions share port pins with the discrete digital I/O capabilities.

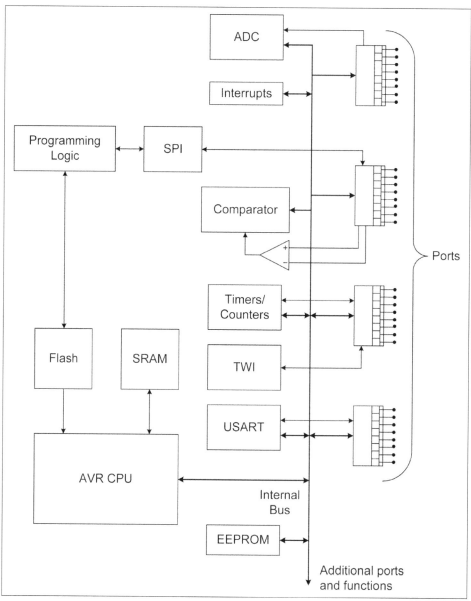

Figure 2-2. Generic AVR microcontroller block diagram

Atmel makes many different types of AVR microcontrollers, which allows hardware designers to pick the part that meets their specific needs and reduce the number of unused pins and wasted space on a printed circuit board. Some, like the tinyAVR parts, come in small surface-mount packages with as few as six pins. Each has one or more discrete digital I/O ports, which can be programmed to perform multiple functions (see "Peripheral Functions" on page 17).

For example, the ATTINY13-20SQ comes in an eight-pin DIP (dual in-line pin) or SOIC (small outline IC) surface-mount package. Six of the device's pins are connected to an internal 8-bit I/O port (port B). The other two are VCC (power) and ground. The six port B pins can be configured as analog inputs, oscillator outputs, interrupt inputs, SPI signals, or discrete digital inputs or outputs. Internally, the device—even one this small—is still an AVR microcontroller, and it has 1K of built-in flash memory for programs, and 64 bytes of RAM for variables.

On the other end of the spectrum there are AVR parts like the ATmega649, with nine 8-bit ports (A through J, but no I, since I can be confused for the numeral 1), 64K of flash memory, 4K of RAM, 2K of EEPROM, 54 general-purpose I/O pins, and an integrated LCD interface. The AVR32 series of parts are 32-bit AVR processors with up to 256K of flash memory, 32K of RAM, an integrated DSP (digital signal processing) unit, protected memory, and 36 general-purpose I/O pins.

No Arduino boards use an AVR part as small as a tinyAVR (it would be a real challenge to squeeze the Arduino bootloader into something with only 1K of flash and still have room left for a useful program), or anything like the ATmega649 or an AVR32, but the point here is that the AVR family offers many choices, and the parts that have been selected for use in Arduino devices aren't the only AVR parts that could be used.

Internal Memory

AVR devices all contain various amounts of three types of memory: flash, SRAM (static random-access memory), and EEPROM. The flash memory is used to store program code, the SRAM is used to hold transient data such as program variables and the stack, and the EEPROM can hold data that needs to persist between software changes and power cycles. The flash and EEPROM can be loaded externally, and both will retain their contents when the AVR is powered off. The SRAM is volatile, and its contents will be lost when the AVR loses power.

Peripheral Functions

The heart of an AVR microcontroller is the 8-bit CPU, but what makes it a truly useful microcontroller is the built-in peripheral functions integrated into the IC with the CPU logic. The peripheral functions of an AVR device vary from one type to another.

Some have one timer, some have two or more (up to six for some types). Other parts may have a 10-bit A/D converter (ADC), whereas others feature a 12-bit converter. All AVR parts provide bidirectional I/O pins for discrete digital signals. Some versions also support a touchscreen and other types of interactive interfaces.

This section contains general descriptions of the peripheral functions that are used with the various types of AVR devices found in Arduino products, with the ATmega168 serving as a baseline example. This section does not attempt to provide an exhaustive reference for each type of AVR microcontroller, but instead covers the general functionality of each type of peripheral function. Refer to Chapter 3 for specific information regarding the processors used in the Arduino boards described in this book, and also see the Atmel technical documentation for low-level details not provided here.

Control Registers

In addition to 32 general-purpose registers in the CPU, AVR devices also have a number of control registers that determine how the I/O ports, timers, communications interfaces, and other features will behave. The control register set will vary from one type of device to another, since different types may have more or less ports than others, and different peripheral function configurations. The control registers for the AVR parts used in the Arduino boards covered in this book can be found in Appendix B. They are also described in detail in the documentation available directly from Atmel.

Even a modest AVR part like the ATmega168 has far more internal functionality than it has pins available to dedicate to each function. For this reason, most of the pins on an AVR microcontroller can be configured to perform specific functions based on the settings contained in the control registers. Because the pin functions are dynamically configurable, it is possible to have a pin perform one type of function at one point in time, and then perform a different function once the control register value has been modified.

For example, pin 12 of an ATmega168 in a 28-pin DIP package is connected to PD6 (Port D, bit 6), but it can also be configured to act as an interrupt source (PCINT22), as the positive input for the AVR's internal analog comparator (AIN0), or as the output of a timer comparison logic circuit (the Timer/Counter0 output compare match A), which can be used to generate a PWM (pulse width modulation) signal.

Digital I/O Ports

AVR microcontrollers use bidirectional I/O ports to communicate with the external world. A port is an 8-bit register wherein some or all of the bits are connected to physical pins on the AVR device package. Different types of AVR devices have different numbers of ports, ranging from one for the ATTINY13-20SQ up to nine for the ATmega649. Ports are labeled as A, B, C, and so on.

Each pin of a port is controlled by internal logic that manages the signal direction, the state of an internal pull-up resistor, timing, and other functions. A simplified schematic is shown in Figure 2-3. The Px in Figure 2-3 refers to port bit/pin x (0 through 7). For a detailed description of the AVR port logic, see the AVR technical references.

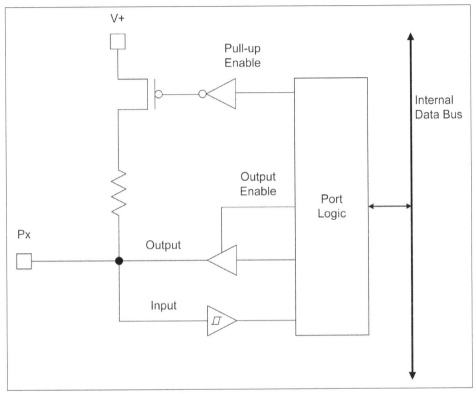

Figure 2-3. AVR I/O port block diagram

Because of the sophisticated logic used to control functionality, an AVR port can perform many different functions—some of them simultaneously. When a port is configured as an output it is still possible to read data from it, and an output can be used to trigger an interrupt (discussed in "Interrupts" on page 26).

8-Bit Timer/Counters

There are two forms of 8-bit timer/counter available in AVR microcontrollers. In the first type the clock input is derived from the primary system clock, and hence the timer/counter is synchronous. The second form has the ability to operate in an asynchronous mode using an external clock source. Figure 2-4 shows a simplified schematic of an AVR timer. The control registers for the timer/counter are defined in Appendix B and described in detail in the Atmel technical documentation.

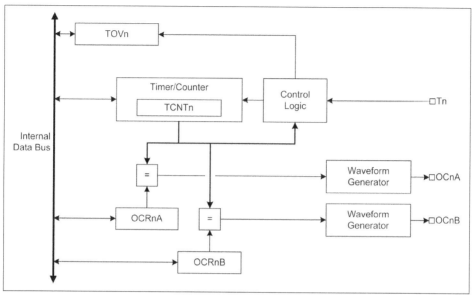

Figure 2-4. AVR timer/counter block diagram

The Timer/Counter0 module in an AVR 8-bit timer/counter peripheral function is a general-purpose timer and/or counter that features two independent output comparison circuits with four modes of operation. The timer/counter modes of operation are as follows:

Normal mode
> This is the simplest mode of timer/counter operation. The count always increments and no counter clear is performed when the counter reaches its maximum 8-bit value. When this occurs, the counter overflows and returns to zero. When the counter wraps back to zero, the Timer/Counter Overflow Flag (TOV0) is set. The TOV0 flag can be viewed as a ninth bit, but it is only set, not cleared, by a timer overflow. The timer overflow interrupt will automatically clear the overflow flag bit, and the interrupt can be used to increment a second software-based counter in memory. A new counter value can be written to the TCNT0 register at any time.

Clear Timer on Compare (CTC) mode
> In the Clear Timer on Compare mode, the OCR0A register is used to manipulate the counter resolution by defining the maximum value of the counter. This results in greater control of the compare match output frequency and helps to simplify external event counting.

Fast PWM mode

The fast pulse width modulation mode supports high-frequency PWM waveform generation.

Phase correct PWM mode

The phase correct PWM mode provides a high-resolution phase correct PWM waveform generation option.

In addition, some AVR devices contain an 8-bit timer/counter with the ability to operate asynchronously using external clock inputs (the TOSC1 and TOSC2 clock input pins). It is functionally equivalent to the synchronous 8-bit timer/counter circuit described previously.

16-Bit Timer/Counters

The 16-bit timer/counter is similar to the 8-bit version, but with an extended count range. It is true 16-bit logic, which allows for 16-bit variable period PWM generation. The module also features two independent output comparison circuits, double-buffered output comparison registers, and an input capture circuit with noise canceling. In addition to PWM generation the 16-bit timer/counter can be used for high-resolution external event capture, frequency generation, and signal timing measurement. It has the ability to generate four different interrupts (TOV1, OCF1A, OCF1B, and ICF1).

Timer/Counter Prescaler

In an AVR device one or more counters may share the same prescaler logic, but with different settings. The prescaler is essentially a divider circuit that generates a derivative of the system I/O clock at f/8, f/64, f/256, or f/1024, which are referred to as *taps*. One timer/counter might use the f/64 tap, whereas another might use the f/1024 tap. The use of a prescaler allows the range of a timer/counter to be extended to more closely match the rate at which an external event occurs, and also increases the time in between timer/counter overflows and resets.

Analog Comparator

The analog comparator section of an AVR microcontroller is used to compare the input voltages on the AIN0 and AIN1 pins. Although AIN0 is defined as the positive input and AIN1 as the negative, this refers to the relationship between them, not the actual polarity of the input voltages. Figure 2-5 shows a simplified schematic of the analog comparator circuit of an AVR.

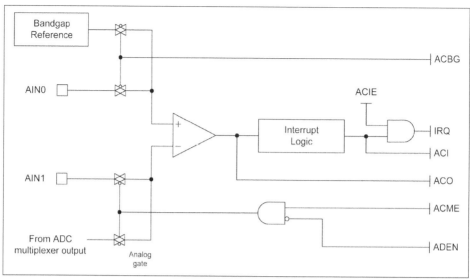

Figure 2-5. AVR analog comparator block diagram

When AIN0 is greater than AIN1, the comparator logic sets the comparator flag ACO. The output of the comparator can be configured to trigger the input capture function of one of the timer/counter modules, and it can also generate an interrupt specific to the comparator. The interrupt event can be configured to trigger on comparator output rise, fall, or toggle.

The analog comparator circuit can do more than just compare the voltages on the AIN0 and AIN1 inputs. The input of the analog comparator may also be configured such that the AIN1 input can be compared to the internal bandgap reference voltage, or AIN0 can be compared to the output of the ADC multiplexer (and this voltage is still available to the input of the ADC). The unusual symbols with four arrows are analog gates. How a gate will respond to a control input is indicated by the inversion circle—when the inverting control input is used it will pass an analog signal when the control is low, and otherwise it will pass a signal when it is high.

Analog-to-Digital Converter

Most AVR microcontrollers contain an 8-bit, 10-bit, or 12-bit analog-to-digital converter. The 8-bit converters are found in the ATtiny6 and ATtiny10 parts. Some of the automotive versions of AVR microcontrollers have no ADC.

When an ADC is part of the AVR design, it will have anywhere from 4 to 28 inputs. The actual number of available inputs depends largely on the physical package. Each input is selected one at a time via an internal multiplexer—they are not all active simultaneously. In addition, some of the I/O pins used by the ADC input multiplexer may also be assigned to other functions.

ATmega168 devices have either six or eight ADC input channels, depending on the package type. The PDIP (plastic DIP) package has a 10-bit ADC with six input channels. The TQFP and QFN/MFL surface-mount packages have a 10-bit ADC with eight input channels. Figure 2-6 shows a block diagram of the AVR ADC peripheral function.

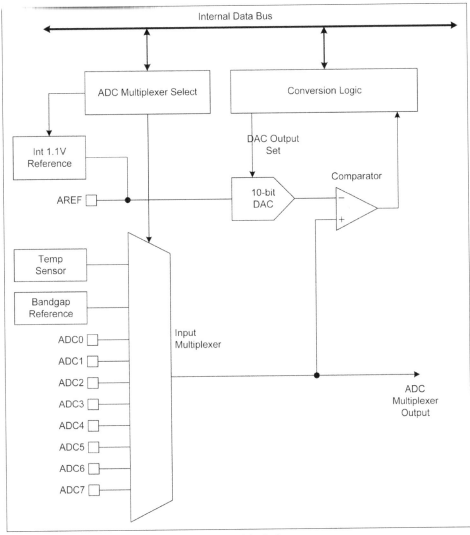

Figure 2-6. AVR analog-to-digital converter block diagram

Notice in Figure 2-6 that the AVR employs what is called a "successive approxima-tion" converter. This type of converter isn't particularly fast, but it is simple to imple-ment, requiring only a DAC (digital-to-analog converter) and a comparator. The typical conversion time for a 10-bit AVR ADC in free-running mode, while still maintaining full resolution, is around 65 microseconds (µs) per sample.

Serial I/O

The ATmega168 provides three primary forms of serial interface: synchronous/asyn-chronous serial, SPI master/slave synchronous, and a byte-oriented two-wire inter-face similar to the Philips I2C (Inter-Integrated Circuit) standard.

USART

A common component of many AVR parts is a built-in USART (universal synchro-nous/asynchronous receiver-transmitter), also referred to as a UART (universal asyn-chronous receiver-transmitter). This function can be used to implement an RS-232 or RS-485 interface, or used without external interface logic for chip-to-chip communi-cations. The baud rate is determined by the frequency of the clock used with the microcontroller, with 9,600 being a typical speed. Higher rates are possible with a fast external crystal. The USART can also be used in SPI (serial peripheral interface) mode, in addition to the dedicated SPI logic found in AVR devices. Figure 2-7 shows the basic internal components of the AVR USART peripheral function.

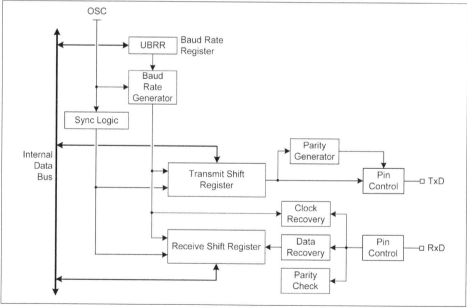

Figure 2-7. AVR USART block diagram

SPI

The SPI peripheral logic of the AVR supports all four standard SPI modes of operation. I/O pins on the AVR device may be configured to act as the MOSI, MISO, and SCK[1] signals used by SPI. These pins are different from the RxD and TxD (recieve data and transmit data) pins used by the USART. Figure 2-8 shows a high-level view of the SPI logic.

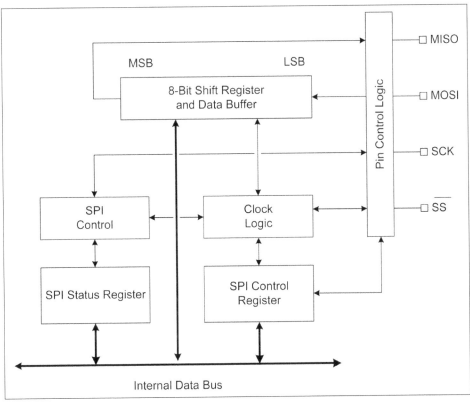

Figure 2-8. AVR SPI block diagram

TWI

The third form of serial I/O supported by many AVR devices is the two-wire interface (TWI). This interface is compatible with the Philips I2C protocol. It supports both master and slave modes of operation, and a 7-bit device address. The TWI interface can achieve transfer speeds of up to 400 kHz with multimaster bus arbitration and has the ability to generate a wakeup condition when the AVR is in sleep mode. Inter-

---

1 Master out, slave in; master in, slave out; and serial clock.

nally, the TWI peripheral is rather complex—much more so than either the USART or SPI peripherals. Figure 2-9 shows an overview of the TWI interface.

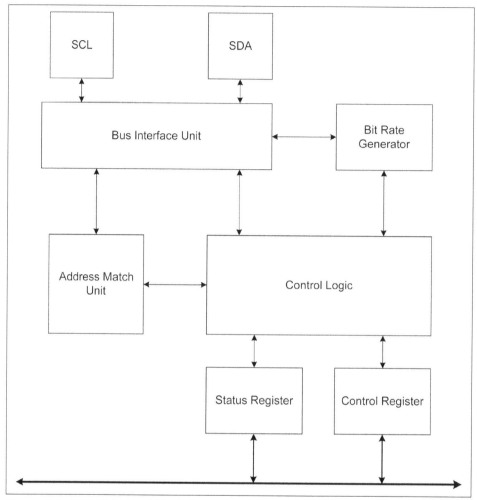

Figure 2-9. AVR TWI (I2C) block diagram

Interrupts

Interrupts are an essential function of a modern processor. They allow the processor to respond to events, either internal or external, by switching to a special block of interrupt handler code to deal with the interrupt. Once the block of code has been executed, control returns to the program that was interrupted at the place where the interrupt occurred. In the AVR an interrupt response may be enabled or disabled via bits in the control registers. This section is specific to the ATmega168. For other microcontroller types, refer to Appendix A or the official Atmel documentation (*http://www.atmel.com/design-support/documentation/*).

The ATmega168 has two external interrupt inputs: INT0 and INT1. These inputs can be configured to trigger on a falling edge, a rising edge, or a low level. The EICRA control register (see Appendix B) is used to configure the behavior. INT0 and INT1 require the presence of an I/O clock. The low-level interrupt mode will generate interrupts as long as the input is held low.

The ATmega168 I/O pins can also serve as interrupt sources. The port-change interrupts are defined as PCINT0 through PCINT23, each of which is associated with one of the device's I/O port pins. When enabled, an interrupt will be generated whenever the state of a port pin changes, even if the pin is configured to act as an output. This allows a pin to generate an interrupt under software control when a program toggles the state of the pin while the port change interrupt detect is enabled.

When any pin in the range from PCINT0 to PCINT7 changes, it will trigger the PCI0 interrupt. Pins in the range PCINT8 to PCINT14 will trigger PCI1, and pins in the range PCINT16 to PCTIN23 trigger the PCI2 interrupt. The registers PCMSK0, PCMSK1, and PCMSK2 control which pins contribute to pin change interrupts.

When an enabled interrupt occurs, the CPU will jump to a location in a vector table in memory that has been assigned to that particular interrupt. The address contains a jump instruction (RJMP) that points to the actual block of code for that interrupt. When the interrupt code is done, execution returns to the point in the original program where the interrupt occurred. Figure 2-10 shows how the interrupt vector table is used to switch execution to the interrupt code block, and then return control to the main program once the interrupt code is finished.

The ATmega168, for example, has a vector table with 26 entries, as shown in Table 2-1. For other processor types, refer to the Atmel documentation for more information on interrupts and how they are managed in the AVR devices.

Table 2-1. ATmega168 interrupt vectors

Vector	Address	Source	Definition
1	0x0000	RESET	External pin, power-on, brown-out, and watchd
2	0x0002	INT0	External interrupt request 0
3	0x0004	INT1	External interrupt request 1
4	0x0006	PCINT0	Pin change interrupt request 0
5	0x0008	PCINT1	Pin change interrupt request 1
6	0x000A	PCINT2	Pin change interrupt request 2
7	0x000C	WDT	Watchdog time-out interrupt
8	0x000E	TIMER2	COMPA Timer/Counter2 compare match A
9	0x0010	TIMER2	COMPB Timer/Counter2 compare match B

Vector	Address	Source	Definition
10	0x0012	TIMER2	OVF Timer/Counter2 overflow
11	0x0014	TIMER1	CAPT Timer/Counter1 capture event
12	0x0016	TIMER1	COMPA Timer/Counter1 compare match A
13	0x0018	TIMER1	COMPB Timer/Counter1 compare match B
14	0x001A	TIMER1	OVF Timer/Counter1 overflow
15	0x001C	TIMER0	COMPA Timer/Counter0 compare match A
16	0x001E	TIMER0	COMPB Timer/Counter0 compare match B
17	0x0020	TIMER0	OVF Timer/Counter0 overflow
18	0x0022	SPI, STC	SPI serial transfer complete
19	0x0024	USART, RX	USART, Rx complete
20	0x0026	USART, UDRE	USART, data register empty
21	0x0028	USART, TX	USART, Tx complete
22	0x002A	ADC	ADC conversion complete
23	0x002C	EE READY	EEPROM ready
24	0x002E	ANALOG COMP	Analog comparator
25	0x0030	TWI	Two-wire serial interface
26	0x0032	SPM READY	Store program memory ready

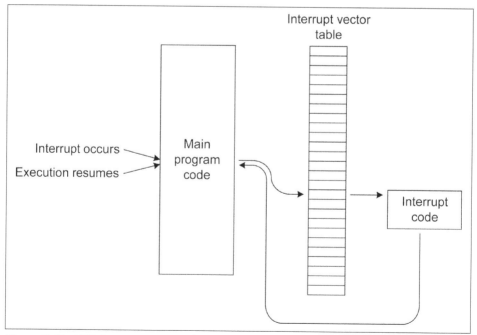

Figure 2-10. AVR interrupt vectors (ATmega168/328)

Watchdog Timer

The AVR provides a watchdog timer (WDT) with a configurable time-out period ranging from 16 ms to 8. If the WDT is enabled it can be used to generate a reset upon time-out, generate an interrupt on time-out, or a combination of both. It uses a separate on-chip oscillator, and because the WDT is clocked from a separate oscillator it will continue to count when the microcontroller is in sleep mode. It can therefore be used to awaken the microcontroller after a specific period of time.

Another common use for the WDT is to force either a reset or an interrupt if the timer expires without a reset action from the software. This is useful for forcing the microcontroller out of deadlocks or detecting runaway code.

Electrical Characteristics

The ATmega168 and ATmega328 AVR microcontrollers can operate with a VCC of 1.8 to 5.5V DC. The ATmega32U4 can utilize a VCC of 2.7 to 5.5V DC.

The current consumption of an AVR device will vary depending on the device type, the active or inactive (sleep) state of the microcontroller, and the clock speed. Values across the ATmega and XMEGA product lines range from 0.55 milliamps (mA) to 30 mA. The total current consumption is also dependent on the amount of current sourced through the I/O pins. Chapter 3 lists specific values for the ATmega168, ATmega328, ATmega1280, ATmega2560, and ATmega32U4 microcontrollers.

For More Information

Chapter 3 describes the pinouts used for the three types of AVR MCUs encountered on Arduino boards, and Chapter 4 shows how the AVR MCU pins are mapped to the I/O pins on various Arduino boards.

Each of the various AVR MCUs may have different combinations of internal peripheral functions, and in some cases the functionality may vary slightly between types. The best source of definitive information for AVR MCUs is the Atmel website (*http://www.atmel.com*).

Arduino-Specific AVR Microcontrollers

This chapter presents technical descriptions of the types of AVR parts used in Arduino models that are based on 8-bit AVR devices. This is intended to build on the functional descriptions presented in Chapter 2 for AVR microcontrollers in general, but with a specific focus on the ATmega168/328, ATmega1280/2560, and ATmega32U4 microcontrollers.

From the perspective of someone programming an Arduino with the IDE, the microcontroller is a simplified abstraction of the actual underlying AVR device. The code necessary to perform operations such as configuring an output pin to generate a PWM signal or internally route an analog voltage into the built-in ADC is straightforward. The internal addresses of the control registers and their control bits are predefined, so the sketch author need not worry about the low-level details.

Because an Arduino board is really nothing more than a carrier for an AVR chip, the electrical characteristics of the Arduino are largely those of the processor. The pins from the chip are connected directly to the pin terminals or solder pads along the edge of the Arduino board. There is no buffering or level-shifting between the chips and the board's connection points.

At the time this book was written Arduino used five basic types of ATmega microcontrollers and three variations, for a total of eight part numbers. These are listed in Table 3-1. The main differences between the various AVR devices lie in the amount of on-board flash memory available, the maximum clock speed, the number of I/O pins on the chip, and of course the available internal peripheral functions. The ATmega32U4 device also has a built-in USB interface, which eliminates the need for a second part to handle the USB communications. The devices all use the same CPU instruction set.

Table 3-1. AVR microcontrollers used in Arduino products

Microcontroller	Flash	I/O pins (max)	Notes
ATmega168	16K	23	20 MHz clock
ATmega168V	16K	23	10 MHz clock
ATmega328	32K	23	20 MHz clock
ATmega328P	32K	23	20 MHz clock, picoPower
ATmega328V	32K	23	10 MHz clock
ATmega1280	128K	86	16 MHz clock
ATmega2560	256K	86	16 MHz clock
ATmega32U4	32K	26	16 MHz clock

This inherent level of compatibility means that programs written for an Arduino Diecimila should compile and run with no changes on an Uno board. The primary difference between a Diecimila and an Uno is that one uses an ATmega168 and the other is based on an ATmega328P. As you can see from Table 3-1, the ATmega328P has twice as much on-board flash memory as the ATmega168. So, programs written for an Uno might not be backward compatible with a Diecimila if the code was designed for use with the larger amount of program memory.

ATmega168/328

The ATmega168 and ATmega328 are basically the same chip with different amounts of on-board memory. The block diagram of an ATmega168 or 328 device is shown in Figure 3-1.

Memory

The ATmega328 has twice the amount of each type of memory as the ATmega168, as shown in Table 3-2. Other than this, the two parts are identical.

Table 3-2. ATmega168/328 on-board memory

	ATmega168	ATmega328
Flash program memory	16K bytes	32K bytes
EEPROM	512 bytes	1K bytes
RAM	1K bytes	2K bytes

Features

These two parts share the following features:

- In-system programming by on-chip boot program

- Two 8-bit timer/counters with separate prescaler and compare mode
- One 16-bit timer/counter with separate prescaler, compare mode, and capture mode
- Real-time counter with separate oscillator
- Six PWM channels
- Six- or eight-channel 10-bit ADC (depends on package type)
- USART
- Master/slave SPI serial interface
- Two-wire serial interface (Philips I2C compatible)
- Programmable watchdog timer
- Analog comparator
- 23 programmable I/O lines

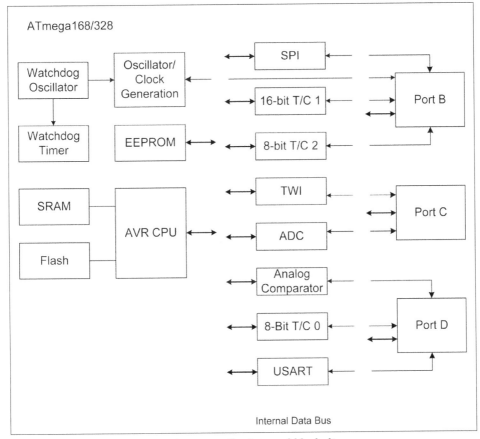

Figure 3-1. ATmega168/328 microcontroller internal block diagram

Packages

The ATmega168 and ATmega384 are available in four different package types: 28-pin DIP, 28-pin MLF surface-mount, 32-pin TQFP surface-mount, and 32-pin MLF surface-mount. The 28-pin DIP is the most commonly used package on Arduino boards, although the Uno SMD uses a 32-pin surface-mount package. This section will focus on the PDIP version of the ATmega168.

Ports

The ATmega168/328 has three ports designated as B, C, and D. Ports B and D are 8-bit ports. Port C has six pins available that can be used as ADC inputs. PC4 and PC5 are also connected to the TWI logic and provide SCL and SDA I2C-compatible signals (clock and data). Also note that PC6 is typically used as the RESET input. There is no PC7 on the ATmega168/328 parts. Also note that there is no Port A in the ATmega168/328 part.

Each port provides bidirectional discrete digital I/O with programmable internal pull-up resistors. The on/off states of the pull-up resistors are selected via control register bits for specific port pins.

The port output buffers have symmetrical drive characteristics with both sink and source capability. As inputs, port pins that are externally pulled low will source current if the internal pull-up resistors are activated. Port pins are placed in a tri-state (high impedance) mode when a reset condition becomes active, even if the clock is not running.

Pin Functions

Figure 3-2 shows the pin functions for the 28-pin DIP package. Refer to the Atmel documentation (*http://bit.ly/atmel-docs*) for pinout information for the surface-mount packages.

Analog Comparator Inputs

Figure 3-3 shows the locations of the AIN0 and AIN1 pins for an ATmega168 or ATmega328 in a PDIP package. Note that AIN0 shares a pin with the OC0A timer/counter output (PD6). So, if PD6 is being used as a PWM output, it cannot be used as the AIN0 input unless it is reconfigured each time its function needs to change.

Analog Inputs

Except for the SCL and SDA pins used for TWI serial communications, the analog input pins are available for use as either discrete digital I/O or analog inputs without conflicts with other AVR peripheral functions. Figure 3-4 shows the pins of the PDIP package for an ATmega168 that are used for analog inputs.

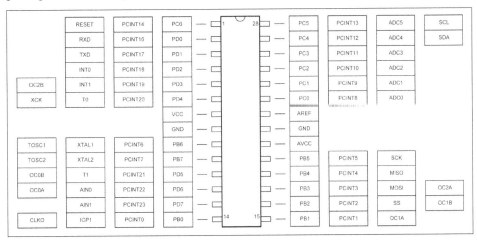

Figure 3-2. ATmega168/328 microcontroller DIP package pin functions

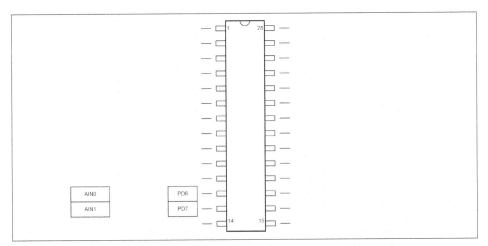

Figure 3-3. ATmega168/328 microcontroller analog comparator input pins

Serial Interfaces

Figure 3-5 shows the I/O pins of an ATmega168 (PDIP package) used by the serial interface functions. Because none of the serial functions share port pins, it is possible to use all three forms without port conflicts.

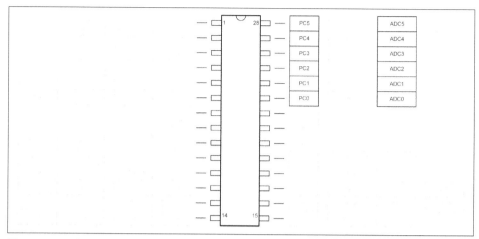

Figure 3-4. ATmega168/328 microcontroller ADC input pins

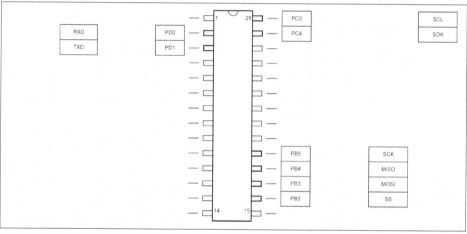

Figure 3-5. ATmega168/328 microcontroller serial I/O pins

Timer/Clock I/O

The internal timer/counter logic of an ATmega168/328 AVR is complex, and this is reflected in the chip's pin assignments, shown in Figure 3-6. Note that the OCxn pins (OC0A, OC0B, OC1A, OC1B, OC2A, and OC2B) are available as PWM outputs, and this is how they are labeled on an Arduino board. Also note that T1 and OSC0B share the same pin (PD5), but otherwise the PWM-capable outputs can be used without conflicts with other timer/counter functions.

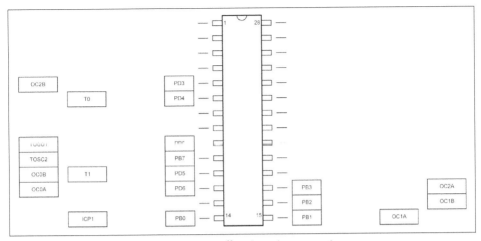

Figure 3-6. ATmega168/328 microcontroller timer/counter pins

External Interrupts

The port D pins PD2 and PD3 are specifically intended for use as external interrupt inputs. Any of the PCINT0 to PCINT23 pins may also be used as external interrupt inputs without interfering with other assigned functions (see "Interrupts" on page 26 for an overview of how these interrupts can be used). Figure 3-7 shows the external interrupt inputs available with the ATmega168/328 devices.

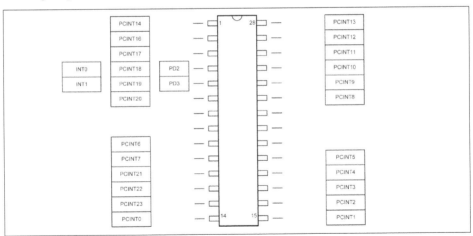

Figure 3-7. ATmega168/328 microcontroller interrupt input pins

Arduino Pin Assignments

The Diecimila, Duemilanove, Uno, and Ethernet boards use the same basic board layout pattern (described in Chapter 4), as shown in Figure 3-8. These Arduino boards employ a nomenclature for the pin socket headers on the board based on the functions of the pins of the ATmega168 or ATmega328 installed on the PCB. The main emphasis is on the functions most commonly used, which are discrete digital I/O, analog inputs, and the PWM output capabilities of the AVR microcontroller.

Basic Electrical Characteristics

Table 3-3 shows some of the basic electrical characteristics of the ATmega168/328 microcontrollers, with an emphasis on power consumption.

With a VCC of between 1.8 and 2.4 volts a low input on an I/O pin is defined as a voltage between –0.5 and two-tenths (0.2) of VCC. For a VCC between 2.4 and 5.5 volts a low input is defined as a voltage between –0.5 and three-tenths (0.3) of VCC.

With a VCC of between 1.8 and 2.4 volts a high input is defined as a voltage between seven-tenths (0.7) VCC and VCC + 0.5 volts. For a VCC between 2.4 and 5.5 volts it is defined as a voltage between six-tenths (0.6) VCC and VCC + 0.5 volts.

Table 3-3. ATmega168/328 power consumption characteristics

Device	Parameter	Condition	VCC	Typical	Max
ATmega168	Power supply current	Active 8 MHz	5V	4.2 mA	12 mA
		Idle 8 MHz	5V	0.9 mA	5.5 mA
	Power-save mode	32 MHz TOSC	1.8V	0.75 uA	
		32 MHz TOSC	5V	0.83 uA	
	Power-down mode	WDT enabled	3V	4.1 uA	15 uA
		WDT disabled	3V	0.1 uA	2 uA
ATmega328	Power supply current	Active 8 MHz	5V	5.2 mA	12 mA
		Idle 8 MHz	5V	1.2 mA	5.5 mA
	Power-save mode	32 MHz TOSC	1.8V	0.8 uA	
		32 MHz TOSC	5V	0.9 uA	
	Power-down mode	WDT enabled	3V	4.2 uA	15 uA
		WDT disabled	3V	0.1 uA	2 uA

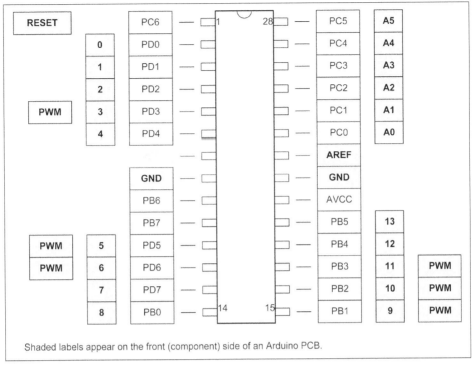

Shaded labels appear on the front (component) side of an Arduino PCB.

Figure 3-8. Arduino labels for ATmega168/328 pins

ATmega1280/ATmega2560

As with the ATmega168/328, the primary difference between an ATmega1280 and an ATmega2560 is in the amount of on-board memory. Otherwise, these devices are identical. A simplified block diagram of an ATmega1280 or 2560 device is shown in Figure 3-9.

For the sake of clarity there are some minor details missing from Figure 3-9, but the essential parts are there. For a more detailed diagram, refer to the Atmel documentation for the ATmega1280 and ATmega2560 parts. Also note that Figure 3-9 shows the internal functions available with the 100-pin package. The 64-pin package supports a subset of what is described here.

Memory

The ATmega2560 has twice the amount of each type of memory as the ATmega1280, and eight times as much flash memory as the ATmega328 MCU. The available memory is shown in Table 3-4. Other than this, the ATmega1280 and ATmega2560 in the same package types are identical.

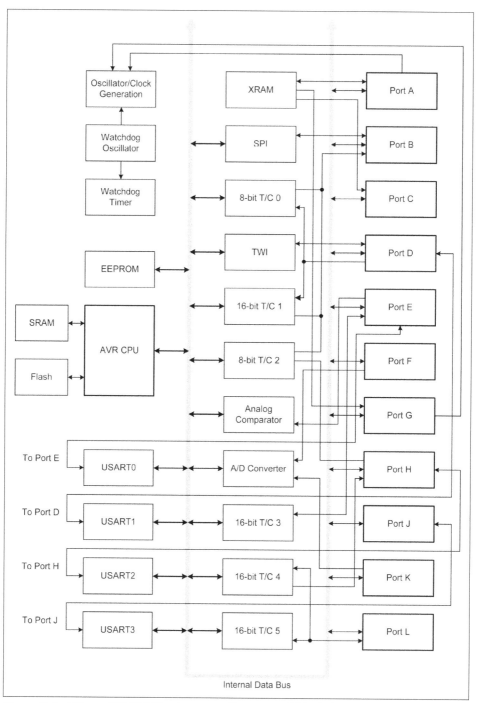

Figure 3-9. ATmega1280/2560 microcontroller internal block diagram

Table 3-4. ATmega1280/2560 on-board memory

	ATmega1280	ATmega2560
Flash program memory	128K bytes	256K bytes
EEPROM	4K bytes	4K bytes
RAM	8K bytes	8K bytes

Features

These two parts share the following features:

- In-system programming by on-chip boot program
- Two 8-bit timer/counters with separate prescaler and compare mode
- Four 16-bit timer/counters with separate prescaler, compare mode, and capture mode
- Real-time counter with separate oscillator
- 12 PWM channels
- Output compare modulator
- Six-channel, 10-bit ADC
- Four USART functions
- Master/slave SPI serial interface
- Two-wire serial interface (Philips I2C compatible)
- Programmable watchdog timer
- Analog comparator
- 86 programmable I/O lines

Packages

The ATmega1280 and ATmega2560 devices are available in a 100-pin TQFP package, a 100-pin BGA (ball grid array) package, and a 64-pin TQFP package. Figure 3-10 shows the relative sizes and pin spacing (pitch) of the three available package types. Only the 100-pin packages provides all of the functionality shown in Figure 3-9.

The Arduino Mega and Mega2560 boards both use the 100-pin version of the TQFP package.

Ports

The 100-pin versions of the ATmega1280 and ATmega2560 parts have 11 ports, labeled A through L. Note that there is no I port, since the letter I can be mistaken for the numeral 1.

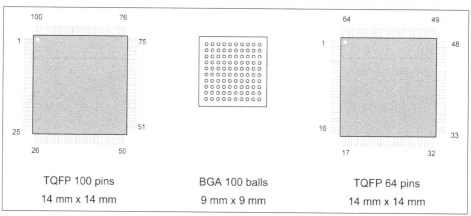

Figure 3-10. A Tmega1280/2560 packages

Ports A, B, C, D, and E are bidirectional 8-bit ports. Port B has better drive capabilities than the other ports. Ports F and K are used as inputs to the internal A/D converter, but can also serve as bidirectional ports. Port G is a 6-bit port, and ports H, J, and L are 8-bit bidirectional interfaces. Each port provides bidirectional discrete digital I/O with programmable internal pull-up resistors. The on/off states of the pull-up resistors are selected via control register bits for specific port pins.

The port output buffers have symmetrical drive characteristics with both sink and source capability. As inputs, port pins that are externally pulled low will source current if the internal pull-up resistors are activated. Port pins are placed in a tri-state (high impedance) mode when a reset condition becomes active.

Pin Functions

The diagrams in this section refer to the 100-pin version of the ATmega1280/2560 devices. For details regarding the BGA and 64-pin parts, refer to the Atmel technical documentation.

Analog Comparator Inputs

Like the smaller ATmega168/328 devices, the ATmega1280/2560 parts have two analog comparator inputs, as detailed in Table 3-5 and shown in Figure 3-11.

Table 3-5. Analog comparator inputs

Pin	Port	Function
4	PE2	AIN0
5	PE3	AIN1

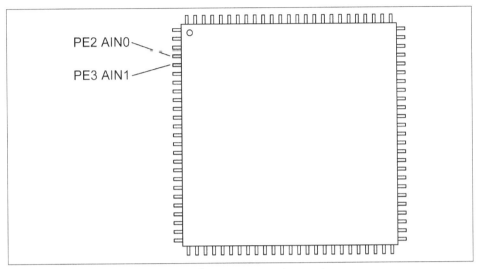

Figure 3-11. ATmega1280/2560 analog comparator input pins

Analog Inputs

The ATmega1280/2560 devices support up to 16 A/D converter inputs. These are located on ports F and K, and are connected to pins 82 through 97 on the 100-pin package (see Table 3-6). Port F also has alternate functions TCK, TMS, TDI, and TDO, and port K pins are connected to port interrupts PCINT16 through PCINT23. The physical pin locations are shown in Figure 3-12.

Table 3-6. Analog inputs

Pin	Port	Function	Pin	Port	Function
82	PK7	ADC15	90	PF7	ADC7
83	PK6	ADC14	91	PF6	ADC6
84	PK5	ADC13	92	PF5	ADC5
85	PK4	ADC12	93	PF4	ADC4
86	PK3	ADC11	94	PF3	ADC3
87	PK2	ADC10	95	PF2	ADC2
88	PK1	ADC9	96	PF1	ADC1
89	PK0	ADC8	97	PF0	ADC0

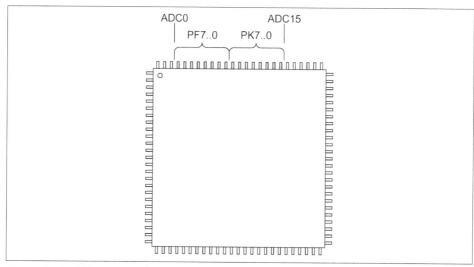

Figure 3-12. ATmega1280/2560 ADC input pins

Serial Interfaces

The ATmega1280/2560 devices have four internal USART functions. These are brought out as four pairs of pins, one serving as TXD and the other as RXD. The SPI interface is available on port B on pins 19 through 22 of the 100-pin package. The two-wire interface (I2C) is connected to port D via pins 43 and 44. The pin assignments are listed in Tables 3-7 through 3-12. Figure 3-13 shows the locations of the serial I/O pins on the 100-pin package.

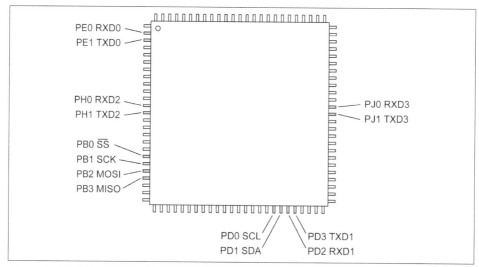

Figure 3-13. ATmega1280/2560 serial I/O pins

Table 3-7. USART 0

Pin	Port	Function
2	PE0	RXD0
3	PE1	TXD0

Table 3-8. USART 1

Pin	Port	Function
45	PD2	RXD1
46	PD3	TXD1

Table 3-9. USART 2

Pin	Port	Function
12	PH0	RXD2
13	PH1	TXD2

Table 3-10. USART 4

Pin	Port	Function
63	PJ0	RXD3
64	PJ1	TXD3

Table 3-11. SPI

Pin	Port	Function
19	PB0	SS (active low)
20	PB1	SCK
21	PB2	MOSI
22	PB3	MISO

Table 3-12. TWI

Pin	Port	Function
43	PD0	SCL
44	PD1	SDA

Timer/Clock I/O

There are five timer/counter functions in an ATmega1280/2560 device, as shown in Figure 3-14. Table 3-13 lists the pins. Note that there is no T2 pin.

Table 3-13. Atmega1280/2560 timer/counter pins

Pin	Port	Function	Pin	Port	Function
1	PG5	OC0B	50	PD7	T0
5	PE3	OC3A	15	PH3	OC4A
6	PE4	OC3B	16	PH4	OC4B
7	PE5	OC3C	17	PH5	OC4C
8	PE6	T3	18	PH6	OC2B
9	PE7	ICP3	27	PH7	T4
23	PB4	OC2A	35	PL0	ICP4
24	PB5	OC1A	36	PL1	ICP5
25	PB6	OC1B	37	PL2	T5
26	PB7	OC0A/OC1C	38	PL3	OC5A
47	PD4	ICP1	39	PL4	OC5B
49	PD6	T1	40	PL5	OC5C

External Interrupts

The ATmega1280/2560 devices support eight external interrupt inputs in addition to the port interrupt functions available on ports B, J, and K. The pins are listed in Table 3-14.

Table 3-14. ATmega 1280/2560 interrupt pins

Pin	Port	Function	Pin	Port	Function
6	PE4	INT4	43	PD0	INT0
7	PE5	INT5	44	PD1	INT1
8	PE6	INT6	45	PD2	INT2
9	PE7	INT7	46	PD3	INT3

Arduino Pin Assignments

The Mega and Mega2560 Arduino boards use the board layout pattern described in Chapter 4. These Arduino boards employ a nomenclature for the pin socket headers on the board based on the functions of the pins of the ATmega1280 or ATmega2650 installed on the PCB. The main emphasis is on the functions most commonly used, which are discrete digital I/O, analog input, and the PWM output capabilities of the AVR microcontroller.

Table 3-15 lists the pins on a Mega or Mega2560 PCB and the connections to the AVR ATmega1280/2560 device on the board. Functions in parentheses are external memory addressing pins, and a tilde (~) is used to indicate an active-low (true when low) signal. Unlike the smaller Arduino boards, the Mega boards are capable of using external SRAM.

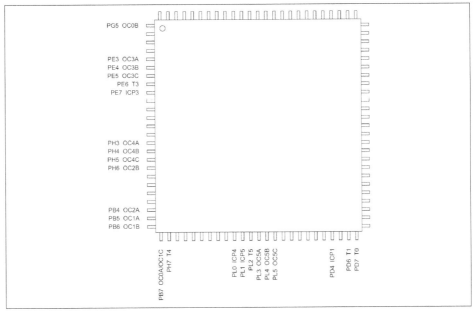

Figure 3-14. ATmega1280/2560 timer/counter pins

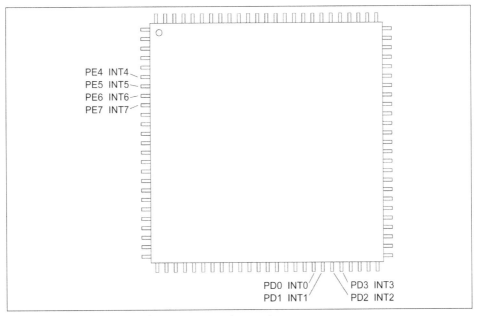

Figure 3-15. ATmega1280/2560 interrupt input pins

Table 3-15. Arduino Mega and Mega2560 pin assignments

Mega board pin	Chip pin #	Function	Mega board pin	Chip pin #	Function
0	2	RXD0	35	55	I/O (A10)
1	3	TXD0	36	54	I/O (A9)
2	5	OC3B [PWM]	37	53	I/O (A8)
3	6	OC3C [PWM]	38	50	T0
4	1	OC0B [PWM]	39	70	I/O (ALE)
5	4	OC3A [PWM]	40	52	I/O (~RD)
6	15	OC4A [PWM]	41	51	I/O (~WR)
7	16	OC4B [PWM]	42	42	PL7
8	17	OC4C [PWM]	43	41	PL6
9	18	OC2B [PWM]	44	40	OC5C [PWM]
10	23	OC2A [PWM]	45	39	OC5B [PWM]
11	24	OC1A [PWM]	46	38	OC5A [PWM]
12	25	OC1B [PWM]	47	37	T5
13	26	OC0A [PWM]	48	36	ICP5
14	64	TXD3	49	35	ICP4
15	63	RXD3	50	22	MISO
16	13	TXD2	51	21	MOSI
17	12	RXD2	52	20	SCK
18	46	TXD1	53	19	~SS
19	45	RXD1	54	97	A0 (analog in)
20	44	SDA	55	96	A1 (analog in)
21	43	SCL	56	95	A2 (analog in)
22	78	I/O (AD0)	57	94	A3 (analog in)
23	77	I/O (AD1)	58	93	A4 (analog in)
24	76	I/O (AD2)	59	92	A5 (analog in)
25	75	I/O (AD3)	60	91	A6 (analog in)
26	74	I/O (AD4)	61	90	A7 (analog in)
27	73	I/O (AD5)	62	89	A8 (analog in)
28	72	I/O (AD6)	63	88	A9 (analog in)
29	71	I/O (AD7)	64	87	A10 (analog in)
30	60	I/O (A15)	65	86	A11 (analog in)
31	59	I/O (A14)	66	85	A12 (analog in)
32	58	I/O (A13)	67	84	A13 (analog in)
33	57	I/O (A12)	68	83	A14 (analog in)
34	56	I/O (A11)	69	82	A15 (analog in)

Note that pins 22 through 37 and pins 39, 40, and 41 may be used to access external memory. Otherwise, they can be used as normal discrete digital I/O pins.

Electrical Characteristics

Table 3-16 shows some of the basic electrical characteristics of the ATmega1280/2560 microcontrollers, with an emphasis on power consumption.

Table 3-16. ATmega1280/2560 power consumption characteristics

Device	Parameter	Condition	VCC	Typical	Max
ATmega1280	Power supply current	Active 8 MHz	5V	10 mA	14 mA
ATmega2560	Power supply current	Idle 8 MHz	5V	2.7 mA	4 mA
both	Power-down mode	WDT enabled	3V	<5 uA	15 uA
		WDT disabled	3V	<1 uA	7.5 uA

With a VCC of between 1.8 and 2.4 volts a low input on an I/O pin is defined as a voltage between –0.5 and two-tenths (0.2) of VCC. For a VCC between 2.4 and 5.5 volts a low input is defined as a voltage between –0.5 and three-tenths (0.3) of VCC.

With a VCC of between 1.8 and 2.4 volts a high input is defined as a voltage between seven-tenths (0.7) VCC and VCC + 0.5 volts. For a VCC between 2.4 and 5.5 volts it is defined as a voltage between six-tenths (0.6) VCC and VCC + 0.5 volts.

ATmega32U4

The ATmega32U4 is a member of Atmel's XMEGA microcontroller line. It has 32 KB of flash program memory, 2.5 KB of SRAM, and 1 KB of EEPROM. I/O functions are accessed through ports B through F. There is no port A in this particular device. Figure 3-16 shows a block diagram of the main internal components of an ATmega32U4.

The ATmega32U4 features an integrated USB 2.0 full-speed interface. This eliminates the need for a separate outboard interface chip. It also includes a 1149.1-compliant JTAG interface for on-chip debugging. The ATmega32U4 can operate with supply voltages ranging from 2.7 to 5.5 volts.

Memory

As shown in Table 3-17, the ATmega32U4 has the same amount of flash memory and EEPROM as an ATmega328, but comes with 2.5 KB of RAM instead of 2 KB.

Table 3-17. ATmega32U4 on-board memory

	ATmega32U4
Flash program memory	32K bytes
EEPROM	1K bytes
RAM	2.5K bytes

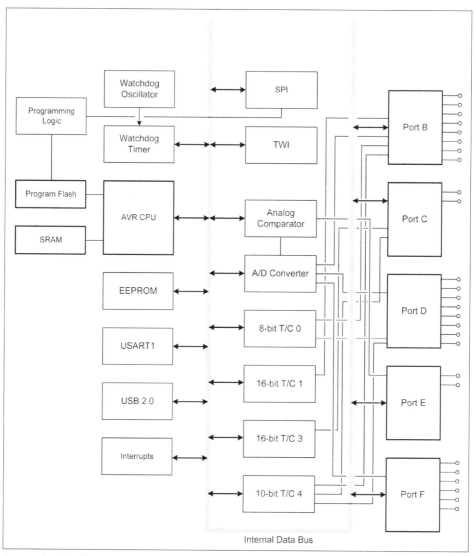

Figure 3-16. ATmega32U4 microcontroller internal block diagram

Features

The ATmega32U4 has the following features:

- In-system programming by on-chip boot program
- One 8-bit timer/counter with separate prescaler and compare mode
- Two 16-bit timer/counters with separate prescaler, compare mode, and capture mode
- One 10-bit high-speed timer/counter with phase-locked loop (PLL) and compare mode
- Four 8-bit PWM channels
- Four PWM channels with 2- to 16-bit programmable resolution
- Six high-speed PWM channels with 2- to 11-bit programmable resolution
- Output compare modulator
- 12-channel, 10-bit ADC
- USART functions with CTS/RTS handshake
- Master/slave SPI serial interface
- Two-wire serial interface (Philips I2C compatible)
- Programmable watchdog timer
- Analog comparator
- On-chip temperature sensor
- 26 programmable I/O lines

Packages

The ATmega32U4 is available in either TQFP44 or QFN44 surface-mount packages. The pin designations, shown in Figure 3-17, are the same for both package types.

Ports

The ATmega32U4 has five ports labeled B through F. In the 44-pin QFN/TQFP surface-mount packages only ports B and D have all eight bits present on the package pins. Ports C, E, and F are represented internally as 8-bit registers, but only bits PC6 and PC7 of port C are available externally. For port E only bits PE2 and PE6 are available externally, and for port F bits PF2 and PF3 are not present on the package pins.

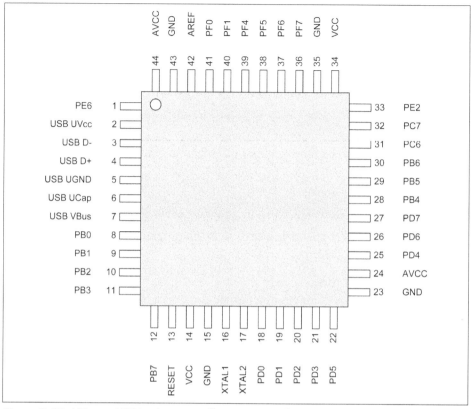

Figure 3-17. ATmega32U4 microcontroller 44-pin package

In addition to various peripheral functions ports B, C, D, E, and F are bidirectional discrete digital I/O ports. Each port provides bidirectional discrete digital I/O with programmable internal pull-up resistors. The on/off states of the pull-up resistors are selected via control register bits for specific port pins.

The port output buffers have symmetrical drive characteristics with both sink and source capability. As inputs, port pins that are externally pulled low will source current if the internal pull-up resistors are activated. Port pins are placed in a tri-state (high impedance) mode when a reset condition becomes active.

Pin Functions

Table 3-18 lists the pin assignments of an ATmega32U4 in a 44-pin package. The mapping of the chip pins to the pin headers on an Arduino Leonardo board can be found in "Arduino Pin Assignments" on page 59.

Table 3-18. ATmega32U4 TQFP/QFN-44 pin assignments

Pin	Functions	Port	Pin	Functions	Port	Pin	Functions	Port
1	INT6/AIN0	PE6	16	XTAL2	n/a	31	OC3A/OC4A	PC6
2	USB UVCC	n/a	17	XTAL1	n/a	32	ICP3/CLK0/OC4A	PC7
3	USB D−	n/a	18	OC0B/SCL/INT0	PD0	33	HWB	PE2
4	USB D+	n/a	19	SDA/INT1	PD1	34	VCC	n/a
5	USB UGnd	n/a	20	RXD1/INT2	PD2	35	GND	n/a
6	USB UCap	n/a	21	TXD1/INT3	PD3	36	ADC7/TDI	PF7
7	USB VBus	n/a	22	XCK1/CTS	PD5	37	ADC6/TD0	PF6
8	PCINT0/SS	PB0	23	GND	n/a	38	ADC5/TMS	PF5
9	PCINT1/SCLK	PB1	24	AVCC	n/a	39	ADC4/TCK	PF4
10	PDI/PCINT2/MOSI	PB2	25	ICP1/ADC8	PD4	40	ADC1	PF1
11	PDO/PCINT3/MISO	PB3	26	T1/OC4D/ADC9	PD6	41	ADC0	PF0
12	PCINT7/OC0A/OC1C/RTS	PB7	27	T0/OC4D/ADC10	PD7	42	AREF	n/a
13	RESET	n/a	28	PCINT4/ADC11	PB4	43	GND	n/a
14	VCC	n/a	29	PCINT5/OC1A/OC4B/ADC12	PB5	44	AVCC	n/a

Analog Comparator Inputs

There is only one external input to the analog comparator in an ATmega32U4. AIN0 is located on pin 1. The other input to the analog comparator comes from the input multiplexer to the on-chip ADC. The internal circuitry is identical to that shown in "Analog Comparator" on page 21 but without the AIN1 pin connection.

Analog Inputs

The ATmega32U4 provides 12 A/D converter inputs, as listed in Table 3-19 and shown in Figure 3-18. These are located on ports B, D, and F. Note that there are no external pins for ADC2 and ADC3.

Table 3-19. ATmega32U4 ADC inputs

Pin	Port	Function	Pin	Port	Function	Pin	Port	Function
41	PF0	ADC0	37	PF6	ADC6	27	PD7	ADC10
40	PF1	ADC1	36	PF7	ADC7	28	PB4	ADC11
39	PF4	ADC4	25	PD4	ADC8	29	PB5	ADC12
38	PF5	ADC5	26	PD6	ADC9	30	PB6	ADC13

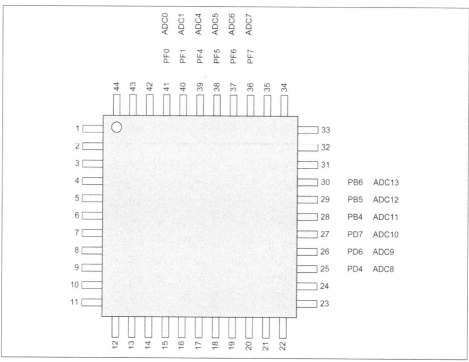

Figure 3-18. ATmega132U4 ADC input pins

Serial Interfaces

The ATmega32U4 has one internal USART function with hardware handshake lines, one SPI interface, and one two-wire interface (TWI) that is I2C compatible (see Tables 3-20 through 3-22). The USB interface is covered later, in Figure 3-22. The pin assignments are shown in Figure 3-19.

Table 3-20. USART

Pin	Port	Function	Pin	Port	Function
20	PD2	RXD1	22	PD5	CTS
21	PD3	TXD1	12	PB7	RTS

Table 3-21. SPI

Pin	Port	Function	Pin	Port	Function
8	PB0	SS	10	PB2	MOSI
9	PB1	SCLK	11	PB3	MISO

Table 3-22. TWI

Pin	Port	Function
18	PD0	SCL
19	PD1	SDA

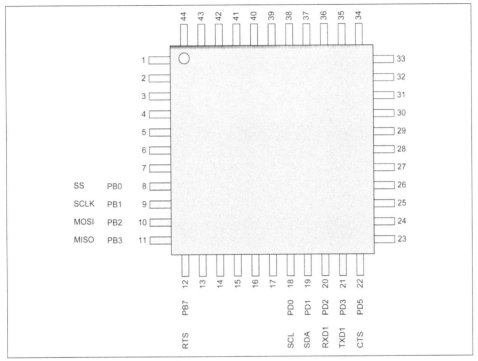

Figure 3-19. ATmega32U4 serial I/O pins

Timer/Clock I/O

The ATmega32U4 has four on-chip timer/counter functions. These consist of one on-chip 8-bit timer/counter, two 16-bit timer/counters with separate prescalers and a comparison modes, and a high-speed 10-bit timer/counter with phase-locked loop (PLL) and a compare mode. Note that the 16-bit timer/counters also support a capture mode.

The timers are numbered from 0 to 4, but there is no timer/counter 2.

The OC1A, OC1B, and T0 pins for timer/counter 0 are bought out on pins shared with timer/counter 4 (the high-speed timer/counter). Likewise, the pins for timer/counters 1 and 3 are also shared with pins available to timer/counter 4.

Table 3-23 lists the pin assignments for the timer/clock functions of the ATmega32U4 MCU. A tilde (~) indicates an active-low (low = true) pin. There are no T3 or T4 pins on the ATmega32U4. Only timer/counters 0 and 1 have T0 and T1 pins available. The pin assignments are illustrated in Figure 3-20.

Table 3-23. Timer/clock pins

Pin	Port	Function	Pin	Port	Function
12	PB7	OC0A/OC1C	29	PB5	OC1A/~OC4B
18	PD0	OC0B	30	PB6	OC1B/OC4B
25	PD4	ICP1	31	PC6	OC3A/OC4A
26	PD6	T1/~OC4D	32	PC7	ICP3/~OC4A

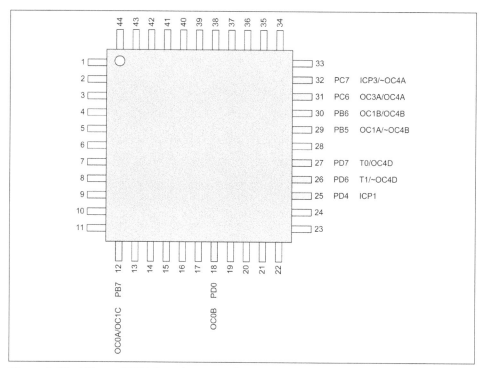

Figure 3-20. A Tmega32U4 timer/counter pins

External Interrupts

The interrupt pin assignments for the ATmega32U4 MCU are listed in Table 3-24 and shown in Figure 3-21. The port D pins PD0 through PD3 and the port E pin PE6 are specifically intended for use as external interrupt inputs. Any of the PCINT0 to PCINT7 pins may also be used as external interrupt inputs without interfering with other assigned functions (see "Interrupts" on page 26 for an overview of how these interrupts can be used).

Table 3-24. External interrupt pins

Pin	Port	Function	Pin	Port	Function
8	PB0	PCINT0	12	PB7	PCINT7
9	PB1	PCINT1	18	PD0	INT0
10	PD2	PCINT2	19	PD1	INT1
11	PB3	PCINT3	20	PD2	INT2
28	PB4	PCINT4	21	PD3	INT3
29	PB5	PCINT5	1	PE6	INT6

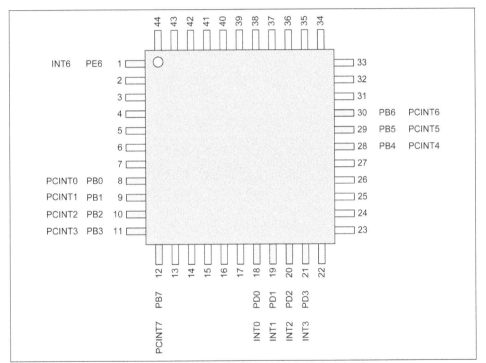

Figure 3-21. ATmega32U4 external interrupt pins

USB 2.0 Interface

The ATmega32U4 incorporates a full-speed USB 2.0 interface, but it cannot be a USB host, only a device. It provides multiple internal endpoints with configurable FIFO buffers. An on-chip PLL generates a 48 MHz clock for the USB interface, and the PLL input can be configured to use an external oscillator, an external clock source, or an internal RC clock source. The 48 MHz output from the PLL is used to generate either a 12 MHz full-speed or 1.5 MHz low-speed clock.

Note that the USB pins on the ATmega32U4 are not associated with a port, only with the internal USB logic and voltage regulator circuitry. The USB I/O pin assignments are listed in Table 3-25 and shown in Figure 3-22.

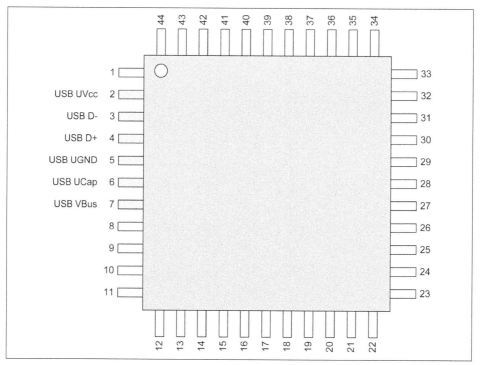

Figure 3-22. ATmega32U4 USB pins

Table 3-25. USB I/O

Pin	Function	Pin	Function
2	USB UVCC	5	USB UGnd
3	USB D−	6	USB UCap
4	USB D+	7	USB VBus

Electrical Characteristics

Table 3-26 shows the essential maximum electrical ratings for an ATmega32U4, and Table 3-27 shows the basic current consumption parameters.

Table 3-26. ATmega32U4 maximum ratings

Parameter	Value(s)	Units
I/O pin voltage	−0.5 to VCC + 0.5	V
Reset pin voltage	−0.5 to +13.0	V
VBUS pin voltage	−0.5 to +6.0	V
Device VCC	6.0	V
I/O pin DC current	40.0	mA
VCC current	200.0	mA

Table 3-27. ATmega32U4 power consumption characteristics

Parameter	Condition	VCC	Typical	Max
Power supply current	Active 8 MHz	5V	10 mA	15 mA
	Idle 8 MHz	5V		6 mA
Power-down mode	WDT enabled	3V	<10 uA	12 uA
	WDT disabled	3V	1 uA	5 uA

Arduino Pin Assignments

The Leonardo Arduino board uses the extended version of the baseline layout, which includes additional pins. The USB interface pins are connected directly to the USB connector on the PCB, and the USB signals do not appear on the pin socket headers on the sides of the PCB. The Leonardo board layout is described in Chapter 4.

The chip-to-Leonardo pin mapping is given in Table 3-28. Note that not all of the ATmega32U4 pins are brought out, and some of the ADC input pins are pressed into service as discrete digital I/O pins.

Table 3-28. Leonardo ATmega32U4 pin mapping

Leonardo pin	Chip pin	Functions	Leonardo pin	Chip pin	Functions
0	20	RXD1/INT2	11	12	PCINT7/OC0A/OC1C/RTS
1	21	TXD1/INT3	12	26	T1/OC4D/ADC9
2	19	SDA/INT1	13	32	ICP3/CLK0/OC4A
3	18	OC0B/SCL/INT0	A0	36	ADC7/TDI
4	25	ICP1/ADC8	A1	37	ADC6/TD0
5	31	OC3A/OC4A	A2	38	ADC5/TMS
6	27	T0/OC4D/ADC10	A3	39	ADC4/TCK
7	1	INT6/AIN0	A4	40	ADC1
8	28	PCINT4/ADC11	A5	41	ADC0
9	29	PCINT5/OC1A/OC4B/ADC12	AREF	42	AREF

Fuse Bits

The AVR MCUs use a set of so-called fuses to set various parameters such as clock source, timing divisors, memory access locks, and so on. You can think of it as a big panel of switches. In this section we'll take a look at the fuse bits for the ATmega168/328. This is the MCU used in the Duemilanove, Mini, Nano, and Uno boards, and the DIP (dual in-line pin) version is commonly used in AVR projects. The general concepts apply to the ATmega1280/2560 and the ATmega32U4 as well.

There are three bytes in the ATmega168/328 for fuse bits: low, high, and extended. A fuse bit is set to zero when it is programmed (i.e., they are active-low logic). Tables 3-29, 3-30, and 3-31 lists the AVR fuse bits and their functions.

Table 3-29. Low fuse byte

Bit name	Bit no.	Description	Def. value	Bit name	Bit no.	Description	Def. value
CKDIV8	7	Clock div by 8	0	CKSEL3	3	Clock source	0
CKOUT	6	Clock output	1	CKSEL2	2	Clock source	0
SUT1	5	Start-up time	1	CKSEL1	1	Clock source	1
SUT0	4	Start-up time	0	CKSEL0	0	Clock source	0

Table 3-30. High fuse byte

Bit name	Bit no.	Description	Def. value	Bit name	Bit no.	Description	Def. value
RSTDISBL	7	External reset disable	1	EESAVE	3	EEPROM preserved	1
DWEN	6	debugWIRE enable	1	BODLEVEL2	2	BOD trigger level	1
SPIEN	5	Enable SPI download	0	BODLEVEL1	1	BOD trigger level	1
WDTON	4	Watchdog timer on	1	BODLEVEL0	0	BOD trigger level	1

Table 3-31. Extended fuse byte

Bit name	Bit no.	Description	Def. value	Bit name	Bit no.	Description	Def. value
-	7	-	1	-	3	-	1
-	6	-	1	BOOTSZ1	2	Boot size	0
-	5	-	1	BOOTSZ0	1	Boot size	0
-	4	-	1	BOOTRST	0	Reset vector	1

The AVR employs logic for clock input and output routing called the "AVR Clock Control Unit," as shown in Figure 3-23.

The timing route "clk" names refer to the timing function of each path. See the Atmel datasheet for the ATmega328 (*http://bit.ly/mc-atmega328*) for details. The clock source multiplexer can be configured with the fuse bits CKSEL(3:0), or the least significant 4-bit nibble. Table 3-32 lists the possible clock source options and bit values for the CKSEL bits.

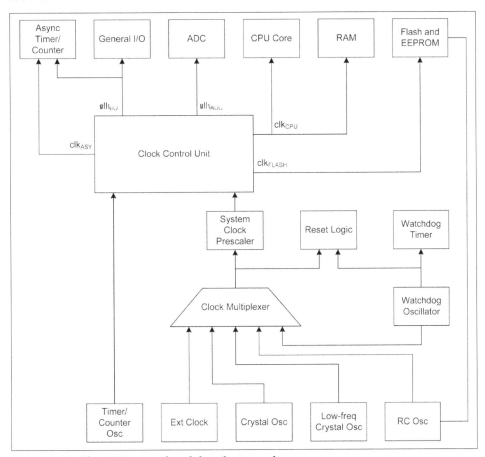

Figure 3-23. The AVR control and distribution subsystem

Table 3-32. Clock source selection using CKSEL fuse bits

Clock source	CKSEL(3:0)	Clock source	CKSEL(3:0)
Low power crystal	1111 - 1000	Calibrated internal	0010
Full swing crystal	0111 - 0110	External clock	0000
Low frequency crystal	0101 - 0100	Not used (reserved)	0001

Note that the crystal clock sources have multiple suboptions, depending on which bits are set. In Chapter 10 these fuse bit are used to configure a brand-new AVR on a custom PCB.

In an Arduino-based device or system the CKSEL bits probably shouldn't be altered, but other fuse bits can be set to provide useful functions. In the high fuse byte, the WDTON and BODLEVEL fuse bits could be used to enable the AVR's watchdog timer and set the BOD (brown-out detection) response level. Both the watchdog timer and the brown-out detection circuit will cause an MCU reset.

The watchdog is actually quite useful, particularly in situations where a fault of some sort might cause bigger problems. So long as the watchdog control register is reset before time-out, usually in the main loop, it will not trigger an interrupt. If some part of the code becomes nonresponsive the main loop will halt, and the watchdog timer will run out. *avr-libc* provides the functions `wdt_enable()`, `wdt_disable()`, and `wdt_reset()` for watchdog control.

The brown-out detection generates an interrupt if the VCC DC supply voltage drops to within a specific range. This may provide enough time to save critical data to attached flash storage or change some discrete digital outputs.

Fuse bits can be set using programming tools like those described in "In-System Programming" on page 147 in Chapter 6. The sidebar "Setting the AVR MCU Fuse Bits for a 16 MHz Crystal" on page 418 describes how to set the fuse bits for a new AVR MCU IC to run with a 16 MHz clock.

This is just a high-level look at fuse bits. As always, refer to the Atmel documentation for detailed descriptions of the watchdog timer and brown-out detection levels.

For More Information

Chapter 2 contains overviews of the internal peripheral functions found in AVR microcontrollers. Chapter 4 contains Arduino board pinouts for the AVR MCU functions.

For detailed information about the internal peripheral functions of AVR microcontrollers, refer to the following Atmel documents (available from Atmel's website, under Support→Datasheets):

- Datasheet: ATmega48PA/ATmega88PA/ATmega168PA/ATmega328P
  8-bit Microcontroller with 4/8/16/32K Bytes In-System Programmable Flash

- Datasheet: Atmel ATmega640/V-1280/V-1281/V-2560/V-2561/V
  8-bit Atmel Microcontroller with 16/32/64KB In-System Programmable Flash

- Datasheet: ATmega16U4/ATmega32U4
  8-bit Microcontroller with 16/32K Bytes of ISP Flash and USB Controller

CHAPTER 4

Arduino Technical Details

This chapter describes the general physical and electrical characteristics of specific Arduino boards, from the Diecimila through recent types like the Leonardo, Esplora, and Micro. Topics covered include pinout descriptions and the physical dimensions of most current Arduino models, from the so-called baseline types like the Uno, to the large form-factor Mega boards and the unique Esplora, to the small-outline boards such as the Mini, Micro, and Nano models.

Arduino Features and Capabilities

Table 4-1 is a comparison of the most common Arduino board types. If you compare this table with the tables in Chapter 1 it is obvious that the basic capabilities of an Arduino board are the capabilities supplied by its microcontroller. However, because the Arduino designs allocate certain pins on the AVR processors to specific functions, or don't bring out all of the processor's pins, not all of the capabilities of the microcontrollers are available at the terminals of an Arduino.

The term "pin" is used in this and other sections when referring to the pin sockets on an Arduino. This is mainly to maintain consistency with the terminology encountered elsewhere, but it's not completely technically correct. The connection points on an Arduino board are sockets, and the jumpers and shields that plug into these sockets are the actual pins. You can think of a "pin" as a connection point of some sort, be it a lead on an IC package, a position on a 0.1 inch (2.54 mm) socket header, or the pins extending from the bottom of a shield PCB.

Table 4-1. Arduino hardware features

Board name	Processor	VCC (V)	Clock (MHz)	AIN pins	DIO pins	PWM pins	USB
ArduinoBT	ATmega328	5	16	6	14	6	None
Duemilanove	ATmega168	5	16	6	14	6	Regular
Duemilanove	ATmega328	5	16	6	14	6	Regular
Diecimila	ATmega168	5	16	6	14	6	Regular
Esplora	ATmega32U4	5	16	-	-	-	Micro
Ethernet	ATmega328	5	16	6	14	4	Regular
Fio	ATmega328P	3.3	8	8	14	6	Mini
Leonardo	ATmega32U4	5	16	12	20	7	Micro
LilyPad	ATmega168V	2.7–5.5	8	6	14	6	None
LilyPad	ATmega328V	2.7–5.5	8	6	14	6	None
Mega	ATmega1280	5	16	16	54	15	Regular
Mega ADK	ATmega2560	5	16	16	54	15	Regular
Mega 2560	ATmega2560	5	16	16	54	15	Regular
Micro	ATmega32U4	5	16	12	20	7	Micro
Mini	ATmega328	5	16	8	14	6	None
Mini Pro	ATmega168	3.3	8	6	14	6	None
Mini Pro	ATmega168	5	16	6	14	6	None
Nano	ATmega168	5	16	8	14	6	Mini-B
Nano	ATmega328	5	16	8	14	6	Mini-B
Pro (168)	ATmega168	3.3	8	6	14	6	None
Pro (328)	ATmega328	5	16	6	14	6	None
Uno	ATmega328	5	16	6	14	6	Regular
Yún	ATmega32U4	5	16	12	20	7	Host (A)

[a] Analog inputs.

[b] Digital I/O.

[c] Pulse-width modulation outputs (alternate DIO pin functions).

Arduino USB Interfaces

Starting with the Leonardo board (2012), the ATmega32U4 XMEGA microcontroller has been used as the primary processor. This part has a built-in USB interface, which eliminates the need for the additional chip seen on earlier Arduino models with a USB interface. The Leonardo (2012), Esplora (2012), Micro (2012), and Yún (2013) all use the ATmega32U4 processor.

The older Arduino models with USB used an FTDI interface chip (the FT232RL), an ATmega8 (Uno), or an ATmega16U2 (Mega2560 and Uno R3). The FT232RL converts between standard serial (such as RS-232) and USB. In the Uno, Uno R3, and

Mega2560 the additional small ATmega processors are preprogrammed to serve as a USB interface. The operation of these parts is transparent when using the Arduino IDE to create and load program sketches.

Those boards that do not have a USB interface must be programmed using an external adapter.

Arduino types that use the FTDI FT232RL serial-to-USB interface chip are essentially identical internally, and consist of a DC voltage regulation circuit and two ICs. Figure 4-1 shows a block diagram of the Diecimila and Duemilanove models with an FTDI interface chip.

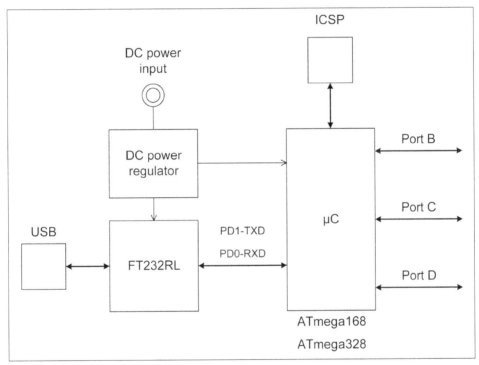

Figure 4-1. FTDI USB interface

Since around 2010, the Uno R2 and Uno SMD boards have employed the ATmega16U2 part instead of the FTDI FT232RL for the USB interface. The Uno R3 also has the ATmega16U2 serving as the USB interface. The ATmega16U2 incorporates a built-in USB 2.0 interface and is basically the same as the ATmega32U4, just with less memory. Figure 4-2 shows a block diagram of the Uno R2 with an ATmega16U2 providing the USB interface. The Uno, with an ATmega8, has the same internal functional arrangement as the Uno R2, just with a different MCU serving as the USB interface.

Arduino Physical Dimensions

Arduino boards come in a variety of shapes and sizes, but generally they can be organized into four groups: full-size or baseline boards, mega-size boards, small form-factor boards, and special-purpose boards.

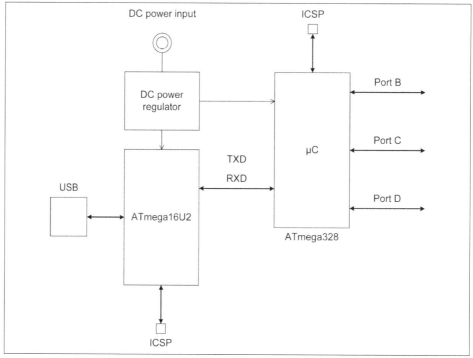

Figure 4-2. A Tmega16U2 USB interface

 The board dimensions given in this section, while generally close, are approximate, as there may be some slight variations between boards from different sources. Refer to the PCB layout from Arduino.cc, which is available for each board, if you need accurate dimensions. Or better yet, just take the measurements from an actual board.

To give some idea of scale, Figure 4-3 shows a lineup of several common Arduino boards. Shown here in clockwise order from the lower left are a Duemilanove, a Leonardo, a clone Mega2560 with an extended I/O pin layout from SainSmart, and an official Arduino Mega2560, with an Arduino Nano sitting in the center.

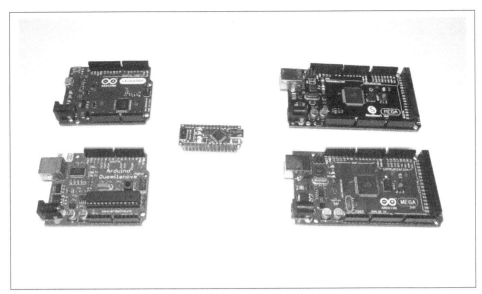

Figure 4-3. Comparison of Arduino and clone boards

Full-Size Baseline Arduino PCB Types

Figure 4-4 shows the physical layouts for six different baseline Arduino boards, from the Diecimila to the Leonardo. In between there are the Duemilanove and Uno variants. Baseline, in this context, refers to the "classic" Arduino PCB layout that determines the physical design of most shields and other add-on components. The functions of the I/O and other pins on each PCB are described in "Arduino Pinout Configurations" on page 73.

With the Diecimila, the Duemilanove, the Uno R2 (revision 2), and the Uno SMD the arrangement of the I/O socket headers along the edges of the PCBs is unchanged. This book refers to this as the baseline Arduino form factor. Also, starting with the Uno R2 a new block of six pins appeared on the PCB, in addition to the block that already existed on earlier boards. This is the ICSP (In-Circuit Serial Programming) interface for the ATmega16U2 processor that is used for the USB interface. The Uno SMD also has this new ICSP interface.

The Uno R3 introduced the new extended I/O pin configuration. This is a backward-compatible extension, meaning that a shield intended for an older model like a Duemilanove will still work with the newer boards. The extension only adds new signal pin sockets, but no new signals, and it doesn't alter any of the pin functions found in the baseline layout. The Leonardo PCB uses the ATmega32U4 processor, which has built-in USB support, so there is only one microcontroller IC on a Leonardo PCB and only one ICSP port. It has the same I/O pin layout as earlier boards, although the actual microcontroller ports used are different.

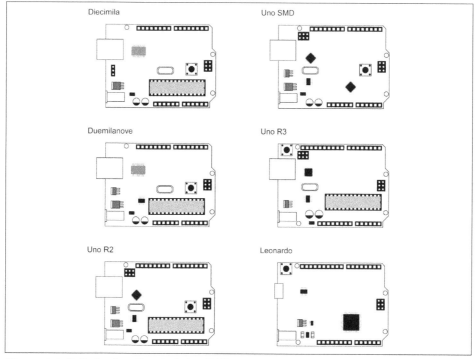

Figure 4-4. Full-size baseline Arduino PCB types

All full-size baseline Arduino boards have the same physical dimensions, as shown in Figure 4-5. The locations of the mounting holes for the PCB vary slightly between models depending on the version of the board, but the overall exterior dimensions are consistent.

Mega Form-Factor Arduino PCB Types

The Mega form-factor boards incorporate the baseline pinout along with additional pins to accommodate the extended capabilities of the ATmega1280 and ATmega2560 microcontrollers (these devices are described in "ATmega1280/ATmega2560" on page 39 in Chapter 3).

Mega and Mega2560

The Mega and Mega2560 are essentially the same layout, with the primary difference being the type of AVR device on the boards. The Mega2560 replaces the Mega, which is no longer produced by Arduino.cc, although some second-source clone boards are still available. Considering the enhanced memory of the Mega2560, there really isn't any reason to purchase a Mega.

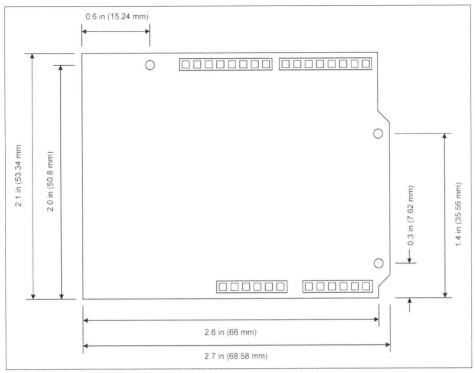

Figure 4-5. Baseline and extended board dimensions

Figure 4-6 shows the overall dimensions of a Mega or Mega2560 board. Note that a baseline (Uno, Leonardo, etc.) shield will work with a Mega board. The I/O pins on the Mega are arranged such that the basic digital I/O and the A/D inputs 0 through 5 are compatible with the baseline pin layout.

Mega ADK

The Mega ADK is based on the Mega2560, but features a USB host interface that allows it to connect to Android phones and similar devices. Other than an additional USB connector located between the B type USB connector and the DC power jack, it is identical to the Mega2560 in terms of dimensions. Like with the Mega2560, a standard baseline-type shield can be used with the Mega ADK.

Small Form-Factor Arduino PCB Types

The full-size Arduino boards were the first to make an appearance around 2005, and by 2007 the layout had settled into the baseline and extended forms seen today. But the Arduino team and their partners realized that the full-size board just wouldn't work for some applications, so they came up with the miniature formats.

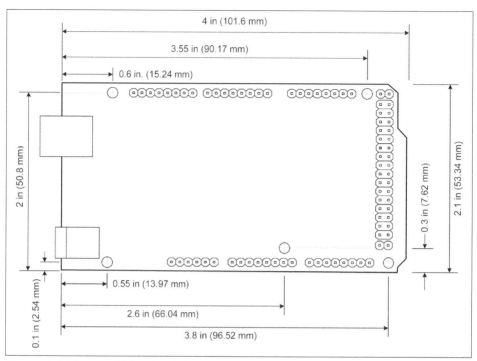

Figure 4-6. Mega and Mega2560 dimensions

The miniature boards include the Mini, Micro, Nano, and Fio layouts. These PCBs are smaller in both width and length, but still have the same AVR processors as the full-size types.

Mini

The Mini is intended for use on breadboards or in other applications where space is limited. It does not have a USB connector, and an external programmer interface must be used to transfer executable code to the microcontroller. Its dimensions are shown in Figure 4-7.

Pro Mini

The Pro Mini is similar to the Mini with regard to pin layout and form factor, but unlike the Mini it is intended for permanent or semipermanent installation. The Pro Mini was designed and manufactured by SparkFun Electronics. Its dimensions are shown in Figure 4-8.

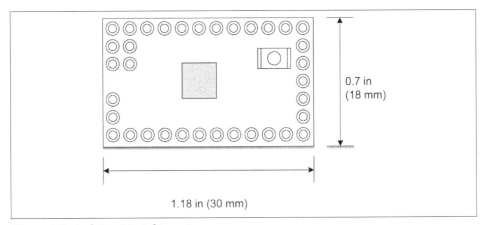

Figure 4-7. Arduino Mini dimensions

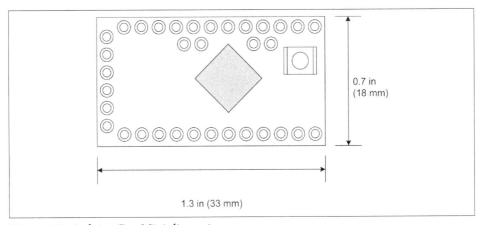

Figure 4-8. Arduino Pro Mini dimensions

Nano

Similar to the Mini, the Nano is a small form-factor board suitable for use with solderless breadboards and as a plug-in module for a larger PCB. It was designed and produced by Gravitech. Its dimensions are given in Figure 4-9.

Fio

The Fio is intended for wireless applications, primarily XBee, and as such it lacks some of the direct connection programmability of other Arduino types. A Fio can be programmed using a serial-to-USB adapter or wirelessly using a USB-to-XBee adapter. It was designed and manufactured by SparkFun Electronics; its dimensions are shown in Figure 4-10.

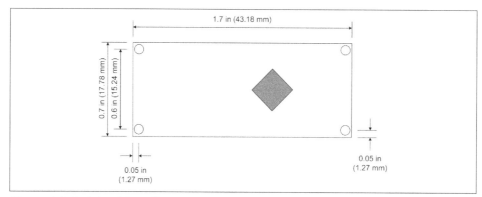

Figure 4-9. Arduino Nano dimensions

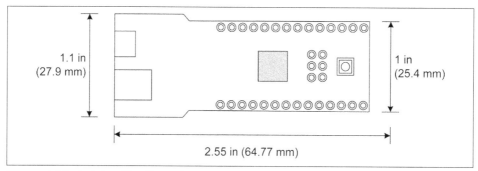

Figure 4-10. Arduino Fio dimensions

Micro

The Micro employs a DIP (dual in-line pin) form factor and uses an ATmega32U4 processor, which is identical to the Leonardo board. Like the Nano, the Micro is suitable for use with a solderless breadboard and as a plug-in module using a conventional IC socket. It was developed in conjunction with Adafruit. The Micro's dimensions are shown in Figure 4-11.

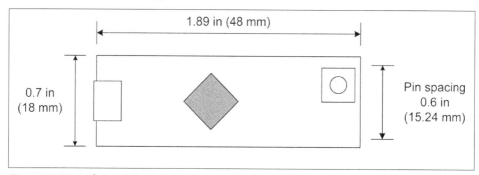

Figure 4-11. Arduino Micro dimensions

Special-Purpose PCB Types

Arduino boards aren't limited to simple shapes like rectangles. The LilyPad is a small disk with connection points arranged around the edge. It can be integrated into clothing to build wearable creations. The Esplora is physically configured like a conventional game controller, although as it is an Arduino it can be programmed to do much more than just play games.

LilyPad

The LilyPad and its variations are intended for wearable applications. The board itself measures about 2 inches (50 mm) in diameter, as shown in Figure 4-12.

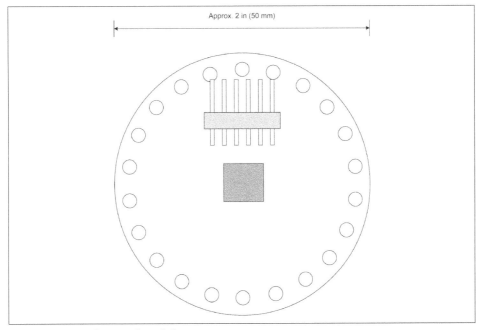

Figure 4-12. Arduino LilyPad dimensions

Esplora

The Esplora is supplied with four pushbuttons, a switch-type joystick, and a micro USB connector. Four mounting holes are available to affix the board to a chassis or panel. The Esplora PCB dimensions are shown in Figure 4-13.

Arduino Pinout Configurations

When creating a shield board for the Arduino, the convention is to follow the common baseline pin layout pattern described here. This configuration is found on the "standard" baseline Arduino boards built between 2007 and 2012. Boards using the newer "extended" pin layout (the Uno R3 and Leonardo), as well as the "Mega" boards, also support the baseline connections, but add new capabilities by extending the rows of terminals along the sides of the PCBs.

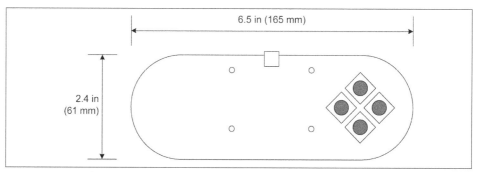

Figure 4-13. Arduino Esplora dimensions

The Baseline Arduino Pin Layout

The baseline Arduino pin layout as it exists today appeared with the Diecimila model. Over the years it has become a de facto standard upon which numerous shield boards have been based. The Arduino boards that utilize the baseline pin layout are listed in Table 4-2.

Table 4-2. Baseline layout Arduino boards

Board name	Year	Microcontroller
Diecimila	2007	ATmega168
Duemilanove	2008	ATmega168/ATmega328
Uno (R2 and SMD)	2010	ATmega328P

Figure 4-14 shows the pinout of a full-size baseline Arduino board. This includes the Diecimila, Duemilanove, Uno R2, and Uno SMD models. The gray boxes in Figure 4-14 give the chip pin number and port designations for the ATmega168/328 parts.

The common baseline I/O and power pin layout for the Arduino consists of 14 discrete digital I/O pins, an analog reference, 3 ground pins, 6 analog input pins, pins for 3.3V and 5V, and a reset line. As shown in Figure 4-14, these pins are arranged as two eight-position connectors and two six-position connectors along the sides of the PCB.

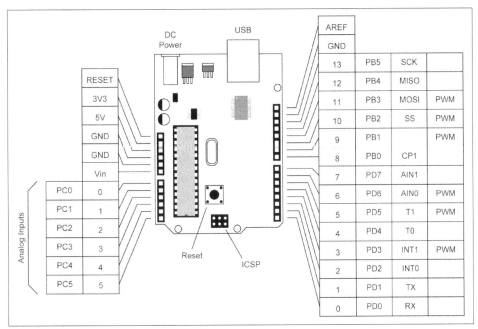

Figure 4-14. Pin functions of standard baseline Arduino boards

From a programming perspective, each interface pin on a Diecimila, Duemilanove, Uno R2, or Uno SMD PCB has a predefined name used to identify it in software. These names are reflected by the labels screened onto the Arduino PCB. Table 4-3 lists the pin assignments for a baseline or R2 Arduino with an ATmega168 or ATmega328 MCU. See the pin assignments for the Arduino Ethernet (Table 4-3) for the Uno SMD board.

Table 4-3. Arduino ATmega168/328 pin assignments

Digital pin (Dn)	Analog pin (An)	AVR pin	AVR port	AVR function(s)	AVR PWM
0	2	PD0	RxD		
1	3	PD1	TxD		
2	4	PD2	INT0		
3	5	PD3	INT1, OC2B	Yes	
4	6	PD4	T0, XCK		
5	11	PD5	T1	Yes	
6	12	PD6	AIN0	Yes	
7	13	PD7	AIN1		
8	14	PB0	CLK0, ICP1		
9	15	PB1	OC1A	Yes	

Digital pin (Dn)	Analog pin (An)	AVR pin	AVR port	AVR function(s)	AVR PWM
10		16	PB2	OC1B, SS	Yes
11		17	PB3	OC2A, MOSI	Yes
12		18	PB4	MISO	
13		19	PB5	SCK	
14	0	23	PC0		
15	1	24	PC1		
16	2	25	PC2		
17	3	26	PC3		
18	4	27	PC4	SDA	
19	5	28	PC5	SCL	

The Extended Baseline Pin Layout

Starting with the R3 version of the Uno, four additional pins appeared on the Arduino PCB. Two of these are near the relocated reset button and provide additional connections for I2C (the SCL and SDA lines). The other two appeared next to the reset connection on the opposite side of the board. One is designated as IOREF (the nominal I/O voltage, may be either 3.3V or 5V depending on board type) and the other is not presently connected. Table 4-4 lists the extended baseline layout boards.

Table 4-4. Extended layout Arduino boards

Board name	Year	Microcontroller
Uno R3	2010	ATmega328
Ethernet	2011	ATmega328
Leonardo	2012	ATmega32U4

Uno R3

Like the Uno R2 and Uno SMD, the Uno R3 utilizes a second microcontroller to handle USB communications. The Arduino Ethernet does not have built-in USB. Figure 4-15 shows the block diagram for the Uno R3 and Uno SMD boards.

The pin functions for the Uno R3 are shown in Figure 4-16.

The extended baseline (R3) Arduino boards with the ATmega328 MCU have the same pin assignments as given in Table 4-3, but with the additional pins for ADC4 and ADC5 (A4 and A5). The Leonardo pin functions are defined in "Leonardo" on page 79.

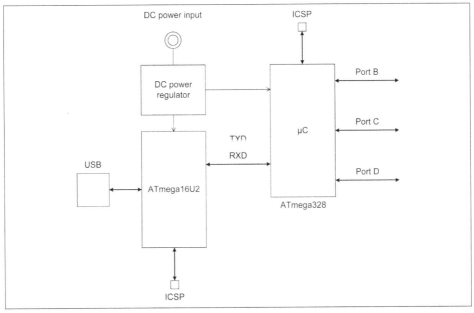

Figure 4-15. ATmega16U2 USB interface

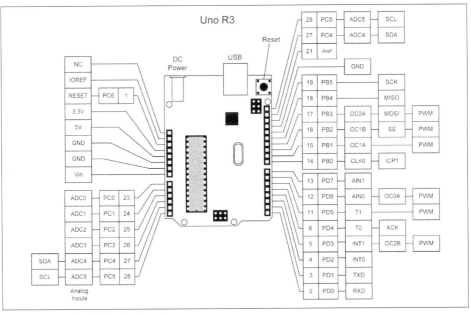

Figure 4-16. Uno R3 pin functions

Ethernet

The Ethernet deviates from the Arduino conventions seen up through the Uno R3 with its inclusion of a 100Mb Ethernet interface and an RJ45 jack. It has no USB interface. The MCU is a surface-mount version of the ATmega328, with different pin functions and numbering from the ATmega328. A WIZnet W5100 chip is used for the Ethernet interface. Figure 4-17 shows a block diagram of the Ethernet board.

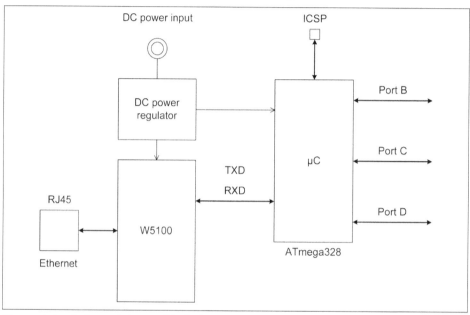

Figure 4-17. Arduino Ethernet block diagram

An FTDI-type interface is used to program the Ethernet with an adapter, like the SparkFun or Adafruit FTDI-type devices. This interface is brought out on a right-angle six-pin header along one edge of the PCB next to the microSD carrier. Figure 4-18 shows the pinouts of the Ethernet board.

This product has been retired by Arduino.cc, but it is still available from multiple sources. Ethernet connectivity can be obtained by using an Ethernet shield (see Chapter 8 for more details on shields).

Table 4-5 lists the pin assignments for the Arduino Ethernet. Note that pins 10, 11, 12, and 13 are reserved for the Ethernet interface and are not available for general-purpose use.

Table 4-5. Arduino Ethernet pin assignments

Digital pin (Dn)	Analog pin (An)	AVR pin	AVR port	AVR function(s)	AVR PWM
0		30	PD0	RxD	
1		31	PD1	TxD	
2		32	PD2	INT0	
3		1	PD3	INT1, OC2B	Yes
4		2	PD4	T0, XCK	
5		9	PD5	T1, OC0B	Yes
6		10	PD6	AIN0, OC0A	Yes
7		11	PD7	AIN1	
8		12	PB0	CLK0, ICP1	
9		13	PB1	OC1A	Yes
10		14	PB2	OC1B, SS	Yes
11		15	PB3	OC2A, MOSI	Yes
12		16	PB4	MISO	
13		17	PB5	SCK	
14	0	23	PC0		
15	1	24	PC1		
16	2	25	PC2		
17	3	26	PC3		
18	4	27	PC4	SDA	
19	5	28	PC5	SCL	

Leonardo

The Leonardo introduced the ATmega32U4 processor, which contains a built-in USB interface and enhanced functionality. This simplified the PCB layout, as can be seen in Figure 4-19. Also, note that the Leonardo uses a mini-USB connector instead of the full-size type B connector found on older Arduino boards. This was a much-needed change, and it allows the Leonardo to work with shields that would have interfered with the B type USB connector on the older models.

The Uno R3 and the Leonardo both use the same PCB pin layout, but some of the microcontroller functions are different. In the Arduino IDE this is handled by using a set of definitions specific to each board type to map functions to specific pins.

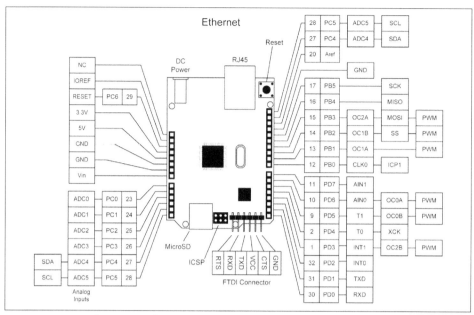

Figure 4-18. Pin functions of Arduino Ethernet board

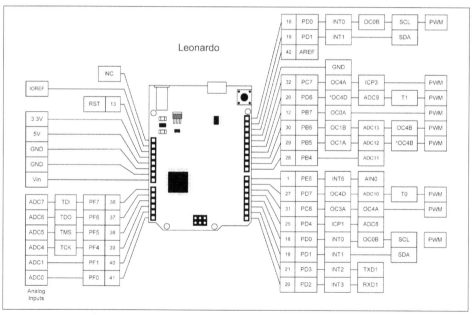

Figure 4-19. Pin functions of Arduino Leonardo board

Table 4-6 lists the pin assignments for an extended or R3 Arduino with an ATmega32U4 MCU.

Table 4-6. Arduino ATmega32U4 pin assignments

Digital pin (Dn)	Analog pin (An)	AVR pin	AVR port	AVR function(s)	AVR PWM
0		20	PD2	INT3, RxD1	
1		21	PD3	INT2, TxD1	
2		19	PD1	INT1, SDA	
3		18	PD0	INT0, OC0B, SCL	Yes
4		25	PD4	ICP1, ADC8	
5		31	PC6	OC3A, OC4A	Yes
6		27	PD7	OC4D, ADC10, T0	Yes
7		1	PE6	INT6, AIN0	
8		28	PB4	ADC11	
9		29	PB5	OC1A, ADC12, *OC4B	Yes
10		30	PB6	OC1B, ADC13, OC4B	Yes
11		12	PB7	OC0A	Yes
12		26	PD6	*OC4D, ADC9, T1	Yes
13		32	PC7	OC4A, ICP3	Yes
14	0	36	PF7	TDI	
15	1	37	PF6	TDO	
16	2	38	PF5	TMS	
17	3	39	PF4	TCK	
18	4	40	PF1		
19	5	41	PF0		

The Mega Pin Layout

The Mega series (which use the ATmega1280 and ATmega2560 processors) also incorporate the standard pinout pattern, but include additional pins to accommodate the extended I/O capabilities of the larger processors. The Mega pin layout is shown in Figure 4-20. The boards that utilize this layout are listed in Table 4-7. Most common shields will work with the Mega boards.

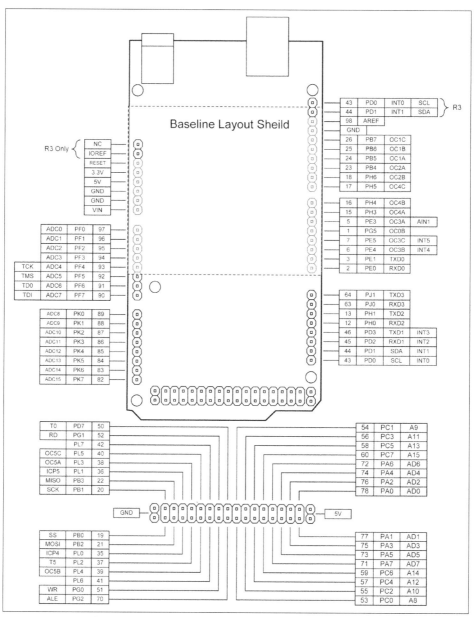

Figure 4-20. Pin functions of the Arduino Mega series boards

In Figure 4-20, the PCINT pin functions are not shown for the sake of clarity. Also note that the R3 version of the Mega2560 contains pins not found on earlier versions, but these do not interfere with baseline layout–style shields.

Table 4-7. Mega pin layout boards

Board name	Year	Microcontroller
Mega	2009	ATmega1280
Mega2560	2010	ATmega2560
Mega ADK	2011	ATmega2560

Nonstandard Layouts

In terms of nonstandard pinout configurations (nonstandard in the sense of being physically incompatible with conventional Arduino shields), the most radical is the LilyPad, with its circular form factor and use of solder pads for connections. The small form-factor Nano, Mini, Mini Pro, and Micro have pins soldered to the under-side of the board, and are suitable for use with a solderless breadboard block or as a component on a large PCB. The Fio uses solder pads with spacing compatible with standard header pin strips, and the Esplora has a game controller–type form factor. None of the nonstandard layout boards can be used directly with a standard shield. The boards that fall into this category are listed in Table 4-8, and the pin functions for these boards are shown in Figures 4-21 through 4-27 in the following sections.

Table 4-8. Nonstandard pin layout boards

Board name	Year	Microcontroller
LilyPad	2007	ATmega168V/ATmega328V
Nano	2008	ATmega328/ATmega168
Mini	2008	ATmega168
Pro Mini	2008	ATmega328
Fio	2010	ATmega328P
Esplora	2012	ATmega32U4
Micro	2012	ATmega32U4

LilyPad

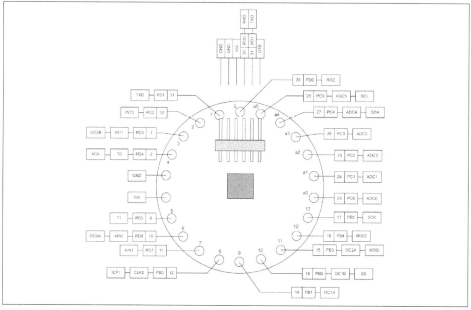

Figure 4-21. Pin functions of the Arduino LilyPad

Nano

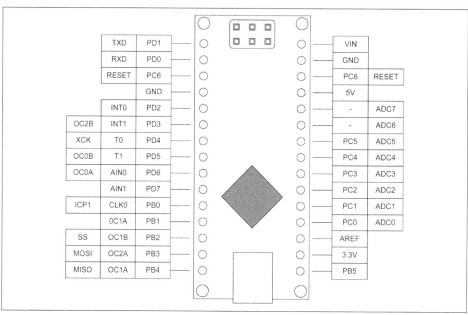

Figure 4-22. Nano pin functions

Mini

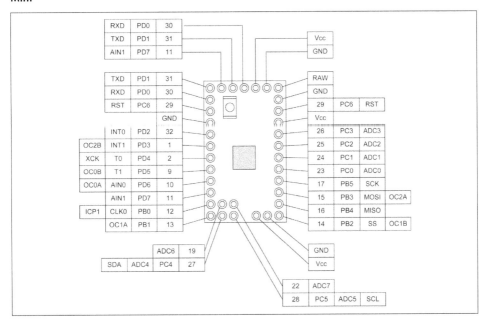

Figure 4-23. Mini pin functions

Pro Mini

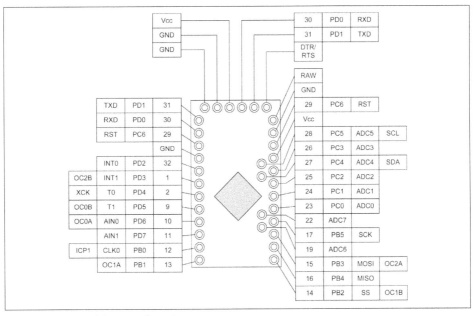

Figure 4-24. Pro Mini pin functions

Fio

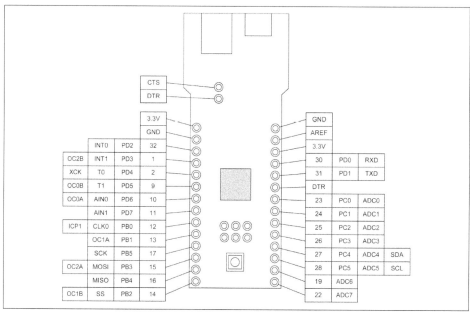

Figure 4-25. Fio pin functions

Esplora

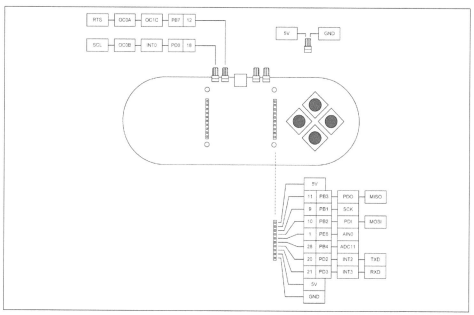

Figure 4-26. Esplora pin functions

Micro

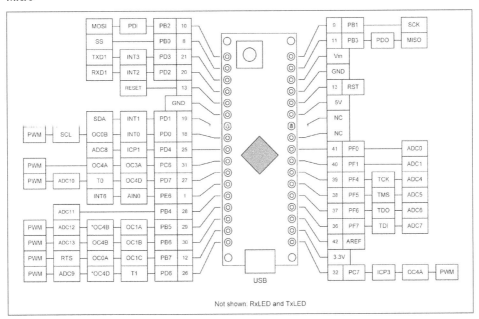

Figure 4-27. Micro pin functions

For More Information

The Atmel website (*http://www.atmel.com/avr*) offers a selection of datasheets, example software, and other resources for working with the AVR microcontrollers. Note that these are only for the AVR, not the Arduino.

Additional information about the various Arduino boards can be found on the Arduino website (*http://www.arduino.cc*).

CHAPTER 5

Programming the Arduino and AVR Microcontrollers

This chapter is an ambitious high-level tour of the tools, concepts, and techniques that you can use to create, compile, assemble, and load software onto an Arduino. There are many deep subjects covered here in broad strokes, and it would be impossible to do any of them real justice in the space of a single chapter. The goal is to provide you with enough information to get previous experience with microcontrollers, you may find out some things about the Arduino environment that you were not aware of before.

 This chapter does not describe the C or C++ languages. Those topics are covered elsewhere in great detail (refer to Appendix D for some suggested books). The intent here is to impart an understanding of how the contents of a program or a sketch are converted into binary codes that the AVR MCU, on an Arduino board or wherever it might be, can execute, and what is involved in making that happen.

This chapter starts with a short overview of cross-compiling, the technique of using a compiler and other tools on one computer system to create executable programs that can be transferred to another computer, perhaps with a completely different architecture. This is exactly what the Arduino integrated development environment is designed to do. Chapter 6 provides a more detailed look at the low-level development tools and techniques that the Arduino IDE utilizes, but here the focus is on what the Arduino IDE can do and how you can use it effectively.

Bootloaders are introduced next, with emphasis on AVR microcontrollers. The bootloader is an essential feature of the AVR family of parts used on Arduino boards, and understanding how it works can help to reduce frustration and lay the foundation for installing your own bootloader should the need arise. In Chapter 6 we will look at methods for installing a new bootloader, perhaps of your own design.

The Arduino IDE is introduced next, along with guidance on how to install the IDE on various host computers and how to configure it for your own preferences. We will also take a quick look at what is going on under the hood of the IDE, and how it evolved from earlier tools. A walk-through of a simple program illustrates the key points in the creation of an Arduino program, known as a sketch.

The chapter wraps up with an overview of the Arduino source code, and where you can get your own copy to examine. Having the source code available, both for the supplied libraries and for the IDE and its components, can help to answer questions about why something happened the way it did, and help you determine if you really want to try an alternate approach to programming an Arduino.

 Remember that since the main emphasis of this book is on the Arduino hardware and related modules, sensors, and components, the software shown is intended only to highlight key points, not to present complete ready-to-run examples. The full software listings for the examples and projects can be found on GitHub (*https://www.github.com/ardnut*).

Cross-Compiling for Microcontrollers

Like for most single-board microcontroller systems, programs for an Arduino and its AVR MCU are developed on another system of some type (Windows, Linux, or Mac) and then transferred, or uploaded, to the AVR microcontroller on the Arduino board. The development system is referred to as the *development host*, and the Arduino or other MCU-based device is called the *target*. This type of development process has been around for quite a long time, and has been used whenever it was necessary to create software for a target machine that didn't have the capability to compile code for itself.

The technique of creating software for one type of processor on a different type of system is referred to as *cross-compiling*, and it is really the only way to create compiled software for microcontroller targets. Small microcontrollers like the AVR (or any small 8-bit device, for that matter) simply don't have the resources to compile and link something like a C or C++ program. A more capable machine with resources such as a fast CPU, large-capacity disk drives, and lots of memory is used to do the compiling and linking, and then the finished program is transferred to the target for execution.

In some cases the host system might even have an emulator available for the target that allows the developer to load and test programs in a simulated environment. An emulator might not provide a 100% perfect simulation of the target MCU and its actual environment, but it can still be a useful tool for checking the basic functionality of the software before it is actually uploaded to the real MCU. There are several AVR emulators available for the Linux operating system, including *simavr* and the GNU AVR Simulator (see Appendix E for a list of Arduino and AVR software tools, with links).

Objects, Images, and Source Code

You may notice that the terms "object," "image," "executable," and "source code" get tossed around quite a bit when talking about compiling, linking, and loading software. These are all very old terms, going back to the days of punched cards, memory drums and magnetic tape, and computers that filled whole rooms.

Source code, as you might expect, refers to the input fed into a compiler or an assembler. Source code is human-readable and created using some type of text editor. For the Arduino this usually means C or C++, created using the Arduino IDE or some other editor or IDE.

The output of the compiler or assembler is referred to as *object code*, and it is really just machine code for a particular CPU. In other words, it consists of the binary values of the CPU's operation codes and any associated data in the form of so-called literal values, also in binary form. This is typically referred to as *machine language*. With modern compilers and assemblers this is what is inside the **.o* and **.obj* files found after the assembly or compilation is complete.

Object code may not be immediately executable. If a program is comprised of two or more modules, or requires object code from an external library, then the output from the compiler or assembler will contain placeholders that refer to external software.

In some cases the external code may be in source code form (such as with most of the Arduino-supplied code) and will be compiled at the same time as your sketch, or it might already be compiled into an *object library* containing one or more separate code modules. Precompiled object libraries typically have an extension of **.a*.

A tool called a *linker* is used to fill in the blanks and connect the various parts into an *executable image*. In the case of Arduino programs this usually involves the inclusion of runtime support in the form of the `main()` function and the basic AVR runtime library functions for input/output and other low-level operations.

The end result is an executable image with everything the program needs to run, all in one bundle. After conversion into an ASCII file comprised of strings of hexadecimal characters, it is uploaded into the AVR MCU, converted back into binary, and stored in the on-board flash memory for execution.

Bootloaders

Getting a program into a modern microcontroller might entail any one of several different methods, but the easiest is to let the MCU itself assist with the process. This is accomplished with a small preloaded bit of firmware called a *bootloader* (see the sidebar "Origins of Firmware" on page 92 for a quick discussion of firmware).

The AVR family of microcontrollers provides reserved space in the on-board flash memory space for a bootloader. Once the MCU has been configured to use the bootloader, the address of this special memory space is the first place the AVR MCU will look for instructions when it is powered up (or rebooted). So long as the bootloader is not overwritten with an uploaded program, it will persist in memory between on-off power cycles.

Reserving the bootloader location typically involves enabling an internal switch, or *fuse*, with a special programming device that communicates with the MCU through an ICSP or JTAG interface. The MCU examines the fuse configuration (nonvolatile configuration bits) at startup to determine how the flash memory is organized and if space has been reserved for some type of bootloader or other startup code. (For more details on the fuse bits used in the AVR, refer to "Fuse Bits" on page 60).

A key feature of the AVR devices is their ability to load program code into their internal flash memory via a bootloader and a serial interface. In the case of the ATmega32U4 (described in Chapters 2 and 3), the USB interface can be used to upload the program code into the MCU; no special programming device or auxiliary MCU is necessary.

Arduino boards—both the official products and the software-compatible boards—come with an Arduino-type bootloader already loaded into the MCU. The Arduino bootloader implements a specific protocol that allows it to recognize the Arduino IDE and perform a program data transfer from the development host to the target board. The Arduino bootloader is described in detail in "Bootloader Operation" on page 154, and some techniques for replacing the bootloader with one of your own choosing are discussed in "Replacing the Bootloader" on page 156.

Origins of Firmware

The term *firmware* is a holdover from the days when embedded computers came preloaded with programs that could not be modified in the field. In some cases a microcontroller or microprocessor might be fabricated with software already incorporated into the silicon of the chip. The developer supplied the chip manufacturer with the binary machine code, which was then incorporated into the process of creating the actual silicon chip. This was called *mask programming*, and while it did result in low-cost parts (if purchased in large enough quantities), it wasn't the most practical way to develop software. It usually required a fair amount of development support equipment, along with special versions of the target microcontroller. In some cases that meant using an in-circuit emulator (ICE), which was basically the functional equivalent of the target processor implemented using discrete logic ICs contained in a metal box with a ribbon cable to plug into the socket where the processor would eventually be placed on the circuit board. Once the code was working correctly and thoroughly tested, it could be implemented on the chip with some degree of confidence—that is, assuming that the ICE was an accurate simulation of the actual microprocessor or microcontroller. With some early versions of the tools, that wasn't always the case.

In other cases the firmware might be loaded onto a chip using a special one-shot programming device similar to what was used to program a read-only memory (ROM) memory chip. In fact, these programming devices could usually handle both ROM and MCU devices. The program was stored in the chip in the form of memory bits that could be set just once to represent values of one or zero. These were known as OTP, or one-time programmable, parts. A one-time programmable part was just that; once it was programmed that was how it was going to stay.

If the estimated production quantities of chips could not justify the development and setup costs of the mask approach, then OTP was a reasonable alternative. In small production environments it was not uncommon to use devices called *gang programmers* with rows of special sockets to program multiple OTP processors at the same time.

However, OTP still required a lot of up-front development and testing effort using special tools. While loading buggy firmware into OTP parts wasn't as financially disastrous as ordering a production run of 10,000 defective mask-programmed parts, it was still a bad day when OTP parts ended up in the trash because of a subtle error in the code.

As technology progressed, newer devices incorporated UV-erasable programmable read-only memory (UV-EPROM, usually found in parts in a DIP, or dual in-line pin, package with a clear glass window for UV light to shine on the chip and reset the memory bits) and electrically erasable programmable read-only memory (EEPROM). Both of these types of parts required special tools to erase the contents of the read-only memory and load new data, but loading bad code only meant going through the erase-load-test-grumble cycle again instead of tossing the device into the trash.

Most of the latest generations of microcontrollers now use flash memory, but there are still OTP parts available. One of the major advantages of flash memory is that it is read/write: it can be written to, as well as read from, while the processor is running. Another major advantage is that it is nonvolatile, meaning that the contents will persist between power cycles. The concept of firmware based on nonvolatile read-only memory that required special tools to modify is becoming a thing of the past, but anything loaded into the flash memory of a microcontroller is still referred to as firmware.

The Arduino IDE Environment

As of the date of this writing, the most recent version of the Arduino IDE is 1.6.4. It looks pretty much the same on each platform. Figure 5-1 shows the initial screen that appears when the IDE is launched.

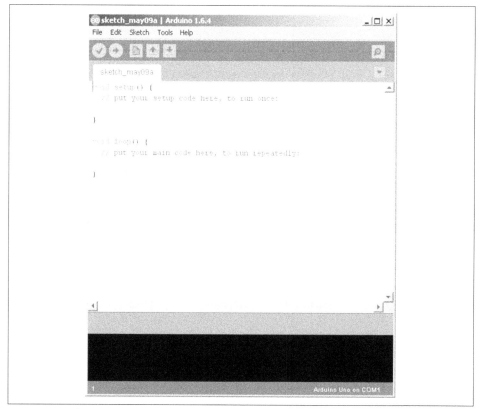

Figure 5-1. The Arduino IDE main screen

The latest version of the IDE incorporates some fixes and updates to some of the menus, adds a command-line interface for managing boards and libraries, and resolves some code highlighting issues. Support for the new Gemma board has been added. It also supports non-Arduino (that is, unofficial) boards by allowing you to enter a URL that will allow the IDE to retrieve information about a board from the supplier's website and integrate it into the IDE's board manager.

If you have an older Arduino board (something like a Diecimila, Duemilanove, or Uno), then not having the latest version of the IDE really isn't that big of an issue. Most people using an older version of the IDE with an older type of Arduino board probably won't even notice that they don't have the latest changes. If this applies to you, unless you plan to work with one of the newer Arduino boards (such as the Yún, Zero, or Gemma), you can probably put off getting the latest version of the IDE, at least for now.

Installing the Arduino IDE

The procedure for installing the Arduino IDE and libraries varies from extremely easy to very involved, depending on your platform and how much effort you want to put into it. For most people using Linux the easiest approach is to download an installation package via a package manager and it let it deal with also installing the necessary support packages.

The downside to this is that what is available from a package repository may not be the latest available version. The Arduino main website (*http://arduino.cc*) will always have the latest version, but gathering all the necessary bits and then getting them up and running on Linux can involve multiple manual steps. Windows and Mac OS X are easier to deal with in this regard, as discussed in the following sections.

After installation is complete, it's a good idea to take a few minutes and look through what has been installed on your system. In the Arduino installation directory you will find the source code for numerous examples, sources and examples for the stock libraries, hex (binary loadable image) files for various versions of the bootloader, and documentation in the form of a suite of HTML pages that can be viewed with a web browser or opened using the Help menu item in the IDE.

Downloads for supported host systems are available from Arduino.cc (*http://www.arduino.cc/en/Main/Software*). Just select the one that it is appropriate for you.

Windows

Installing the Arduino IDE on a Windows system is straightforward. The installation package, *arduino-1.6.4-windows.exe*, takes care of all the details. If you happen to have an older version of the IDE already installed it will remove it for you before installing a newer version (but it won't remove your existing sketches). The Arduino executables and libraries are placed in *Program Files\Arduino*. You can also find the examples here. A directory called *Arduino* is created in your home directory. This is where all user-created sketches and user-supplied libraries will be located.

Linux

Most Linux distributions provide the Arduino IDE in prepackaged form, be it a *deb*, *rpm*, *ymp*, or other package type. This is the preferred way to get the IDE and libraries quickly and correctly installed on a Linux system. As mentioned earlier, the downside to this approach is that the package maintainers for the various Linux distributions might not have the latest version from Arduino.cc available in distribution package format. For example, the latest version of Arduino available for Kubuntu is 1.0.4.

While a distribution package is the easiest way to get the Arduino IDE installed on a Linux system, the various components can also be downloaded directly from Arduino.cc and installed manually. For some distributions this might be the only option. You can read about the supported Linux distributions on the Arduino website (*http:// bit.ly/apg-linux*).

Package installation on an Ubuntu-type system (Ubuntu, Kubuntu, or some other Ubuntu-derived distribution) is done using *apt-get* or a similar tool. OpenSuse systems use the *yast* tool to install the IDE package, and other systems might use rpm or some other package manager. After installation, the various examples, hardware, library files, and tools are located in */usr/share/arduino*. On my Kubuntu and Xubuntu systems my sketches and custom libraries are placed in a default location in my home directory, in a subdirectory called *sketchbook*.

Mac OS X

The Mac OS X version of the Arduino IDE and libraries is supplied in a ZIP file with the name *arduino-1.6.4-macosx.zip*. According to the Arduino website it is intended for version 10.7 (Lion) or newer. After unzipping the archive, copy *Arduino.app* to the appropriate location on your system (the *Applications* folder is one possibility).

For additional information on installing the Arduino IDE on a Mac OS X system, refer to the Arduino.cc website (*http://www.arduino.cc/en/Guide/MacOSX*).

Configuring the Arduino IDE

You can tailor the IDE to suit your needs by using the Preferences dialog, found under File→Preferences on the main menu (under "Arduino" on a Mac). This dialog allows you to specify the sketch directory, specify an external editor (if you don't like the one that comes with the IDE), and modify various behaviors of the IDE. Figure 5-2 shows the Preferences dialog for an older version of the IDE on a Linux system.

The preferences file, which the dialog refers to as */home/jmh/.arduino/preferences.txt* in Figure 5-2, contains many more settings that are not shown in the dialog. But, as it states, don't edit this file while the IDE is active.

Figure 5-3 shows the Preferences dialog for version 1.6.4 of the IDE. It has many more options than the older version, but it is still just a graphical version of the preferences file.

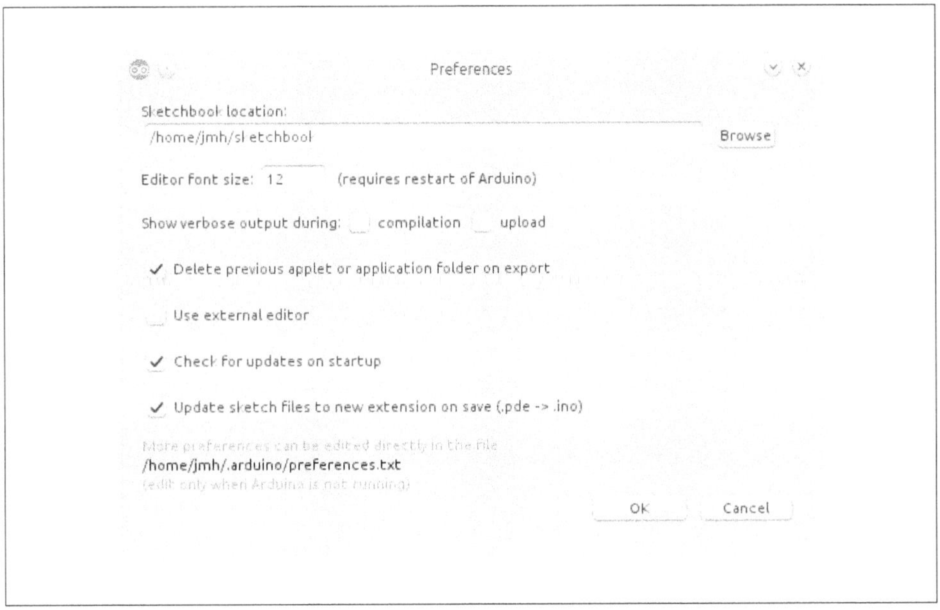

Figure 5-2. Old-style IDE preferences dialog

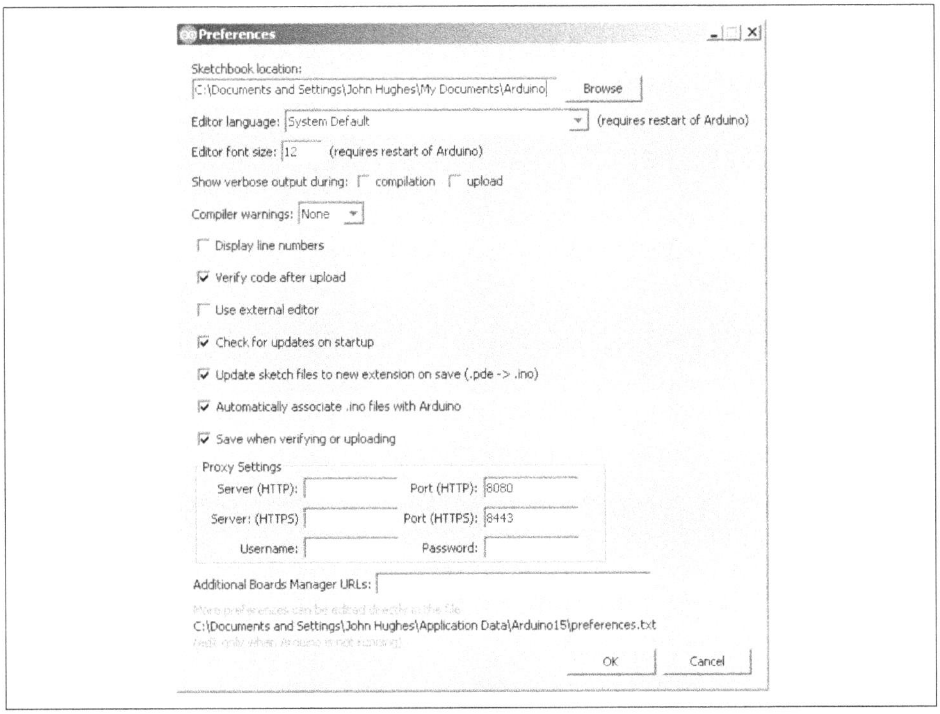

Figure 5-3. New-style IDE Preferences dialog

The preferences file is an ASCII KVP (key/value pair) data file that contains settings for the editor, application initial display geometry (the size of the Arduino IDE window), serial interface parameters (baud rate, data size, etc.), and the browser to use for viewing the supplied HTML help files, among other things.

The preferences file also contains information about the last sketch that was edited, and the last size of the IDE window. Because it is dynamically modified by the IDE while it is running, it should only be manually edited when the IDE is not active.

For the older version of the IDE, this may be the only way to modify parameters such as the indent size (the default is 2 spaces, but I prefer 4), the size and type of font used in the editor (the default is Monospaced with a size of 12, but I usually set it to 11 so I see more in the editor window), and tab expansion. The newer versions of the IDE offer more options in the Preferences dialog, but there may still be a few things you might want to tweak in the *preferences.txt* file itself.

Cross-Compiling with the Arduino IDE

What makes an Arduino special, instead of just being yet another PCB with an AVR soldered onto it, is the Arduino IDE, the Arduino firmware, the runtime code, the software libraries developed by Ardunio.cc, and of course the programs supplied by you, the developer. The Arduino IDE is a quick and easy way to build and load software for an AVR chip. It accomplishes this by effectively hiding much of what goes on when code is compiled, linked, and transferred to an AVR target.

The programming language used by the Arduino IDE is typically C++, although it can also use C, as the AVR-GCC toolchain will accept either. For this reason you will see C/C++ when this text refers to both languages, and C or C++ when discussing a particular language.

Individual source code files are displayed in the IDE in what are referred to as "tabs." You can switch from one file to another by simply selecting the appropriate tab at the top of the editor window. When the source code is compiled the IDE will step through each tab, creating an object file for each one.

The Arduino IDE uses the *avr-gcc* compiler and related tools (referred to as the *toolchain*) to build the binary executable code for an AVR device. The libraries supplied by Arduino.cc have functions available for things like time delays, serial data output, and other capabilities (see Chapter 7).

Sometimes the term "Arduinoese" has been used to imply that the Arduino uses its own unique dialect of C/C++. *This is incorrect.* The language used is real C or C++, with some limitations on what can be done with C++ source code. Recall that the

Arduino IDE is really nothing more than a "wrapper" around the AVR-GCC tool-chain (covered in "The AVR Toolchain" on page 125).

Specifically, in *avr-libc* there is no support for the C++ new and delete operators, but they are provided by the Arduino team. You can find the Arduino definitions in */usr/share/arduino/hardware/arduino/cores/arduino/new.h* on a Linux system, in *C:\Program Files\Arduino\hardware\arduino\avr\cores\arduino* on a Windows system, and in */Applications/Arduino.app/Contents/Java/hardware/arduino/avr/cores/arduino* on the Mac.

Instantiating Class Objects with new

Dynamic memory allocation in an embedded system is generally considered to be a bad idea. Dynamically allocated memory is typically taken from the SRAM, and most MCUs, the AVR included, don't have a lot of SRAM to start with. *avr-libc* supports malloc(), but not new—but the Arduino folks thought it would be a good idea to include it, so they did.

Note that when a class object is instantiated in AVR C++ code it is *not* created in the SRAM space. The 8-bit AVR MCUs are Harvard architecture processors, and they will not execute code from SRAM, only from flash memory space. The magic here is that the object is being created at compile time in flash memory and the new operator returns a pointer to the object.

As shown in Example 5-1, you can use new with version 1.0.5 or later of the IDE in a source file containing global variables.

Example 5-1. Instantiating a global object with new

```
// global_objs.cpp
#include "Arduino.h"
#include <LiquidCrystal.h>
#include <DDS.h>

LiquidCrystal *lcd = new LiquidCrystal(LCD_RS, LCD_E, LCD_D4, LCD_D5,
                                       LCD_D6, LCD_D7);
DDS *ddsdev = new DDS(DDS_OSC, DDS_DATA, DDS_FQ_UP, DDS_W_CLK, DDS_RESET);
```

If *lcd* and *ddsdev* are declared as exports in a corresponding include file, as shown in Example 5-2, they can be accessed from any other source module that includes the global include file.

Example 5-2. Declaring global objects as extern

```
// global_objs.h
#include <LiquidCrystal.h>
#include <DDS.h>
```

```
extern LiquidCrystal *lcd;
extern DDS *ddsdev;
```

Alternatively, you can instantiate the objects in the global variables source file and then specifically assign the pointers (the same include file could be used here as well). This is shown in Example 5-3.

Example 5-3. Assigning objects to global pointers

```
// global_objs.cpp
#include "Arduino.h"
#include <LiquidCrystal.h>
#include <DDS.h>

LiquidCrystal lcdobj(LCD_RS, LCD_E, LCD_D4, LCD_D5, LCD_D6, LCD_D7);
DDS ddsobj(DDS_OSC, DDS_DATA, DDS_FQ_UP, DDS_W_CLK, DDS_RESET);

LiquidCrystal *lcd = &lcdobj;
DDS *ddsdev = &ddsobj;
```

An interesting observation is that the executable objects these statements are part of have different compiled build sizes, depending on which method is used. The new approach results in an executable object that is 662 bytes larger than the pointer assignment method. It's not much, to be sure, but when you only have 32K of flash to work with, every little bit counts.

These code snippets were taken from the source code for the signal generator presented in Chapter 11.

In addition, C++ templates are not available, and exceptions are not supported. Classes, including constructors and destructors, are supported, however. Pure C code has few limitations. Even `malloc()` and `free()` can be used, with some cautions (see the *avr-libc* user manual (*http://bit.ly/avr-libc-malloc*) for details and guidance). But bear in mind that using dynamic memory allocation with an embedded MCU is technically questionable and can lead to some difficult problems and clumsy code. Like the old saying goes: just because you can, doesn't mean you should. There needs to be a very compelling reason to use dynamic memory with a memory-constrained microcontroller.

The source of confusion over the language used with the Arduino IDE may be the predefined macro definitions and functions that the Arduino environment uses to simplify access to the various I/O functions of the AVR MCU. An Arduino sketch may have an odd name and an unusual file extension (.*ino*), and it might look like a stripped-down version of C, but that is intentional. It makes perfect sense when you consider the original intended audience for the Arduino—namely, people who might not have any significant programming experience (or even none at all). But all the messy complexity of C and much of C++ is still there, if you want to use it. Chapter 6 describes the process of creating executable code for an Arduino (or any AVR MCU, for that matter) using the underlying AVR-GCC toolchain components, a text editor, and makefiles.

Microcontrollers Are Not Microprocessors

It is important to bear in mind that a microcontroller is not a general-purpose CPU. A typical microcontroller is a self-contained device with only a limited amount of program memory and very little RAM, but it does have lots of I/O functions already built in. Microcontrollers are typically used for a limited set of specific tasks, such as sensing keypress actions in a keyboard, controlling some motors, monitoring temperatures, or providing the "smarts" for a dynamic art project—or all of that at the same time. With the possible exception of the ATmega2560, with its ability to access external memory, an AVR microcontroller isn't really a suitable platform for a word processor or a complex video game. That being said, clever folks have used tight and efficient code to make an AVR microcontroller do some very interesting and complex things.

The Arduino Executable Image

Executable Arduino software (the executable image described earlier) typically consists of three primary components: the sketch created by a developer (you, perhaps), the libraries used to access the various functions of an AVR, and the runtime code that contains a `main()` function and a loop for the application software to execute within. If you look back at Figure 5-1, you can see that the IDE has helpfully supplied a minimal template for you to fill in with your own code.

The tools and software components involved in creating executable code for an AVR microcontroller can be divided into two primary categories: host development tools and runtime compilation sources, and the target-side executable binary code for the AVR microcontroller. Figure 5-4 shows a block diagram of the primary Arduino software components, and includes the user-supplied program (sketch) to show how things fit into a complete executable binary image.

The Arduino Software Build Process

There are five main steps in the Arduino build process when using the IDE:

Source preparation
> The sketch is modified slightly by the IDE, which adds the `#include` statement `"WProgram.h"` for version 1.0 or greater of the IDE, or `"Arduino.h"` for older versions of the IDE. Tabs (source files) that do not have an extension are concaten-

ated with the main sketch, creating a large single source code file (tabs with *.c* or *.cpp* extensions are compiled separately).

The IDE also attempts to create function prototypes for any functions other than setup() and loop() found in the sketch. These are placed at the top of the sketch file, immediately after any comments or preprocessor statements (#include and #define) but before any other types of statements. In a conventional C or C++ source file these would be placed in a *.h* header file to be included when the code is compiled, but the Arduino IDE takes care of this for you by dynamically creating and inserting them into the sketch source file. The last part of the source preparation entails appending the contents of the standard *main.cxx* file to the sketch (this is where main() is defined).

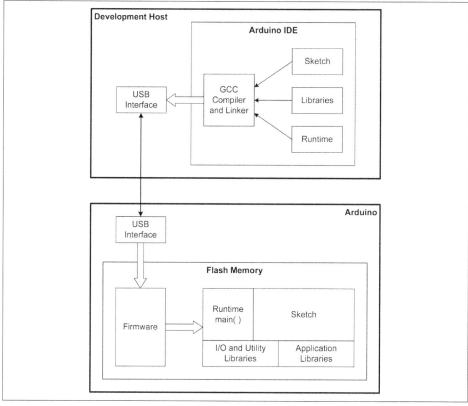

Figure 5-4. Arduino software organization

Compilation

Arduino uses the AVR-GCC compiler suite to translate the source code into binary files called *object files*. These are not immediately executable, but must be processed by the tool called the linker before the Arduino hardware can deal with them (refer to the next step). The AVR-GCC compiler is a form of the base GCC compiler built specifically for use with Atmel's line of AVR devices. Due to the limitations imposed by the target hardware (memory, mainly) not all C++ capabilities are supported, but C support is complete.

The compiler accepts a large number of command-line arguments, also called switches. The Arduino IDE takes care of providing the correct switches to the compiler. These include specifying where standard include files are located, optimization options, target-specific options, and warning message levels, among other things.

Linking

Linking is the process of connecting object files and library modules. The basic idea is to fill in the "holes" in the object files where the original source code referred to a data object or function in an external library or object, but the compiler couldn't resolve the address at the time the code was compiled. The linker's job is to locate the missing references and write them into a final executable binary file, along with the binary code for the referenced data or functions.

Conversion

The binary file created by the linker must be converted to what is called *Intel hex format*, which is what the bootloader firmware on the AVR device expects. The utility *avr-objcopy* can be used to do this, and it is described in Chapter 6.

Uploading

The final step involves transferring the completed executable binary file to the Arduino hardware. This is accomplished using a USB link (usually) and the bootloader firmware in the microcontroller, along with a utility called AVR-DUDE (it's actually an acronym). It is also possible to load executable code into an AVR device directly via the ICSP interface (such as when there is no bootloader), and the Arduino IDE supports this.

Sketch Tabs

It is possible, and actually quite convenient, to divide a large sketch into smaller source code modules (i.e., files), each with its own include file. When the IDE loads a sketch it looks in the sketch's directory for any additional files. Auxiliary files may have no extension, or a *.c*, *.cpp*, or *.h* extension. Files with a *.c* or *.cpp* extension appear in a tab but are compiled into object files to be linked with the main sketch code. Files without an extension are included into the sketch. In order to use a tab file with a *.h* extension, it must be referenced via an #include statement using double quotes (not angle brackets).

If your auxiliary files have their own include files they must all be included in the main sketch, even if it doesn't reference any of the code itself. The same applies to external library modules. If an auxiliary file uses a class in a library but the main sketch does not, both it and the main sketch still need to reference the library's include file.

A technique I like to use involves putting all the global variables (including class objects, as mentioned earlier) into a separate source file. This allows any other modules that need to access the variables to do so. It also makes the main sketch a lot easier to read, as it could end up with just the setup() and loop() functions. You can read more about multifile sketches at arduino.cc (*http://bit.ly/arduino-build-process*), and Chapter 11 describes a working example created for the DDS signal generator (the full source code is available on Github at *https://github.com/ardnut*).

Arduino Software Architecture

Regardless of how big or small it might be, a sketch always consists of at least two required functions: setup() and loop(). The setup() function is called once when the sketch starts. The loop() function executes continuously until the power is disconnected or the Arduino board is reset. loop() is called repeatedly by the main() function that is automatically supplied by the Arduino IDE. A sketch may also have additional functions. For example, the sketch for the thermostat shown in Chapter 12 has multiple functions in addition to the mandatory setup() and loop(). The main function runtime code is described in "Runtime Support: The main() Function" on page 106.

A sketch will also have statements at the start of the file that define other files to include, constants for I/O pins and other values, and global variables. You can see how a typical sketch is organized in "An Example Sketch" on page 107. The sections "Constants" and "Global Variables" on page 111 discuss constants and global variables, respectively.

A never-ending main() loop, sometimes called an *event loop*, is a common way to program microcontrollers. A small device like an AVR doesn't load program files from disk, and any program executing on it is designed to perform a specific function (or range of functions) continuously. So, using the concept of a run-forever primary loop makes perfect sense. Because the program code is stored in flash memory, it will be ready to run again after the Arduino is powered up or reset.

To help put things into perspective, Figure 5-5 shows how the main() function supplied by the Arduino IDE calls the setup() and loop() functions in the sketch.

In Figure 5-5, an external library class object is instantiated (i.e., initialized) outside of both setup() and loop() so it will be available to all the functions in the sketch. In addition to setup(), loop(), and any other functions in the main sketch, there can also be functions in other files, and additional files of source code included in the sketch will be handled by the Arduino IDE as tabs.

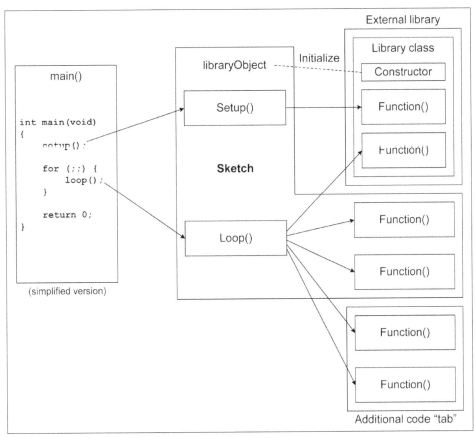

Figure 5-5. Arduino program structure

Note that a tab is not a true library; it's just more sketch source code kept in a separate file. The code used in a tab could be reused in other sketches, but unlike with a real library, you need to manually specify the tab when a sketch is created—just providing an include directive in the main sketch source file is not sufficient. To add code to a sketch in the form of a tab, use Sketch→Add File from the IDE menu bar. The source file for the tab does not need to reside in the same directory as the main sketch file.

External libraries are integrated with the final executable image after the sketch code, along with any tab files, is compiled. The Arduino IDE will generate the correct linker options based on the include statements placed at the start of the sketch. See "Using Libraries in Sketches" on page 112 for more on incorporating libraries into a sketch.

Runtime Support: The main() Function

Compilers used with Linux, Windows, and Mac OS X systems provide a platform-specific library of functions called the *runtime library*. On a GCC-based system this is typically called *libgcc.a* or *libgcc_s.so.1*, and on Windows the runtime libraries usually go by the names *MSVCRT.DLL* for C and *MSVCPP.DLL* for C++. The AVR version is *avr-libc*, which is discussed in Chapter 6. The runtime library will contain commonly used functions specific to a particular platform. These can be common math operations, low-level I/O functions, system timers, support for printf(), and so on.

The Arduino also has its own additional runtime support module, but with a twist. By itself, a program sketch won't do much. It has no main() function to get it started and no way to continuously execute once it does start. The Arduino runtime support includes the necessary main() function, along with other startup configuration functions. As can be seen from Example 5-4, it's actually very simple, and is a typical design for small microcontrollers.

Example 5-4. Arduino main() function

```
int main(void)
{
    init();
    initVariant();

    #ifdefined(USBCON)
    USBDevice.attach();
    #endif

    setup();
    for (;;) {
        loop();
        if (serialEventRun) serialEventRun();
    }
    return 0;
}
```

The calls to init(), initVariant(), and (if applicable) USBDevice.attach() are determined at compile time based on the type of Arduino target hardware selected in the IDE. These comprise the primary components of the runtime support code specific to each type of Arduino board. In a larger *realtime operating system* (RTOS) these might be part of the board support package (BSP) supplied by the RTOS vendor or created by a developer. The functions are described here:

init()

    Located in *Arduino/hardware/arduino/avr/cores/arduino/wiring.c*, this function initializes various AVR peripheral components such as timers, timer/counter prescaling, A/D converter prescaling, and PWM output modes according to the AVR part used with a particular Arduino board.

`initVariant()`
> This provides a "hook" in the form of a so-called weak function declaration. Typically used to provide additional runtime initialization for hardware not covered by the definitions and code supplied with a standard Arduino development environment.

`USBDevice.attach()`
> This refers to the `attach()` method of the `USBDevice` class found in *Arduino/hardware/arduino/avr/cores/arduino/USBCore.cpp*. This class provides the functionality necessary to communicate with a USB interface.

For many people, what these initialization functions actually do is irrelevant. However, if you want to really understand what the Arduino IDE does and how it does it, then reviewing the source code for these support functions is a worthwhile endeavor. Obtaining the source code is described in "Arduino Source Code" on page 119.

As described previously, the `setup()` and `loop()` functions are provided by you in the program sketch. The `setup()` function is typically used to define specific I/O ports and starting states, initialize some parts of the AVR peripheral functions, or perform other one-time operations. It is largely optional and can even be an empty function, but it must exist in the sketch or the linker will generate an error message.

The `loop()` function is where the main activity occurs in a program sketch. As can be seen from Arduino main it is called repeatedly until power to the Arduino board is removed. Interrupts can and do occur while `loop()` is executing, and some applications may incorporate a timer interrupt or a delay to give `loop()` some degree of definite periodicity (as opposed to simply free-running).

An Example Sketch

As we've already seen, a basic Arduino program sketch consists of two parts: the `setup()` function and the `loop()` function. Since the `main()` function is provided by the IDE, we don't need to worry about it when working with simple sketches.

The simple sketch in Example 5-5 will monitor several inputs and produce an output if any of the inputs change from low to high. It could be the basis for a simple burglar alarm.

Example 5-5. Simple intrusion alarm

```
// CONSTANTS
// Define the digital I/O pins to use
#define LOOP1   2       // sense loop 1
```

```
#define LOOP2    3      // sense loop 2
#define ARM      6      // Arm the alarm
#define RESET    7      // Reset alarm state
#define ALERT    12     // Annunciator output
#define LED      13     // Same LED on board

#define SLEEPTM 250     // loop dwell time

#define SLOWFLASH  10  // unarmed flash divisor
#define FASTFLASH  2   // armed flash divisor

// GLOBAL VARIABLES
// State flags
bool arm_state;         // T = armed, F = unarmed
bool alarmed;           // T = in alarm state, F = all quiet
bool led_state;         // control on-board LED
bool loop1state;        // status of sense loop 1
bool loop2state;        // status of sense loop 2
bool in_alarm;          // T = alarm active, F = quiet

int led_cnt;            // flash cycle counter
int flash_rate;         // count divisor

// We will use Arduino's built-in pinMode() function to set the input or
// output behavior of the discrete digital I/O pins

void setup()
{
    // Initialize pins 2, 3, 6, and 7 as digital inputs
    pinMode(LOOP1, INPUT);
    pinMode(LOOP2, INPUT);
    pinMode(ARM, INPUT);
    pinMode(RESET, INPUT);

    // Initialize pins 12 and 13 as digital outputs
    pinMode(ALERT, OUTPUT);
    pinMode(LED, OUTPUT);

    // Initialize state flags
    arm_state  = false;
    alarmed    = false;
    led_state  = false;
    loop1state = true;
    loop2state = true;
    in_alarm   = false;

    // Set LED start condition
    led_cnt = 0;
    flash_rate = SLOWFLASH;
    digitalWrite(LED, LOW);
}
```

```
void loop()
{
    // Get arm switch value
    arm_state = digitalRead(ARM);

    // If reset is low (true), cancel alarm state
    if (!digitalRead(RESET)) {
        in_alarm = false;
    }

    // If not armed, just loop with a short pause
    if (!arm_state) {
        flash_rate =  SLOWFLASH;
        in_alarm = false;
    }
    else {
        flash_rate =  FASTFLASH;
    }

    // Check the sense loops
    loop1state = digitalRead(LOOP1);
    loop2state = digitalRead(LOOP2);

    // only go into alarm state if armed
    if (arm_state) {
        if ((loop1state) || (loop2state)) {
            in_alarm = true;
        }
    }

    if (in_alarm) {
        digitalWrite(ALERT, HIGH);
    }
    else {
        digitalWrite(ALERT, LOW);
    }

    led_cnt++;
    if (!(led_cnt % flash_rate)) {
        led_cnt = 0;                    // reset flash count
        led_state = !led_state;    // invert LED state
        digitalWrite(LED, led_state);
    }

    delay(SLEEPTM);
}
```

Figure 5-6 shows how an Arduino might be connected to external switches for doors and windows. The reasoning behind connecting switches in series and connecting one end of the chain to ground is simple: if any of the wiring connecting the sensor switches is cut, it will trigger the alarm.

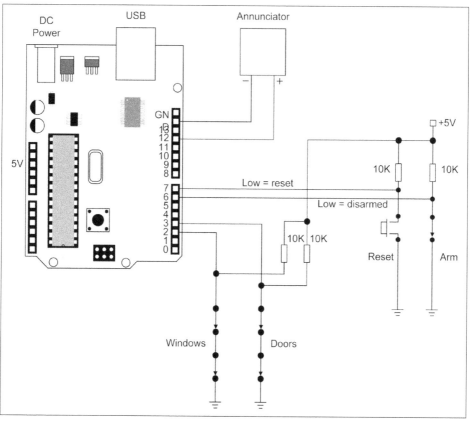

Figure 5-6. Simple Arduino intrusion alarm

In an application like this the sensor switches could be window frame plungers, magnetically actuated reed relays, leaf switches on sliding doors, or even a sensor module such as a sound detector. The annunciator can be anything from a 50-cent piezo buzzer to a relay to control a siren or even a phone autodialer. For more on Arduino-compatible sensors, see Chapter 9. This basic design could be extended in any number of interesting directions, all the way to creating a commercial-quality alarm system. You can find downloadable software from this book on GitHub (*https://www.github.com/ardnut*).

Constants

There are two ways to define constants for I/O pin numbers, time delays, and other values. The first is to use #define statements. The second is to use integers that are initialized to some value and then never modified. For example, a `#define` statement to associate pin 2 with a name might look like this:

```
#define LOOP1 2   // sense loop 1
```

The alternative approach is to create an integer to hold the value:

```
int LOOP1 = 2;    // sense loop 1
```

If you look through example Arduino code you will see both approaches used, and both will work equally well. Although some folks may eschew the `#define` statement, for whatever reasons, there is a memory use trade-off to consider.

The advantage of using `#define` statements, as is done in Example 5-5, is that they result in a smaller compiled program size. With the `#define` statements the final size of the intrusion alarm program is 1,452 bytes. With the `#defines` replaced with `int` declarations, it is 1,540 bytes. That difference of 88 bytes may not seem like much, but in a large sketch with lots of I/O definitions and constants it can add up. If, for example, everything has to fit into 30,720 bytes for an ATmega328, it might make the difference between loading and failing.

Global Variables

One thing you may notice about all sketches is the use of global variables. This is common in Arduino sketches for one very good reason: the `loop()` function is called repeatedly by `main()` when the sketch code is running, and if the variables were defined in `loop()` they would be erased and reloaded each time that `loop()` was called. If `loop()` needs a variable with a persistent value (a counter, perhaps), then it needs to be outside of `loop()` so that it won't be cleared and recreated repeatedly. Global variables also make it convenient for `setup()` to set initial values before `loop()` is first called, and any other functions in the sketch will also have access to the variables.

Global variables have gotten a bad reputation in some circles because of the problems that can arise with unintentional "coupling" between different parts of a program. In large applications that load from disk, execute, and are then terminated by a user, such as word processors, games, and web browsers, this makes sense. It's not a good thing when one part of an application modifies a global variable and another part of the application causes a crash when it also modifies the same global variable. In systems with lots of available memory, dynamic memory allocation, semaphores and mutex locks, and the ability to pass around pointers to things like complex structures, the use of global variables might not be justifiable.

However, in the realm of embedded systems global variables are commonly used as an efficient form of shared memory that is visible to every function in the program. You can think of global variables as a panel full of switches, indicators, knobs, and dials, like in the cockpit of an airplane. Just so long as you follow the "modified by one function, read by many functions" rule things should be fine.

Since an Arduino sketch is not running in parallel with other threads or processes, the chances of an access collision are zero. Interrupts might be a challenge, depending on how they are implemented, but a little forethought can negate potential problems there, as well.

Libraries

The libraries supplied with the Arduino IDE cover a lot of things, but not everything, as can be seen in Chapter 7. Unless someone has put in the time and effort to create a library for a specific sensor or interface, then you will need to supply the code yourself. This is particularly often the case when working with custom or uncommon sensors or shields that require special functions.

With the Arduino IDE the term "library" is used in a way that might seem out of line with what you are used to if you've worked with GCC or Visual Studio on large projects. When building software for a full-size computer like a desktop PC, a library typically refers to a binary object archive. On Linux and Unix systems this is accomplished using the *ln*, *ar*, and *ranlib* tools. The resulting binary file is an indexed collection of object files (see "Objects, Images, and Source Code" on page 91) that the linker can use to fill in the gaps and generate a complete program. If it's a dynamic library, then it will be loaded when the application is started (or even during program execution) and the linking will occur at that time.

In the Arduino IDE environment libraries usually exist as source code until they are needed in a sketch. So what is actually happening when a sketch is compiled is that the sketch (and any tab files), any necessary libraries, and the runtime code are all compiled at the same time and linked into a single binary executable image.

Using Libraries in Sketches

As mentioned earlier, when the Arduino IDE encounters an include statement that refers to a library already registered with the IDE it will generate the necessary build steps and linker options to incorporate the library automatically. However, in order for this to work the IDE must know about the library in advance.

Registering a library with the IDE is described in "Adding a Library to the Arduino IDE" on page 116. The Arduino IDE comes with a selection of libraries for common operations and I/O devices already preloaded. Figure 5-7 shows the available library listing for an older version of the IDE, and Figure 5-8 shows the library listing and library management options available in the latest version (1.6.4) of the IDE.

Figure 5-7. Registered library listing from older Arduino IDE

Example 5-6 shows how a library—in this case, the SoftwareSerial library—is incorporated into a sketch. The first thing to notice is the include statement at the top of the sketch file:

```
#include <SoftwareSerial.h>
```

From this the IDE will determine that it needs to locate a library with the name "SoftwareSerial," and it will expect that library to be in its list of known library modules.

Example 5-6. Library example 1

```
#include <SoftwareSerial.h>

SoftwareSerial softSerial(2, 3);

int analogPin = A0;    // select the input pin for the potentiometer
int analogVal = 0;  // variable to store the value coming from the sensor

void setup()
{
    // set the data rate for the SoftwareSerial port
    softSerial.begin(9600);
}
```

```
void loop() // run over and over
{
    // read some data from an analog input
    analogVal = analogRead(analogPin);

    // write the data to the softSerial port
    if (softSerial.available())
        softSerial.write(analogVal);

    // pause the program for 1 second (1000 milliseconds)
    delay(1000);
}
```

The SoftwareSerial object is instantiated by the line:

```
SoftwareSerial softSerial(2, 3);
```

This also defines the pins to use for the serial I/O function, namely 2 and 3 (digital I/O pins).

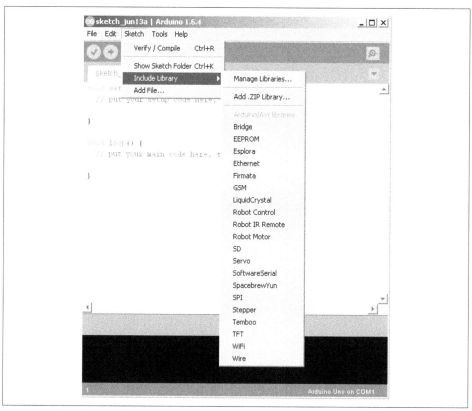

Figure 5-8. Registered library listing from latest version of Arduino IDE

In the setup() function, the serial baud rate is set.

The loop() function reads a value from the A0 analog input pin, and then sends the binary value out via the softSerial object. The loop() function pauses for 1 second after each binary value is sent.

This technique can be used to send and receive binary data between two Arduino boards, but it won't work very well if you want human-readable output. For that you will need to use the print() and println() methods in the library, as shown in Example 5-7.

Example 5-7. Library example 2

```
#include <SoftwareSerial.h>

SoftwareSerial softSerial(2, 3);

int analogPin = A0;     // select the input pin for the potentiometer
int analogVal = 0;  // variable to store the value coming from the sensor

void setup()
{
    // set the data rate for the SoftwareSerial port
    softSerial.begin(9600);
}

void loop()
{
    // read some data from an analog input
    analogVal = analogRead(analogPin);

    // write the data to the softSerial port
    softSerial.print(analogVal);
    softSerial.println();               // print a linefeed character

    // pause the program for 1 second (1000 milliseconds)
    delay(1000);
}
```

It is also possible to use the standard library printf() function to send data out over the serial port, but printf() doesn't come preenabled with the Arduino core functions. If you want to learn how to enable printf(), and get an understanding of how it might impact your sketches, refer to the Arduino web page on enabling printf() in Arduino sketches (*http://playground.arduino.cc/Main/Printf*).

You can learn more about the SoftwareSerial library module from the integrated help in the IDE in the "Libraries" section, and the source code can be found in the Arduino installation directories. SoftwareSerial is also described in Chapter 7.

Adding a Library to the Arduino IDE

User-supplied add-on libraries are the key to using external devices like humidity and temperature sensors, radio frequency (RF) modules, and infrared (IR) remote control components. You can write your own library of functions, or you can elect to install code created by someone else.

Add-on libraries are typically placed in a directory called *libraries* in the *sketchbook* directory. The set of files that comprise the add-on library reside in a subdirectory with an appropriate name. It is the name of this subdirectory that appears in the library list in the IDE. There are two ways to accomplish this:

Method 1, automatic

Recent versions of the Arduino IDE (1.0.5 and later) can automatically install library source code into your *sketchbook* directory for you (the actual location will depend on your operating system). From the IDE's main menu, click Sketch→Import Library→Add Library and then select the directory or ZIP file that contains the library source code. The latest version of the IDE (1.6.4) uses a slightly different series of steps, and the library management functions are grouped under "Include Library."

One nice thing about this method (other than its simplicity) is that the IDE will immediately recognize the new library—there is no need to restart the IDE. Also, with newer versions of the IDE you can download and install libraries from external sources.

Method 2, manual

Manually installing an add-on library for the Arduino IDE involves the following steps:

- First, close the IDE if it is running.
- Unzip the new library files into a temporary directory (most add-on libraries come in the form of ZIP files).
- You should see at least two source code files, a *.c* or *.cpp* file (the program) and a *.h* file (an include or header file).
- Create a subdirectory under *<home>/sketchbook/libraries* with the same name as the two source files.
- Copy the entire contents of the temporary directory to the new library directory you created in the previous step.

  When the IDE is restarted you should see the library listed in the Sketch→Include Library drop-down list.

The convention for organizing the contents of an add-on library is as follows:

```
IRTracker/IRTracker.cpp
         /IRTracker.h
         /keywords.txt
         /README
         /utility/IRCal.cpp
                 /IRCal.h
                 /IREncode.cpp
                 /IREncode.h
                 /IRDecode.cpp
                 /IRDecode.h
         /Examples/ScanIR/ScanIR.ino
                  /SendStop/SendStop.ino
```

Note that this is an example only. So far as I know, there is no IR tracker library available for the Arduino (yet).

The library subdirectory must have the same name as the source files, and this convention is used throughout the Arduino IDE's directories. For example, in the Arduino examples directory (*/usr/share/arduino/examples* in a Linux system) the subdirectories have the same names as the *.ino* files.

As you can see in the example directory structure, there can be additional subdirectories for utility functions, and even more source files than the two files that share the same name as the subdirectory. They won't appear in the libraries list, but the library code will be able to access them when it is compiled.

You might notice a file called *keywords.txt* at the base level of the library directory structure. This is an important file, as it gives the IDE some definitions regarding what things do in the library source code. The *keywords.txt* file for the hypothetical IR tracker library used for our directory structure example might look like this:

```
# IR Tracker
################################
MAXPWR     LITERAL1
MAXTIME    LITERAL1

LastCal    KEYWORD1
TotHours   KEYWORD1

IRState    KEYWORD2
IRSense    KEYWORD2
```

The format is very simple. The # (pound or hash) indicates that everything to the end of the line is a comment and should be ignored. Special keywords, listed in Table 5-1, are used to indicate the type of literal constants, classes, structures, variables, or functions to the IDE.

Table 5-1. Keyword definitions for keywords.txt

Type	Definition
LITERAL1	Literal constants (i.e., macros)
LITERAL2	Literal constants (i.e., macros)
KEYWORD1	Classes, data types, and C++ keywords
KEYWORD2	Methods and functions
KEYWORD3	Structures

You can also find more information in the *keywords.txt* file located in the directory where the Arduino runtime components are installed on your system. On my Linux machine this is */usr/share/arduino/lib*, and on Windows it can be found at *C:\Program Files\Arduino\lib*. You can also look at what others have done by examining the *keywords.txt* files in the library subdirectories supplied with the IDE.

While it is possible to place add-on libraries in the Arduino IDE's predefined set of directories, this is not the recommended approach. The predefined libraries supplied with the IDE are subject to change when the IDE is upgraded to a newer version. The user-contributed libraries and sketches are never altered by an upgrade.

In any case, a library component consists of two basic parts: the source module (*name.c* or *name.cpp*) and the include file (*name.h*). The code in the source module is a set of functions that comprise a simple C++ class. The class itself is defined in the header file. To use the library one need only copy it into the appropriate directory, start the Arduino IDE, and then select the new library from the Sketch→Import Library list. The IDE will examine the header file and place the statement #include *<name.h>* into the sketch.

Creating Custom Libraries

Creating a custom library for use with the Arduino IDE is straightforward. That doesn't necessarily mean it's simple, as libraries can be very complex, but simple or complicated, they all follow the same basic template. The source code is written in the AVR's restricted version of C++ as a class, perhaps with one or more associated classes to help out. Look at the source code for some of the libraries listed in Chapter 7 to get an idea of how library source code is organized.

As earlier noted, a minimal library is a set of at least two files: a *.cpp* source file and a *.h* include file. The source file contains the implementation of the library class, and the include file contains the class definition, type definitions, and macro definitions.

Recall from our earlier sample directory structure that the directory containing the source and include files for the library will have the same name as the primary source (*.cpp*) file. It should also contain the *keywords.txt* and *README* files, and an *examples* directory is always a nice touch (particularly if you intend to release your library for others to use).

Arduino Source Code

The full Arduino source code set contains source files for both AVR-based boards and the ARM-based Due board, which isn't covered in this book. The source code is available from GitHub (*https://github.com/arduino/Arduino.git*) using the URL and the `git clone` command. You can also download a ZIP file with an image of all the repository files from GitHub (*https://github.com/arduino/Arduino/archive/master.zip*).

If you plan to delve deeper into Arduino than just the IDE, then it is helpful to have the source code on hand. Looking at the source files can help to make things much clearer. A tool such as Doxygen (*http://bit.ly/doxygen-main*) can be used to create a linked set of web pages with dependency graphs, call graphs, and an index of classes, source files, and functions. Although the Arduino source doesn't have much in the way of Doxygen-specific tags, it is still possible to generate useful documentation.

At the top level, the Arduino source directory structure contains directories for application source code (*app*), build modules (*build*), hardware-specific files (*hardware*), and the library modules included with a standard Arduino distribution (*libraries*). From a low-level perspective, the *hardware* and *libraries* directories are the most interesting.

In the *hardware* subdirectory you can find the source code for various bootloaders, the runtime support code (which is called "core" in this source file set) that includes *main.cpp*, and a small collection of modules for EEPROM, serial I/O, SPI, and two-wire interfaces (in the *hardware/avr/arduino/libraries* directory, which is different from the *libraries* directory mentioned earlier). The support libraries that come with the Arduino IDE are located in the *libraries* directory. The subdirectories here include source code for audio, Ethernet, liquid crystal displays, SD memory cards, servo motors, stepper motors, TFT displays, and WiFi modules. There are libraries for the Arduino Robot, Esplora, and Yún products as well. Many of the library subdirectories also contain example program sketches, and some have documentation in the form of a text file.

One thing to bear in mind when reading through the Arduino source code is that it makes heavy use of `#if`, `#ifdef`, `#ifndef`, and `#define` statements to determine what will be included and compiled for a specific type of Arduino board. This can be confusing at first, and it might take a little effort to work through what is going on. It is also worth noting that in some cases a function or set of functions is used, while in other cases a class is defined for a specific purpose.

CHAPTER 6

Life Without the Arduino IDE

The Arduino hardware actually isn't all that special; it is just a very basic development board based on the Atmel AVR devices. It is the Arduino IDE and bootloader firmware that make it easier for nonprogrammers to work with it and get things running. It is, however, possible to completely forgo the Arduino IDE. It is a convenient application that takes care of a lot of the messy details of the software build process for the programmer, but those who want to work from the command line with just a text editor can do so without ever using an IDE.

In this chapter we will look at some examples of alternative ways to build programs for an Arduino, and how to use the AVR-GCC toolchain from the command line, without any assistance other than a makefile. We will also see how assembly language can be used "down on the metal" to wring the last bit of performance from an AVR MCU.

Just as there is more than one way to create executable code for an Arduino, there is more than one way to upload software into an AVR. In this chapter we will look at some of the ways to get the job done that don't involve the Arduino IDE.

IDE Alternatives

The Arduino IDE isn't the only way to develop and load programs for an AVR MCU on an Arduino board. One Arduino programming alternative is the PlatformIO tool, which runs under Linux, Mac OS X, and Windows. It is a Python-based code builder and library manager that is executed from the command line. Another Python tool for building Arduino programs is the Ino tool; it works with Linux and Mac OS, but does not currently run in a Windows environment.

PlatformIO

PlatformIO (*http://platformio.org*) is a Python-based command-line tool that supports over 100 different target microcontrollers. It is largely platform independent, requiring only Python 2.6 or 2.7 to run on Windows, Mac OS X, or Linux (I recommend installing Python 2.7). For more information, visit the PlatformIO website.

Based on the type of microcontroller specified, PlatformIO will determine what toolchain components are necessary and then call them in the correct order to complete the compilation, linking, and target upload operations. The boards that PlatformIO supports include the Trinket series from Adafruit, all of the Arduino AVR boards, the BitWizard Raspduino (mentioned in Chapter 1), the Digispark boards from Digistump, the Engduino, the LowPowerLab Mote boards, the Microduino, the Sanguino AVR boards, and the AVR boards from SparkFun. It also supports various ARM-based products such as boards using the Atmel SAM MCU, the STM32 MCU, the LPC from NXP, the Nordic nRF51, as well as the TI MSP430 series and more.

Installing PlatformIO on a Linux or Mac system is straightforward. You can use the command line, download an installation script, or use a Python *pip* tool. You may need to install the cURL data transfer tool first, but this is a common utility and on a Linux system it can be obtained directly from a package repository.

Figure 6-1 shows the console output after successfully downloading and installing the PlatformIO packages.

Once it's installed, you can find out what it can do. If you type in `platformio boards` you will be presented with a long list of currently supported boards.

PlatformIO uses the notion of projects to keep things neat and tidy. Each project starts out with a configuration file and two subdirectories, *src* and *lib*. The configuration file *must* be edited before it can be used. It is a conventional ASCII KVP (key/value pair) INI-type file. You can find detailed documentation on the "Project Configuration File" (*http://bit.ly/platformio-pcf*) reference page.

I recommend creating a subdirectory for your PlatformIO projects separate from the *sketchbook* directory used by the Arduino IDE. In my case, I called it *platformio* (not very creative, but it works for me). To create a new project, enter `platformio init`. Figure 6-2 shows the console output when a project is created.

PlatformIO has many capabilities, including predefined frameworks for various board types, a library manager, and the ability to integrate with an IDE or IDE-like editor such as the Arduino (along with Eclipse, Energia, Qt Creator, Sublime Text, Vim, and Visual Studio).

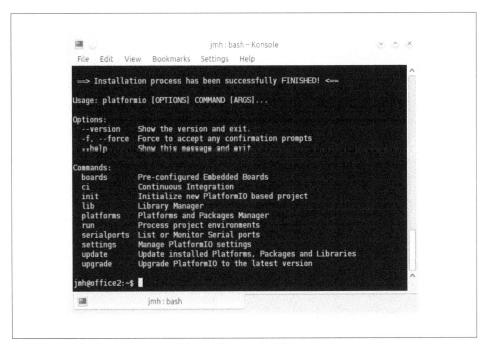

Figure 6-1. PlatformIO successful installation console output

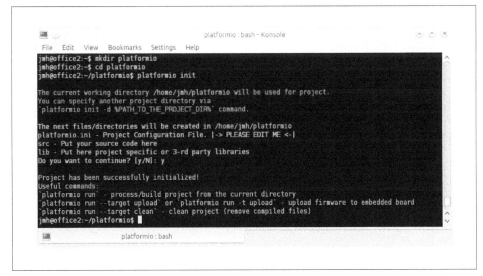

Figure 6-2. Initializing a new PlatformIO project

Ino

Ino is a simple command-line build tool that takes a makefile-based approach to compilation for Arduino targets. Like PlatformIO it is built using Python, but unlike PlatformIO it is specifically intended for use with Arduino boards. Ino supports *.ino* and *.pde* sketch files as well as *.c* and *.cpp* sources, and it claims to support all the boards supported by the Arduino IDE. Note that the current version of Ino only works with Linux and Mac OS X host platforms. Ino will work with Python 2.6 or 2.7.

You can download and install Ino either by downloading a compressed TAR file (*http://inotool.org/#installation*), cloning it from GitHub, or using the Python *pip* or *easy_install* tools. I used *pip* to install Ino and its various components. After installation, running the command ino -h displayed the output shown in Figure 6-3 on the console window.

Figure 6-3. The help output from the Ino tool

Ino creates and uses makefiles, but these are transparent to the user. Like PlatformIO, it uses a directory-based project scheme, and when a project is initialized Ino will create two subdirectories: *src* and *lib*. It also creates a minimal template sketch

(*sketch.ino*) in the *src* directory, just as the latest version of the Arduino IDE does. It is up to you to fill in the blanks and provide the rest of the necessary files.

More information about Ino is available at the official website (*http://inotool.org/*); the tool can be downloaded from the Python Package Index (*http://bit.ly/ino-ppi*).

The AVR Toolchain

The primary means of converting a source file with C or C++ source code into a binary object that can then be incorporated into a finished executable AVR program is the AVR-GCC compiler and its suite of utilities. These are collectively referred to as the "toolchain." As previously mentioned, the primary role of the Arduino IDE is to wrap a user-friendly shell around these tools and hide the messy details as much as possible. PlatformIO and Ino also hide the toolchain behind Python scripts. But it's still there, regardless, and the compiler, linker, assembler, and other tools are available if you want to build your AVR code using a makefile from the command line, or if you just want to perform each step manually.

 The Arduino IDE installation package (in whatever form works for your OS) will take care of installing the AVR-GCC toolchain for you. On a Linux system the Arduino IDE requires the AVR-GCC toolchain and its associated components, so the package manager will install it at the time the Arduino IDE is installed. The main reason for installing the toolchain in addition to what the Arduino IDE provides is to be able to use the latest versions of the tools. On a Windows system the Arduino IDE, tools, and libraries will be placed in a separate directory apart from where something like the WinAVR suite will usually place its toolchain components, so you could have both available if you wanted to do so.

Table 6-1 lists the AVR tools found on a Linux system after installing the Arduino distribution package. The same basic set of programs will be found on a Windows or Apple system after installing the Arduino IDE. Methods for obtaining and installing the GNU AVR tools on various host systems are covered in "Installing the Toolchain" on page 127.

Not all of the tools listed in Table 6-1 are necessary to build executable programs for an AVR chip on an Arduino board. With the exception of AVRDUDE, they are AVR versions of existing GNU tools, with similar or identical functionality.

The important tools in the toolchain, from the perspective of getting something compiled for an Arduino (or any AVR MCU), are *avr-gcc*, *avr-g++*, *avr-ar*, *avr-as*, *avr-ld*, *avr-objcopy*, *avr-ranlib*, and *AVRDUDE*.

Table 6-1. AVR cross-compilation tools

Tool	Description
avr-addr2line	Converts addresses into filenames and line numbers
avr-ar	Creates object code archives, and modifies or extracts code
avr-as	The portable GNU assembler for the AVR
avr-c++filt	Demangles C++ symbols
avr-gcc	The GCC compiler backend to produce AVR object code
avr-g++	The G++ compiler backend to produce AVR object code
avr-ld	The GNU linker for AVR object code
avr-nm	Lists symbols embedded in object files
avr-objcopy	Copies and translates object files
avr-objdump	Displays information from object files
avr-ranlib	Generates an index for a library archive
avr-readelf	Displays information about ELF files
avr-size	Lists object file section sizes and total size
avr-strings	Prints strings of printable characters in binary files
avr-strip	Discards symbols from AVR object files
AVRDUDE	The driver program for various AVR MCU programmers

The heart of the toolchain is the compiler, which is called *avr-gcc*, or *avr-g++* for C++ sources. These are versions of the GNU compiler that have been tailored specifically for use with C or C++ source code and Atmel's line of AVR microcontrollers. They're similar to the full version of *gcc*, with some of the compiler's options preset to make them more convenient to use for AVR MCU cross-compiling.

If you happen to look in the directory where the toolchain components have been installed, you might see the following:

```
avr-c++
avr-cpp
avr-g++
avr-gcc
avr-gcc-4.5.3
```

These are all just variants of the GNU compiler. The *avr-c++* and *avr-g++* files are identical. The *avr-cpp*, *avr-gcc*, and *avr-gcc-4.5.3* files are also identical. If you're curious, you can get the version information by typing avr-gcc -v and avr-g++ -v.

After the source code is compiled, the linker (*avr-ld*) combines all the binary modules into a single executable. The source modules must have already been compiled into object files before they can be processed by the linker.

Compiling and linking isn't the end of the process, however, because the binary executable image file created by the linker must be converted into a so-called Intel Hex file, which contains an ASCII representation of the binary code in the form of ASCII hex characters. Then it can be uploaded to the target board, where the bootloader will translate the ASCII hex back into binary values and write them into the flash memory on the microcontroller.

Other members of the GNU toolchain, such as *ar*, *ranlib*, *nm*, *ld*, and *strip*, also have AVR versions. Figure 6-4 shows a diagram of how the compiler, linker, converter, and uploader all work in sequence to get a program into a compiled form, link in necessary functions from library modules, and then transfer the finished program to an AVR target.

The object files shown in Figure 6-4 might be the compiled code from other modules in your project, or they may be from code supplied with a sensor or other accessory, or they could be runtime AVR-GCC support code such as the `main()` function that the Arduino IDE supplies. The library objects shown could be actual binary libraries such as *avr-libc* or other libraries created with *avr-ar* and *avr-ranlib*, or they might be object files created when an Arduino-style library is compiled prior to linking.

 One component not mentioned yet is the runtime AVR-GCC support library, *avr-libc*, which provides many of the same functions found in a standard C runtime AVR-GCC library. It also includes support functions specific to AVR microcontrollers. *avr-libc* is described in detail in "avr-libc" on page 135.

Installing the Toolchain

Although the toolchain components are typically installed for you when you install the Arduino IDE using a package manager, the Windows installer, or the Mac OS X ZIP file, you can install these components individually if you don't plan on using the Arduino IDE.

At a minimum, you will need the following packages (these are Linux package names):

- *avr-lib*
- *avrdude*
- *avrdude-doc*
- *binutils-avr*
- *gcc-avr*
- *gdb-avr*

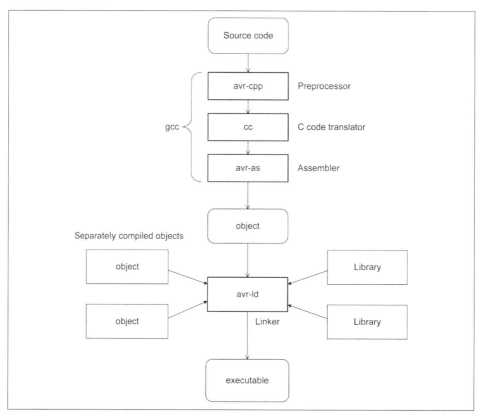

Figure 6-4. The AVR-GCC toolchain

If you are using Windows, then you may also want to consider installing other Unix/Linux-compatible tools such as *grep*, *fgrep*, *ls*, and *diff*. These are already present on Linux and Mac OS X platforms. Although not specifically described here, it is possible to install the AVR GNU toolchain on systems running Solaris or BSD Unix (FreeBSD, for example), or something even more off the beaten path. You just need a really good reason, a lot of patience, and possibly a significant amount of skill in working with source code packages. In general, it is much easier to stick with systems that already have ready-made installation packages available.

Windows installation

As previously mentioned, the Arduino package comes with the necessary compiler and *binutils* programs. Just install the Arduino IDE, and it will install the toolchain components it needs in the same directory where the main executable code for the IDE is located.

The Atmel Corporation provides precompiled binary installation packages, basic documentation, and the source code for AVR *gcc* and *binutils*. You can access the installation packages and other resources from the Atmel website (*http://bit.ly/atmel-avr-win*). Source code packages are also available on the website (*http://bit.ly/atmel-source*).

Another easy way to install the AVR-GCC toolchain on a Windows system is with the package WinAVR (*http://winavr.sourceforge.net*). The install script creates a directory (it's *C:\WinAVR-20100110* on my machine), and the binary executables, libraries, and include files are placed there. There is also a good collection of documentation in both PDF and HTML formats in *C:\WinAVR-20100110\doc*. Be forewarned that WinAVR has not been updated since about 2010, so some of the toolchain components are a bit stale, but they're still usable for most projects. The Atmel version appears to be more recent.

A collection of GNU utilities precompiled for Windows can be found at GnuWin32 (*http://gnuwin32.sourceforge.net*). This is a large collection of system tools and utilities, but it doesn't include the *gcc/g++* compilers or the *binutils* packages. You can obtain those from the Atmel or WinAVR download locations.

You can find more packages and tutorials on Google by searching for "Windows avr binutils." Those who like to use the Eclipse editor/IDE environment and want to try it for AVR software development might want to check out the "AVR Eclipse Environment on Windows" (*http://bit.ly/protostack-eclipse*) tutorial on Protostack.com.

Linux installation

On most Linux systems the installation of the GNU *gcc*, *binutils*, and other AVR-related packages is just a matter of selecting packages using a software package manager. You can have both the "normal" GCC toolchain components and the AVR toolchain components installed at the same time because the AVR versions of the compiler and *binutils* tools will have a prefix of "avr-". I wouldn't recommend attempting to build *avr-gcc* or any of the other members of the toolchain from source unless you have a very compelling reason to do so, and you happen to be very comfortable working with large, complex source code packages.

Mac OS X installation

Because Mac OS X is based on a BSD Unix foundation, a lot of what can be said about Linux can also apply to Mac OS X. The good folks at Adafruit have created a helpful guide (*http://bit.ly/avr-osx*) to installing the AVR GNU toolchain in a Mac OS X environment. That page's link to the Mac package is obsolete, but see the next paragraph for the current link.

The CrossPack development environment from Object Development contains all of the toolchain components needed to develop AVR software in a Mac OS X environment, and it doesn't require Xcode to compile. You can download it from the Object Development website (*http://bit.ly/crosspack-avr*). Note that this is just the toolchain; it doesn't provide a GUI IDE or an editor, so you'll need to install those yourself.

A version of the Eclipse IDE is available for Mac OS X with the AVR toolchain, and of course there is a version of the Arduino IDE available for the Mac OS X platform.

make

For small programs using the toolchain programs from the command line may be fine, but when things start to expand it is handy to have some way to automate the process. The tool of choice for this in a command-line environment is *make*.

make is a type of interpreter, and it processes what are called *makefiles*. The language used by *make* is not a general-purpose programming language, but rather a set of actions, rules, and relationship definitions. You can think of it as a form of a script, and it is often referred to as a macro language because the statements use replacement and substitution to build commands for other tools. *make* does more than this, however, since it can also detect source file changes and track dependencies between source files (for example, if A depends on B and C, and B changes, then B will be recompiled and A will also need to be recompiled to incorporate those changes). *make* was initially created by Richard Stallman, Roland McGrath, and Paul D. Smith.

Many IDEs for large-scale code development utilize *make* as the "backend" for compilation. The PlatformIO and Ino tools also use *make*, but they do it in a way that hides what is going on, and then clean up afterward. There are also tools available that automate the process of creating input for the *make* tool, and if you have ever used the *configure* tool to build a software package, then you've seen this type of utility in action.

The basic idea behind *make* is managing large sets of program source files. It can determine which source files have changed, and what other source files may need to be recompiled if they depend on the files that have changed. The *make* utility can invoke compilers, linkers, automatic documentation generators, and even other makefiles (for situations where the source code may be distributed over multiple subdirectories). *make* can also detect when a tool such as *avr-gcc* or *avr-ld* has encountered an error.

A good place to start in order to get an idea of how *make* is used is looking at the makefiles found in existing projects. Describing all of the capabilities of *make* is far beyond the scope of this book (the official GNU manual is over 200 pages long), and there are many books available that cover *make* and its applications in detail (see

Appendix D). You can download the official user's manual for *make* in PDF format from the GNU website (*http://www.gnu.org/software/make/manual/make.pdf*).

avr-gcc

GCC is an acronym for GNU Compiler Collection. The GCC is based on the concept of utilizing different frontend symbolic processors for specific languages, and then passing the resulting intermediate code to a backend for a specific target platform. The *avr-gcc* cross compiler is built from the GCC source and preconfigured to generate object code specifically for the AVR family of microcontrollers. The GCC can generate object code for many different types of processors. These include the Intel CPUs found in PCs, SPARC RISC CPUs, the 68000 family of large-scale microprocessors, the MSP430 MCUs from Texas Instruments, various ARM-based MCUs, and many others.

avr-gcc and *avr-g++* accept a number of command-line arguments, also called switches. These define things like optimization, compilation mode, pass-through arguments for the linker, paths for include files, and the target processor. Although there are literally hundreds of command-line switches, only a few are necessary to successfully compile code for a specific target processor.

Compilation may involve up to four steps: preprocessing, compilation, assembly, and linking. These always occur in this order. The preprocessor expands all include statements (#include) by loading the text in the named file into the current source. Normally #include statements should only be used to include so-called header files, like *stdio.h* or *stdlib.h*, and not source files. (Although that's generally considered to be a bad idea, I've seen it done.) The preprocessor also strips out any comments and interprets any conditional preprocessor statements, such as #ifdef, #else, and #endif. The output of the preprocessor is a squeaky-clean source file containing pure source code and nothing else. You can use the -E switch to stop the process after the preprocessor is finished and examine the stripped source code.

The GCC is basically a translator for a particular C-like language such as traditional or ANSI C, C++, Objective-C, or Objective-C++. The output from the compiler is an intermediate assembly language file. Usually this would be the input into the assembler (*avr-as*), which in turn will generate an object file. The intermediate assembly language file is deleted after the assembler executes. It is possible to stop the process just before the assembler is run and examine the assembly language output using the -S switch.

The -c switch is used to create an object file without invoking the linker. This is used in makefiles where all the source files are compiled first and then linked in a single step. The compiler also has switches for optimization, warnings (it can generate a lot of warnings), and path specification so that include files can be located. The -o switch specifies the name of a compiled executable image; the default is usually *a.out.*

For more information refer to the GCC manual pages, or you can download a user manual in PDF, PostScript, or HTML format from the GCC online documentation (*https://gcc.gnu.org/onlinedocs*). GCC is not a simple utility, and the number of available options borders on overwhelming. Fortunately, you don't need to use all of them to create working executable code.

binutils

The GNU *binutils* are a collection of programs that handle the various tasks involved with converting the output of the compiler into something that a processor can execute. Table 6-2 lists the contents of the *binutils-avr* package for a Linux system. This suite of tools contains everything needed to assemble, link, convert, and process binary executable files into a form suitable for a target AVR microcontroller. Manuals for the *binutils* tools, and for most all other GNU software as well, can be found on the official web page at GNU Manuals Online (*http://bit.ly/gnu-manuals*).

Table 6-2. AVR binutils collection

Tool	Description
avr-addr2line	Converts addresses into filenames and line numbers
avr-ar	Creates object code archives, and modifies or extracts code
avr-as	The portable GNU assembler for the AVR
avr-c++filt	Demangles C++ symbols
avr-ld	The GNU linker for AVR object code
avr-nm	Lists symbols embedded in object files
avr-objcopy	Copies and translates object files
avr-objdump	Displays information from object files
avr-ranlib	Generates index for a library archive
avr-readelf	Displays information about ELF files
avr-size	Lists object file section sizes and total size
avr-strings	Prints strings of printable characters in binary files
avr-strip	Discards symbols from AVR object files

The essential support utilities needed to build programs for an AVR MCU are *avr-ar*, *avr-as*, *avr-ld*, *avr-objcopy*, and *avr-ranlib*, but the other components in the binutils suite may or may not be of use to you in your software development efforts. The main applications are:

avr-ar

> *avr-ar* is used to create binary object code archives, or static libraries. It can also be used to modify an existing library or extract code from a library. A binary library file (usually with a *.a* extension) is a collection of binary code modules (i.e., object modules) with a master index (created using *avr-ranlib*, described momentarily). An object code library is referred to as a "static" library because any component that is used in another program is incorporated into and becomes a permanent, or static, part of the final executable object. If you're curious, a dynamic shared library is a different sort of thing, and since these aren't typically used with AVR devices (or any small microcontroller, for that matter) they are not covered in this book.

avr-as

> *avr-as* is the portable GNU assembler for the AVR family of MCUs. Although it is often used in conjunction with the GCC, it can also be used as a standalone assembler (as discussed in "AVR Assembly Language" on page 140). There are other assemblers available for AVR microcontrollers, and these are discussed in "AVR Assembly Language" as well, but only *avr-as* is intended to be used with the gcc/g++ compilers.

avr-ld

> *avr-ld* is typically the last step in the process of creating an executable binary object. The primary function of the linker is to combine two or more object files, resolve any address references between them, and relocate data as necessary.
>
> When an executable is built from multiple object files, each of the object files may contain a reference to a function or data that does not exist within a particular object, but does exist in another object. The AVR version of *libc*, discussed next, is an example of this type of situation. For instance, a program may refer to something like `atoi()` (ASCII-to-integer), but not include the source for `atoi()` within itself. When the program is compiled into an object file (a *.o* file) the compiler will leave a hole, so to speak, in the binary code that refers to `atoi()`. The linker detects this empty location, finds the address of the `atoi()` function in a library (i.e., *avr-libc.a*), writes the external address into the code, and then includes the object code for the `atoi()` function in the final binary executable image.

avr-objcopy

> The *avr-objcopy* utility copies the contents of an object file to another format, typically a so-called Intel-format ASCII hex file suitable for uploading to an AVR MCU using the Arduino bootloader. *avr-objcopy* can also generate a type of ASCII hex file called an S-record, which is commonly used with Motorola (Freescale) MCUs.

The Intel hex format file might look something like the following, which shows the beginning and end lines of the Arduino bootloader for an ATmega168 or ATmega328 MCU:

```
:107800000C94343C0C94513C0C94513C0C94513CE1
:107810000C94513C0C94513C0C94513C0C94513CB4
:107820000C94513C0C94513C0C94513C0C94513CA4
:107830000C94513C0C94513C0C94513C0C94513C94
:107840000C94513C0C94513C0C94513C0C94513C84
:107850000C94513C0C94513C0C94513C0C94513C74
:107860000C94513C0C94513C11241FBECFEFD8E036
<em>...more data here...</em>
:107F700009F0C7CF103011F00296E5CF112480919F
:107F8000C00085FFB9CEBCCE8EE10E94C73CA2CD19
:0C7F900085E90E94C73C9ECDF894FFCF0D
:027F9C00800063
:040000030000780081
:00000001FF
```

Many lines have been omitted from the middle part of the listing for the sake of brevity, but you can find the original file, and others, in the directory */usr/share/arduino/hardware/arduino/bootloaders* on a Linux system, or in *C:\Program Files\Arduino\hardware\arduino\avr\bootloaders* on a Windows system.

In this listing, each line contains a start code (the : character), a byte count (which for all but the last four lines in our example is hex 10, or 16), the location where the code is to be written in the MCU's flash memory (the bootloader can alter this if need be), a record type code, the actual code written as up to 32 ASCII hex characters (2 characters per byte, for 16 bytes), and an end-of-line checksum. You can learn more about the Intel hex file format at Wikipedia (*https://en.wikipedia.org/wiki/Intel_HEX*), although it is seldom necessary to examine a hex file directly.

To convert a binary executable to a hex file you could use *objcopy* like so:

```
avr-objcopy -O ihex execpgm execpgm.hex
```

objcopy, like almost all GNU tools, is capable of much more and has a plethora of command-line options, most of which you will probably never find a use for. The online manual for *objcopy* can be found at Sourceware.org (*http://bit.ly/sw-objcopy*).

avr-ranlib

avr-ranlib generates an index for inclusion into a binary object archive file. This helps to speed up the linking process, because this is what the linker will use to locate the address of an object needed to fill in a link "hole" in another object file. If the index is not available, then the linker will have to scan through the library file, object by object, looking for a suitable match.

avr-libc

avr-libc is an AVR version of the C/C++ runtime library. Together with *avr-gcc* and *avr-binutils* it forms the core of the GNU toolchain for AVR microcontrollers.

External libraries supplied with the AVR toolchain, such as *arv-libc*, should be in a standard location, which on a Linux system would be something like */usr/lib/avr/lib/* or */usr/local/avr/lib*, depending on how *avr-gcc* was built and how your system is configured. External libraries can be in any directory, actually, just so long as the linker can find them.

avr-libc is the one critical component not provided with *avr-gcc* and *binutils*. It is a standard C/C++ library that contains AVR equivalents of the same functions found in a regular (i.e., GNU *libc*) standard C library, with some limitations related to the capabilities of AVR MCUs (limited available memory, for example).

Table 6-3 lists the include files available with *avr-libc*. If you are experienced with C or C++ programming on a full-size system, then most of these will look familiar to you.

Table 6-3. Common include files provided by avr-libc

Filename	Description
alloca.h	Allocates space in the stack frame of the caller
assert.h	Tests an expression for false result
ctype.h	Character conversion macros and ctype macros
errno.h	Defines system error codes
inttypes.h	Integer type conversions
math.h	Basic math functions
setjmp.h	Defines nonlocal goto methods `setjmp()` and `longjmp()`
stdint.h	Defines standard integer types
stdio.h	Standard I/O facilities
stdlib.h	General utilities
string.h	String operations and utilities

Some of the include files supplied with *avr-libc* are unique to the AVR target; these are listed in Table 6-4. These include files are located in the */usr/lib/avr/include/avr* directory (on a Linux system). Some define functions and constants for things like boot management, time delays, EEPROM access, fuse settings, and port pin functions. Others define the interrupts and I/O mappings for specific processor types.

Table 6-4. AVR-specific include files provided by avr-libc

boot.h	io90pwm316.h	iom169pa.h	iom32u2.h	iomx8.h	iotn45.h	iox128a3.h
builtins.h	io90pwm3b.h	iom169p.h	iom32u4.h	iomxx0_1.h	iotn461a.h	iox128d3.h
common.h	io90pwm81.h	iom16a.h	iom32u6.h	iomxx4.h	iotn461.h	iox16a4.h
cpufunc.h	io90pwmx.h	iom16.h	iom406.h	iomxxhva.h	iotn48.h	iox16d4.h
crc16.h	io90scr100.h	iom16hva2.h	iom48.h	iotn10.h	iotn4.h	iox192a3.h
delay.h	ioa6289.h	iom16hva.h	iom48p.h	iotn11.h	iotn5.h	iox192d3.h
eeprom.h	ioat94k.h	iom16hvb.h	iom640.h	iotn12.h	iotn84a.h	iox256a3b.h
fuse.h	iocan128.h	iom16hvbrevb.h	iom644.h	iotn13a.h	iotn84.h	iox256a3.h
interrupt.h	iocan32.h	iom16m1.h	iom644pa.h	iotn13.h	iotn85.h	iox256d3.h
io1200.h	iocan64.h	iom16u2.h	iom644p.h	iotn15.h	iotn861a.h	iox32a4.h
io2313.h	iocanxx.h	iom16u4.h	iom6450.h	iotn167.h	iotn861.h	iox32d4.h
io2323.h	io.h	iom2560.h	iom645.h	iotn20.h	iotn87.h	iox64a1.h
io2333.h	iom103.h	iom2561.h	iom6490.h	iotn22.h	iotn88.h	iox64a1u.h
io2343.h	iom1280.h	iom3000.h	iom649.h	iotn2313a.h	iotn9.h	iox64a3.h
io43u32x.h	iom1281.h	iom323.h	iom649p.h	iotn2313.h	iotnx4.h	iox64d3.h
io43u35x.h	iom1284p.h	iom324.h	iom64c1.h	iotn24a.h	iotnx5.h	lock.h
io4414.h	iom128.h	iom324pa.h	iom64.h	iotn24.h	iotnx61.h	parity.h
io4433.h	iom128rfa1.h	iom3250.h	iom64hve.h	iotn25.h	iousb1286.h	pgmspace.h
io4434.h	iom161.h	iom325.h	iom64m1.h	iotn261a.h	iousb1287.h	portpins.h
io76c711.h	iom162.h	iom328p.h	iom8515.h	iotn261.h	iousb162.h	power.h
io8515.h	iom163.h	iom3290.h	iom8535.h	iotn26.h	iousb646.h	sfr_defs.h
io8534.h	iom164.h	iom329.h	iom88.h	iotn28.h	iousb647.h	signal.h
io8535.h	iom165.h	iom32c1.h	iom88pa.h	iotn40.h	iousb82.h	signature.h
io86r401.h	iom165p.h	iom32.h	iom88p.h	iotn4313.h	iousbxx2.h	sleep.h
io90pwm1.h	iom168.h	iom32hvb.h	iom8.h	iotn43u.h	iousbxx6_7.h	version.h
io90pwm216.h	iom168p.h	iom32hvbrevb.h	iom8hva.h	iotn44a.h	iox128a1.h	wdt.h

avr-libc also includes a number of utility and compatibility include files, as shown in Table 6-5. The files *delay.h*, *crc16.h*, and *parity.h* in the *avr/* directory actually point to include files in the *util/* directory, so if you include, say, *<avr/parity.h>* in your code it will actually use *<util/parity.h>*.

Table 6-5. Utility and compatibility include files provided by avr-libc

Filenames	Description
util/atomic.h	Atomically and nonatomically executed code blocks
util/crc16.h	CRC computations
util/delay.h	Convenience functions for busy-wait delay loops

Filenames	Description
util/delay_basic.h	Basic busy-wait delay loops
util/parity.h	Parity bit generation
util/setbaud.h	Helper macros for baud rate calculations
util/twi.h	TWI bit mask definitions
compat/deprecated.h	Deprecated items
compat/ina90.h	Compatibility with IAR EWB 3.x

For information on using *avr-libc* refer to the user manual, and don't forget to check the include files themselves for notes regarding applications and limitations. Bear in mind that things like `malloc()` and `printf()`, while they can be used with an AVR MCU, have limitations in the memory-constrained environment of the AVR. Math is another issue, since the AVR MCUs do not have floating-point math functions. They support integer and fixed-point math, but floating-point math must be done using a floating-point processor simulation. This is slow, so avoid it if at all possible.

The *avr-libc* home page can be found at *http://www.nongnu.org/avr-libc/*.

 A bzip2 compressed version of the PDF user documentation is located at *http://bit.ly/avr-libc-manual*. The documentation for *avr-libc* also covers the AVR toolchain.

Building C or C++ Programs from Scratch

If you want to build your own software without using the Arduino IDE and its standard runtime AVR-GCC and libraries, you can definitely do that. It is, however, often easier to get something up and running using the Arduino tools, and then move on to rewriting the parts that need optimization or customization.

Compiling with avr-gcc or avr-g++

What *avr-gcc* or *avr-g++* does with the source code internally is not something you would normally be concerned about, but you can, of course, read more about the internals of the GCC tools at GNU.org (*https://gcc.gnu.org/onlinedocs/gccint*).

The commands `gcc` and `g++` actually do more than just compile source code. If you use a command like this:

```
avr-gcc -mmcu avr5 -o test test_src.c -L../avrlibs -lruntime
```

gcc will compile *test_src.c* for the ATmega32U4, call *avr-ld* to link it to file called *libruntime.a* located in *../avrlibs*, and then put the final result into a binary executable image called *test*.

If you just want the compiler to compile something into an object file for linking at a later step, you can use the -c (compile only) switch. This is common in makefiles where multiple source files are compiled first, and then linked into an executable file (possibly with external libraries as well) as a final step. The command for this is simple:

```
avr-gcc -mmcu avr5 -c test_src.c
```

avr-gcc also supports a suite of switches for things like optimization, warnings, and alternate paths for include (header) files.

Multiple Source Files and make

When working with multiple source files, the *make* utility is essential. You can write your own makefile (which is what I typically do), or use a tool like *arduino-mk*. This is a predefined makefile that incorporates the necessary logic to build Arduino sketches using the components of the AVR-GCC toolchain.

You can download *arduino-mk* from GitHub. For some Linux platforms it is available as an installable package. If you do install it from a package manager, note that it will most likely put the main file, *arduino-mk*, in the same directory where the Arduino IDE installed its components. The main website is at Hardware Fun (*http://bit.ly/hf-makefile*), and you can download a ZIP archive from GitHub (*http://bit.ly/gh-makefile*). Once *arduino-mk* is installed, you will probably need to install some additional support functions and define some shell environment global variables. Refer to the documentation on the Hardware Fun website for details.

If you are not familiar with the language of makefiles, then using them might involve a steep learning curve, but I think it's well worth the effort if you want to move beyond an IDE and into the realm of hardcore embedded systems development. Not every MCU has an IDE available for it, and not every IDE can do all of the things that can be accomplished when you have direct control over the build process.

We can reuse the concepts from Example 5-5 to create a simple makefile such as the one shown next. In this case we are supplying our own main() function in the source file *main.cpp*, the equivalent of the setup() and loop() functions in *alarmloop.c*, global variables in *globals.c*, and the #define statements in the file *defs.h*. Our makefile looks like this:

```
CC = avr-gcc
LD = avr-ld
OC = avr-objcopy

SRCS    = main.c alarmloop.c globals.c
HDRS    = main.h globals.h defs.h
OBJS    = main.o alarmloop.o globals.o
EXEC    = alarm
TARGET  = alarm.hex
```

```
$(TARGET): $(OBJS)
        $(LD) -o $(EXEC) $(OBJS)
        $(OC) -O ihex $(EXEC) $(TARGET)

main.o: main.c $(HDRS)
        $(CC) -mmcu avr5 -Wall -c main.c

alarmloop.o: alarmloop.c $(HDRS)
        $(CC) -mmcu avr5 -Wall -c alarmloop.c

globals.o: globals.c
        $(CC) -mmcu avr5 -Wall -c globals.c
```

The directory for this simple project would contain, at a minimum:

```
Makefile
main.c
main.h
alarmloop.c
alarmloop.h
globals.c
globals.h
```

Just typing make will start the process. *make* looks in the current directory for a file named *Makefile* (note the capital *M*, that's required) and will load and process it if found. You can use another name for the makefile if you wish by telling *make* what to look for, like so:

```
make -f mymake
```

Our sample makefile has several features that are worth covering briefly. First is the declaration of alias names, such as *CC*, *LD*, and *SRCS*. These are expanded when encountered further on in the file, and save on repetitive typing as well as perform substitutions with conditional statements.

Another thing to note is the rules. In this case I've used explicit rules in the form of:

```
target: dependencies
    action
```

The `action` is indented using a tab character. *make* requires this as part of its syntax, so be careful not to use spaces here. Also note that there can be as many actions under a rule as you like, so long as each is on its own line and has a leading tab.

When *make* evaluates a rule it looks at the dependencies to determine what is necessary to build the target. In the case of *main.o*, the rule specifies *main.c* and the other include (header) files, which are defined by the macro $(HDRS). If the target does not yet exist then it will always be built, but if it does exist then *make* will check to see if any of the dependencies have changed by examining the time and date stamps of the files. Only if there has been a change will the target be recompiled. In the case of $ (TARGET) the rule states that the two subsequent actions (linking and object conversion) must occur if any of the source objects have changed.

So, for example, if you edit *alarmloop.c* to change something and then run *make*, the first thing it will do is recompile *alarmloop.c* to create a new *alarmloop.o*. It will then rebuild $(TARGET) since the *alarmloop.o* object has now changed. If the include file *alarmloop.h* has changed, then it will recompile both *main.c* and *alarmloop.c* before rebuilding $(TARGET).

Creating a makefile can be a bit of a chore, but once it's done you generally don't need to modify it very often, except perhaps to add a new source module or change a compiler or linker switch setting.

AVR Assembly Language

For those who really want to wring out every last bit of performance from an AVR device, there is assembly language programming. This is definitely not for the faint of heart, but it allows you to take full control of what the MCU does and when it does it. In some situations this level of control may be necessary to get the best possible performance with the least amount of program memory. For some of the AVR MCUs, such as the ATtiny series, assembly language is really the only way to go, as a C or C++ program may easily compile down into something that will be too large to fit comfortably in the limited amount of memory.

This section provides a very, very high-level overview of AVR assembly language programming. As stated in the preface, this book is not a programming tutorial; it is a hardware reference and guide to sources of more detailed information. The intent of this section is to give you a sense of what is involved with assembly language programming so that you can decide if it is something you want to pursue. To do the subject justice would require another, much larger, book. Fortunately there are already such books available, and you can also download a lot of useful information from various websites, such as those listed in "AVR Assembly Language Resources" on page 146.

Because assembly language is closer to the hardware than any other language, it is sometimes referred to as *machine language*. Actually, machine language is comprised of the binary codes for various instructions; assembly language is a human-readable form of the machine language. The assembler's job is to translate from human readable to machine readable, and also add some handy features like macros, symbolic names, conditional assembly, and the ability to refer to functions in external libraries.

In assembly language we can write something like MOV R6, R7, which copies the contents of register R7 into register R6, and it will be translated by the assembler into a binary form (machine language) that the MCU can execute. The names used for the various operations carried out by the MCU's process are called mnemonics, and the parameters that follow a mnemonic (if any) are called operands.

Programming in assembly language is the art of providing the detailed step-by-step instructions necessary for a CPU to do something useful. A language like C takes care of the details so the programmer doesn't have to decide which registers to use and what status bits to check. With assembly language there is nothing between the programmer and the fundamental logic of the processor, and even the simplest operation must be explicitly described.

The AVR Programming Model

Internally, an AVR MCU consists of an AVR core, flash and EEPROM memory, and a suite of peripheral functions. The core contains an instruction register and decoder, a program counter, a small amount of RAM, various status bits, 32 general-purpose registers, and an Arithmetic Logic Unit (ALU). Figure 6-5 shows how the various components are organized inside an AVR MCU. Refer to Chapter 2 and Chapter 3 for additional details on the internal functions of AVR MCUs.

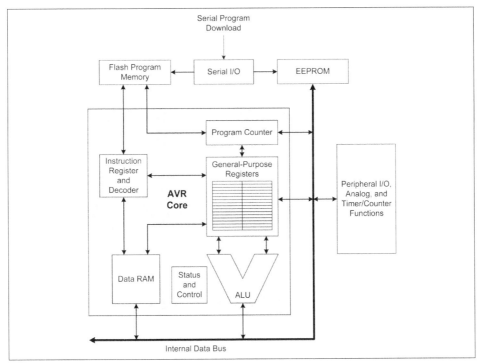

Figure 6-5. AVR CPU block diagram

Instructions operate on registers and memory, in that data can be copied from one register to another; two registers can be compared, swapped, added, subtracted, multiplied, and divided (to name a few of the operations); and data can be read from flash memory, EEPROM, or RAM. Peripheral functions are controlled and accessed via registers as well, but these are not part of the core set of 32 general-purpose registers. The three primary characteristics of an MCU, memory, instructions, and registers, are described here:

Memory organization

As mentioned in Chapter 2, the Atmel AVR MCU uses what is called a Harvard architecture. In a Harvard architecture the program code is stored in read-only (flash) memory and modifiable data (variables) are stored in a separate memory space (the RAM in the AVR core). Other microprocessors may use an alternate scheme called the Von Neumann architecture, in which programs and data share the same memory space.

In an AVR MCU the general-purpose registers and the I/O registers used by the peripheral functions are technically part of the read/write memory space. Figure 6-6 shows how memory is laid out in the AVR MCU.

Figure 6-6 is intentionally generic. The assembler directives CSEG, DSEG, and ESEG (used by the Atmel assembler) refer to the code, data, and EEPROM memory spaces, respectively. The highest addresses for the CSEG, DSEG, and ESEG memory spaces will vary from one AVR type to another. Note that space for an optional bootloader is reserved at the end of the CSEG space.

Instruction processing

An AVR MCU utilizes what is called a *single-level pipeline* to fetch, decode, and execute instructions from flash memory. While one instruction is being executed the next instruction is being prefetched from memory, and it will be ready to be decoded and executed as soon as the current instruction completes. This feature, and the RISC nature of the AVR core, is what allows an AVR MCU to execute most instructions in a single clock cycle.

Registers

Many of the AVR instructions set or clear bits in an 8-bit status register (SREG). Each bit has a specific purpose, as shown in Table 6-6, and the true or false (1 or 0) state of certain bits is checked by instructions such as BREQ (Branch if Equal) or BRNE (Branch if Not Equal) following a CP (Compare) instruction, which can alter the Z, C, N, V, H, and S status bits.

The 32 general-purpose registers are organized as 8-bit registers, with the final three pairs (26-27, 28-29, and 30-31) available as 16-bit index (indirect address) registers referred to as the X, Y, and Z pointers, respectively.

A limited number of instructions can operate on two 8-bit registers as a 16-bit register pair. The lower-numbered register of the pair holds the least significant bits. The least significant register must also be even numbered, so the pair of r0:r1 is valid, but r1:r2 is not.

Table 6-6. AVR SREG status bits

Bit	Symbol	Function
0	C	Carry flag
1	Z	Zero flag
2	N	Negative flag
3	V	Two's complement overflow indicator
4	S	For signed tests
5	H	Half carry flag
6	T	Transfer bit used by BLD and BST instructions
7	I	Global interrupt enable/disable flag

In addition to the status register, the general-purpose registers, and the I/O registers, the AVR has a program counter (PC) register and a stack pointer (SP) register. These are modified by certain instructions (a jump will modify the PC, for example) or a subroutine call, which utilizes both the PC and SP registers. For example, when a CALL instruction is encountered the current value of the PC is adjusted to point to the instruction following the CALL, then it's pushed onto the stack. The PC is then loaded with the start address of the subroutine. When the subroutine returns via a RET instruction, the saved PC is "popped" from the stack and the program resumes at the location immediately following the CALL instruction.

There are, of course, many other details about the inner workings of AVR MCUs and AVR assembly language that have not been covered here. If you want to explore assembly language programming with an AVR, then reading a good book or two is highly recommended. Appendix D contains some suggestions, and the links already mentioned are also very useful.

Creating AVR Assembly Language Programs

Depending on the assembler, single-line comments may begin with a semicolon (;), a hash or pound (#) symbol, or some other character. The GNU assembler (*avr-as*) supports multiline comments with an opening and closing /* and */, like those found in C code. Refer to your assembler's documentation for specifics.

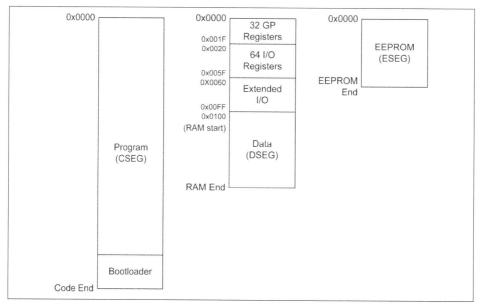

Figure 6-6. AVR memory layout

Comments in assembly language are very important. Say, for example, you have some assembly language that looks like this:

```
LOOP:
LDI             R16, 0X00
OUT             PORTB, R16
LDI             R16, 0XFF
OUT             PORTB, R16
RJMP    LOOP
```

What is it doing? In this case it's fairly easy to see that all it does is load register 16 (R16) with either the value of 0 or 255 (0xFF) and then write that value to PORTB. The pins for PORTB will toggle on and off as fast as the MCU can execute the loop.

We can improve on this and add in the rest of the missing bits for things like program origin and port initialization. The result looks like this:

```
; set power-up/reset vector
.ORG 0X0000
RJMP MAIN

; code entry point
.ORG 0X0100
MAIN:
LDI             R16, 0XFF       ; load R16 with 0b11111111
OUT             DDRB, R16       ; set port B data direction register

; endless loop - toggles port B pins on and off
```

```
LOOP:
LDI         R16, 0X00       ; load R16 with zero
OUT         PORTB, R16      ; write to port B
LDI         R16, 0XFF       ; now load it with 0xFF
OUT         PORTB, R16      ; and write it to port B
RJMP   LOOP             ; jump back and do it again
```

I've gone bit overboard with the comments, but in assembly language it is not uncommon to see a comment on almost every line. Bear in mind that even something as simple as the common "hello world" program requires many more assembly language statements than it does C or C++ statements to achieve the same result, and some of those assembler statements may not be intuitively obvious.

Most assemblers provide a set of predefined keywords called *directives*. The previous example has one, the .ORG directive. Table 6-7 lists a few more useful directives. These keywords are called directives because they direct the assembler to make specific associations or perform certain actions.

Table 6-7. AVR assembler directives

Directive	Operation
BYTE	Reserve one or more bytes as a variable
CSEG	Use the code segment
CSEGSIZE	Configure code segment memory size
DB	Define a constant byte or bytes
DEF	Define a symbolic name for a register
DEVICE	Define which device to assemble for (the target)
DSEG	Use the data segment
DW	Define a constant word or words (16-bit values)
ENDM, ENDMACRO	End of a macro definition
EQU	Assign a symbol to an expression
ESEG	Use EEPROM segment
EXIT	Exit from file
INCLUDE	Read and include source from another file
LIST	Enable list file generation
LISTMAC	Enable macro expansion in list file
MACRO	Start of a macro definition
NOLIST	Disable list file generation
ORG	Set program origin
SET	Assign a symbol to an expression

The following code fragments show how some of the directives can be used.

```
; Disable listing generation when including the external files.
; This helps keep the listing output neat and uncluttered.

.NOLIST
.INCLUDE "macrodefs.inc"
.INCLUDE "mcutype.inc"
.LIST


; Define a macro to do something useful

.MACRO  SUBI16                  ; Define macro
        SUBI    @1,low(@0)      ; Subtract low byte
        SBCI    @2,high(@0)     ; Subtract high byte
.ENDMACRO                               ; End macro definition

SUBI16 0x2200,R16,R17    ; Subtract 0x2200 from R17:R16
```

AVR Assembly Language Resources

There are many good sources of information and useful tutorials available online if you want to delve deeper into AVR assembly:

- The AVR Assembler Site (*http://avr-asm.tripod.com*) has a wealth of information, all neatly organized into various categories.

- AVRbeginners.net (*http://www.avrbeginners.net*) is a very slick website with lots of details on the inner workings of AVR MCUs and some assembly language examples.

- Atmel has the online reference AVR Assembler (*http://bit.ly/avr-assembler*), and a description of Atmel's assembler can be found in the AVR Assembler User Guide (*http://bit.ly/avr-assembler-guide*). It's not the same as the *avr-as* assembler, but the general principles are similar.

- The *avr-as* assembler is described in the documentation found at GNU Manuals Online (*http://bit.ly/gnu-manuals*), and the *avr-libc* documentation also has a brief overview.

- Gerhard Schmidt has created a website (*http://bit.ly/avr-overview*) with a lot of useful information, and it's available in both English and German. The material is also available as a PDF file (*http://bit.ly/avr-overview-pdf*).

- You can download an assembly language summary from the Johns Hopkins University website (*http://bit.ly/jhu-assembler*).

This is just the tip of the iceberg, so to speak. For some book suggestions, refer to Appendix D.

Uploading AVR Executable Code

Compiling or assembling code into an executable file is only half the process. The other half involves uploading the executable code into an AVR MCU for execution. There are several ways to do this, including via the Arduino bootloader, the ICSP interface found on many Arduino boards (at least, the relevant pins are usually available), and a JTAG interface.

In-System Programming

Atmel application note AVR910 (*http://bit.ly/avr-insystem*) describes the In-System Programming interface used on AVR MCUs. This is what is called the ICSP interface on Arduino boards. It is basically an extension to the SPI interface of the AVR MCU. This function (serial I/O) was shown as a separate functional block in Figure 6-5 because it has the ability to communicate directly with the flash and EEPROM memory in the MCU.

Flash Memory Lifetime

All flash memory has a finite limit to the number of write operations it can endure. In modern components this is usually a large number (it can vary from one part to another), with 10,000 or more write cycles not uncommon. This might not sound like a lot when compared to the amount of work that the typical disk drive in a desktop computer performs, but bear in mind that even though it can be viewed as a type of slow disk drive (like a USB memory stick), the flash in a microcontroller is not a disk drive. It is used to load a program for execution, not to store runtime AVR-GCC data during operation (microcontrollers usually have some RAM available for that purpose, and something like a microSD wafer, with a life expectancy of around 100,000 write cycles, can be used to record large amounts of runtime AVR-GCC data). So, even if you were to reload the contents of the flash memory with program code once a day, it would still take 4 or 5 years to wear it out.

Figure 6-7 shows the pinout for the primary ICSP connector on an Arduino Uno (R2). There is a second ICSP connector on the board, which is used for the AVR MCU that handles the USB interface. It is wired the same, but there really isn't a reason to use it unless you need to reprogram that MCU as well (and perhaps lose USB functionality). Note that there is also a 10-pin connector format defined by Atmel, but it is not used with most Arduino-type boards. The 10-pin connector has the same signals as the 6-pin connector, and more ground connections to take up the additional 4 pins.

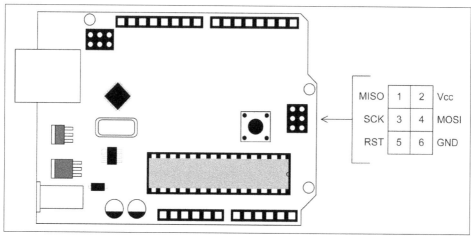

Figure 6-7. The ICSP connector on an Arduino Uno R2

In "binutils" on page 132 the *avr-objcopy* tool was described, with a focus on converting an executable binary image to an ASCII hex file. The reason for converting the binary image to a hex file is the ability of the Intel hex format to define record segments, addresses, and the end of the file, and then send the data over a communication link that might have issues with pure binary. The AVR MCU uses binary data to program the flash memory, but if you look at the AVR910 application note you can see that whatever is doing the programming has a lot of control over the process. The hex file is used by a programming device, not by the AVR MCU.

Programming with the Bootloader

In the case of an Arduino board, the programming device is the bootloader in flash. This allows the MCU to program itself, at the cost of losing some of the flash memory space to hold the bootloader. Without the bootloader it falls on the programming device to deal with final data placement (target addresses), MCU configuration bits (fuses), and other low-level details. (For more details on the fuse bits used in the AVR MCUs, refer to "Fuse Bits" on page 60).

It is important to note that the Arduino bootloader firmware uses the AVR's serial interface pins (RxD and TxD on pins D0 and D1, respectively), so if you want to attach something like an RS-232 converter you can use that to program the AVR, or you can use a USB-to-serial adapter like the SparkFun module shown in Figure 6-8. On Arduino boards that use an FTDI FT232L, an ATmega8, or an ATmega16U2 for the USB interface, a quick look at the schematic will show that the interface chip or MCU is using the D0 and D1 serial pins through 1K resistors, and the DTR signal is used to generate a reset of the primary AVR MCU.

Figure 6-8. SparkFun USB-to-serial adapter

You can still use the D0 and D1 pins, provided that they are isolated correctly. The partial schematic in Figure 6-9 shows how an FTDI FT232L is typically connected to an AVR MCU in an Arduino. Note that this does not apply to the Leonardo and other boards with the ATmega32U4, which has a built-in USB interface.

Uploading Without the Bootloader

If you really need to use the maximum available space in the AVR flash for a program, or you don't want to use the D0 and D1 pins in the standard Arduino way to upload a program, then you can load a compiled program directly using the ICSP interface. This approach works with a "fresh" ATmega MCU directly from Atmel, as well as with an Arduino board.

To upload without the Arduino IDE requires direct interaction with the AVRDUDE utility (described in "AVRDUDE" on page 152). Typically this would also involve the use of makefiles or some other technique to produce compiled code.

If you want the convenience of letting the Arduino IDE handle the compilation chores for you, then you can go that route as well. The Arduino IDE supports direct AVR device programming by allowing you to select a programming device via a menu item under Tools on the main menu. An upload can then be started using the "Upload Using Programmer" function in the File drop-down menu. If you have overwritten the bootloader firmware and then decide to go back to using the bootloader in the conventional manner, then it will need to be reloaded as described in "Replacing the Bootloader" on page 156.

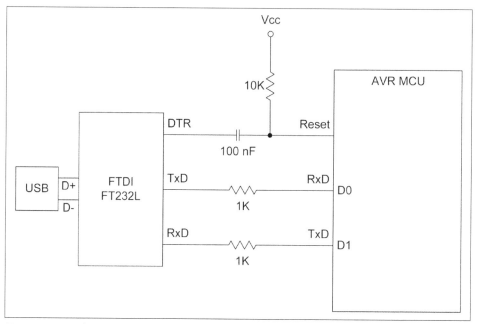

Figure 6-9. Arduino USB interface using an FTDI converter IC

Of course, if memory space is not an issue and the Arduino bootloader has already been installed, then there is no need to remove the bootloader. The ICSP interface works fine with or without the bootloader. If you elect to use the ICSP interface, you can simply ignore the bootloader.

Life without the bootloader requires a special programming device. Atmel tools such as the AVRISP MKII and the new Atmel-ICE are the gold standards because of their capabilities and compatibility. Sadly, Atmel has discontinued the AVRISP MKII in favor of the Atmel-ICE, but there are many compatible devices currently available. The AVRISP MKII did not support JTAG.

However, since the ISP interface is essentially an SPI serial port you can use a variety of devices to get the job done, including another Arduino board (more on this shortly). Some readily available programming devices include the USBtinyISP from Adafruit (Figure 6-10) and the Pocket AVR Programmer from SparkFun (Figure 6-11).

The USBtinyISP programmer is a kit, but it is relatively easy to assemble. You can read more about it on the Adafruit website (*http://bit.ly/usbtinyisp-avr*). The Pocket AVR Programmer comes preassembled, and you can find more information about it at SparkFun (*http://bit.ly/sf-pocketavr*).

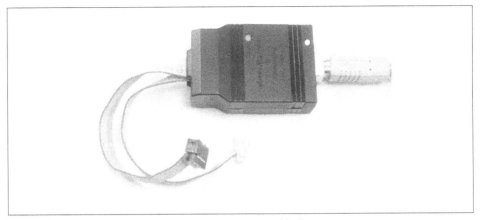

Figure 6-10. The USBtinyISP from Adafruit (assembled)

Figure 6-11. The Pocket AVR Programmer from SparkFun

In addition to loading software onto an AVR, a programmer can also be used to read registers, examine memory, and set fuse bits. The ability to set fuse bits is a compelling reason to have some type of ISP device available. For more information about fuse bits, see "Fuse Bits" on page 60.

JTAG

JTAG, an acronym for Joint Test Action Group, is a low-level interface designed to provide access to debugging facilities incorporated into an MCU or other logic device. The formal definition is found in the IEEE document *Standard Test Access Port and Boundary-Scan Architecture*, IEEE Standard 1149.1-1990. The latest version of this and other standards can be obtained from the IEEE (*https://standards.ieee.org*).

Not all AVR MCUs have JTAG support. As far as I can tell from the Atmel selection guides and datasheets, the XMEGA series devices have it, but the 8-bit Mega series parts (like those used in Arduino boards) do not. But since there are many types of AVR MCUs available, it is entirely possible that there are some XMEGA parts without JTAG, and some Mega parts that have it.

In most cases, though, you don't really need the advanced features of JTAG. It's nice when you want to step through the code with a debugger, or examine register contents on the fly, but it is often the case that just looking at the pins with an oscilloscope or logic analyzer will provide plenty of information.

As for accessing the internal functions in an AVR, you can use something like USBtinyISP to set the internal AVR fuse bits or load the EEPROM. So unless you have a real need for a JTAG tool, you can probably skip the expense.

AVRDUDE

An AVR programmer is good to have, but it needs something that can provide it with the data to be uploaded into the AVR MCU. That something is called AVRDUDE.

AVRDUDE, or the AVR Download UploaDEr, is a utility program for uploading and downloading code and data from the memory spaces of an AVR MCU. Figure 6-12 shows the output AVRDUDE will generate when executed without any arguments.

AVRDUDE can also program the on-board EEPROM memory, and the fuse and lock bits as well. The tool runs in either command-line or interactive mode. The command-line mode is useful when incorporating AVRDUDE into a script or a makefile, while the interactive mode can be used to poke around in the MCU's memory, modify individual bytes in the EEPROM, or fiddle around with the fuse or lock bits.

AVRDUDE supports a variety of programming devices, including the Atmel STK500, the AVRISP and AVRISP MKII, serial bit-bangers, and a parallel port interface. A manual is available in PDF format (*http://bit.ly/avrdude-pdf*).

The Arduino IDE uses AVRDUDE to handle the upload process, and you can see what the command line looks like by enabling the upload output from the Preferences dialog. Here is what it looks like when uploading the simple intrusion alarm sketch from Chapter 5 (note I have wrapped the line in order to fit it on the page):

```
/usr/share/arduino/hardware/tools/avrdude
 -C/usr/share/arduino/hardware/tools/avrdude.conf
 -v -v -v -v -patmega328p -carduino -P/dev/ttyUSB0 -b57600 -D
 -Uflash:w:/tmp/build2510643905912671503.tmp/simple_alarm.cpp.hex:i
```

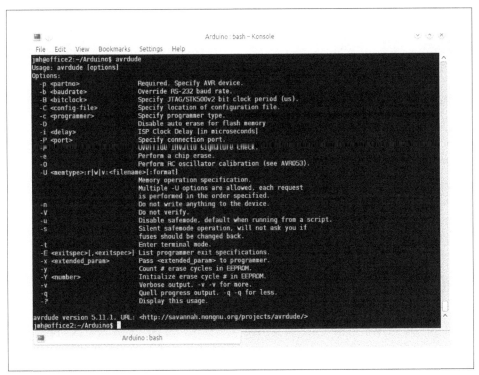

Figure 6-12. AVRDUDE help output

The -p, -c, -P, -D, and -U switches are the key things to note here. These and some of the other available switches are described in Table 6-8. The multiple -v switches just tell AVRDUDE to be as verbose as possible.

Table 6-8. AVRDUDE command-line switches

Switch	Function	Description
-p	Processor ID	Identifies the part connected to the programmer. In our case, the Arduino board being programmed does indeed have an ATmega328p—it's a Duemilanove.
-c	Programmer ID	Specifies the programmer to use. In this case, the Arduino bootloader is the programmer.
-C	Configuration	Specifies a configuration file to use.
-P	Port name	Identifies the port to which the programmer is attached. Since this is a Linux system, the pseudo serial port at */dev/ttyUSB* is used.
-b	Port baud rate	Overrides the default baud rate.
-D	Auto-erase	Disables the flash auto-erase.
-U	Upload	Upload specification.

Note that the -U switch in the command line is comprised of multiple parts. It defines the memory target (flash), the mode (write), and a source file containing the hex form of the executable image. The final i at the end of the argument string specifies that the hex source is in Intel format.

Using an Arduino as an ISP

By loading a utility program onto an Arduino, you can make it become an ISP for another Arduino. You can upload a new bootloader using this technique. The Arduino website (*http://bit.ly/arduino-isp*) provides simple directions.

Essentially all that is necessary is a sketch supplied with the Arduino IDE, and two Arduino boards. Figure 6-13 shows how the boards are connected. In this case a Duemilanove is acting as the ISP device for an Uno R3. It can also work with boards that don't have the same pinouts for the SPI signals, like the Leonardo, but you will need to make adjustments for that. Refer to the Arduino documentation for details.

Bootloader Operation

The purpose of the bootloader is to accept a program for the AVR from the host development system. The microcontrollers installed on Arduino boards come with the bootloader preinstalled, and for most applications there is seldom any need to remove or reload the bootloader software. Unless there is a compelling reason to try to reclaim the few kilobytes of memory (depending on the processor type and the vintage) that the bootloader consumes, the easiest approach is just to leave it alone and take advantage of it.

The operation of the Arduino bootloader is similar to that of any other flash memory–equipped microcontroller. The primary objective is to get the user-supplied program into on-board memory and then transfer control to the new program for execution. When an Arduino is powered on, it begins to execute the bootloader. The bootloader then checks for new incoming program data on the USB interface, and if nothing is detected after some small amount of time, the program loaded into the main section of flash memory is executed.

On newer versions of the Arduino this check for incoming data also occurs even while the AVR processor is running a previously stored program, so that when a new program upload is detected the existing program code is interrupted and the bootloader is given full control of the processor. On older Arduino boards the reset button must be pressed as soon as the IDE starts to upload the compiled program in order to detect the upload. You may need to do this a few times to get the timing just right. Note that some Arduino-compatible boards also behave this way.

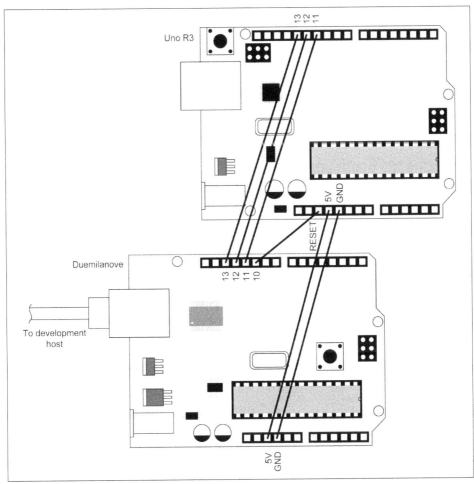

Figure 6-13. Using an Arduino as a programmer for another Arduino

Once the bootloader has determined that the incoming data is a valid program upload, it then unlocks the on-board flash memory and begins reading in the new program and writing it to the flash memory (but not in the location where the bootloader itself resides). Once the upload is complete the memory is once again locked so that it cannot be accidentally modified, and an interrupt is generated that will cause the processor to be directed (or vectored, as it is sometimes called) to the starting address of the new program.

Current versions of the Arduino bootloader can read data at 19,200 baud (19.2 Kbaud), or about 1,900 bytes per second. Older versions of the bootloader listened for incoming data at 9,600 baud, and you can alter the behavior of the Arduino IDE to accommodate this if necessary. At 19.2K it can still take many seconds to transfer a large program sketch to the target processor.

The bootloader source code is available from Arduino.cc, and it's worth reading if you are really curious about how it works. If you do peruse it you will notice that the section that manipulates the flash memory is written in AVR assembly language. At this point the code is "down on the metal" and interacting with the microcontroller's internal control registers, and assembly is the appropriate language for this task. The bootloader source code also gives some insight into what goes on at the lowest levels in a microcontroller. There is a lot happening behind the scenes when a program is loaded and executed.

Replacing the Bootloader

Should you have the need (or the desire) to install a new bootloader on the AVR processor in an Arduino, you will need a device to program the microcontroller's flash memory directly via the ICSP port like the ones shown in Figures 6-10 and 6-11, or you can use the Arduino-to-Arduino trick described in "Using an Arduino as an ISP" on page 154.

The bootloader resides in a particular region of the flash memory in an AVR available for just this purpose, and the available size ranges from 256 bytes to 4 KB, depending on the AVR type. The Arduino IDE supports bootloader uploading via a programming device like those shown earlier. Under Tools on the main IDE menu you must select the type of programmer you have, compile the bootloader, and then upload it to the AVR microcontroller. Refer to the Arduino's built-in help and to the main Arduino website for details.

Summary

This chapter has been a whirlwind tour through multiple topics, from the components of the AVR toolchain to the *make* utility, assembly language programming, and finally the nitty-gritty of AVR bootloaders. Even if you never use any of the tools or techniques covered here, it is still useful to know something about what is going on under the hood of the Arduino IDE. It also serves as a glimpse into what working with embedded systems was like before the Arduino IDE came along.

If you want to explore any of the topics presented here in more detail be sure to avail yourself of the references given, and don't forget to look at Appendix D for even more sources of information. Embedded microcontroller devices are key components of modern civilization, and for every computer you can see there are many, many more that you cannot, hidden in TV remote controls, microwave ovens, your stereo system and DVD player, traffic light controls, your automobile, and even the keyboard for the computer sitting on your desk. At one time all these little devices were programmed using the techniques covered briefly in this chapter, and many of them are still programmed with these same methods today.

CHAPTER 7

Arduino Libraries

The Arduino IDE comes with a collection of libraries that can be used with program sketches. These contain functions to access peripheral devices like an Ethernet interface, a liquid crystal display, a conventional serial interface, and many others.

Note that although the term "library" is used to describe auxiliary code, the modules themselves aren't always what one might think of as a library in the sense of a pre-compiled module, such as the *.a* (archive) or *.so* (shared object) libraries in Unix or Linux. In many cases they're just standard C or C++ source files (with AVR-GCC limitations, of course), but the end result is largely the same. Library code is compiled as necessary along with the sketch code into object files and linked with the sketch (see Chapters 5 and 6). In other cases a library really is a binary object, such as the components supplied with the *avr-libc* library suite. If you want to know where a library or external code module is coming from, check both the *avr-libc* documentation and the Arduino documentation.

After a program sketch and any library modules are compiled, the linker utility resolves the address references between the library components and the user-supplied functions, and then binds all the components into one executable binary image. The AVRDUDE utility employed by the IDE (discussed in Chapter 6) handles the process of interacting with the on-board bootloader (covered in Chapter 5) to transfer the compiled binary code onto an AVR device and save it in the processor's on-board memory.

For an overview of the software development process using the Arduino IDE, see Chapter 5. Chapter 6 covers code development using just the AVR-GCC toolchain.

Library Components

The Arduino development environment comes with a selection of support libraries for things such as serial I/O over the USB connection, EEPROM read/write, Ethernet I/O, an LCD interface, and support for servo and stepper motor actuators, among other things. These are described in the following sections.

 These descriptions are, by necessity, terse. For more details and usage examples refer to the Arduino libraries page (*http://bit.ly/arduino-libraries*) or see the reference documentation supplied in the form of HTML files with the Arduino IDE (the built-in help pages). Bear in mind that the Arduino website will have the latest documentation for the most recent versions of the libraries, but what comes with the Arduino IDE will describe the libraries included with that release.

You can examine a list of available libraries by selecting Sketch→Import Library from the IDE toolbar. This will also show any libraries that you have added to the environment (adding libraries is described in Chapter 5).

The following descriptions cover the libraries supplied with a basic Arduino IDE installation (additional libraries are available from Arduino and from other sources like shield suppliers):

EEPROM
Supports reading and writing to "permanent" storage using an AVR's built-in EEPROM

Ethernet
Used with the Arduino Ethernet shield for Ethernet connectivity

Firmata
Provides communications with applications on the computer using a standard serial protocol

GSM
Used with the GSM shield to connect to a GSM/GPRS network

LiquidCrystal
Contains functions for controlling liquid crystal displays (LCDs)

SD
Provides support for reading and writing SD flash memory cards

Servo
A collection of functions for controlling servo motors

SPI

  Supports the use of the Serial Peripheral Interface (SPI) bus

SoftwareSerial

  Implements serial communication on any of the digital pins

Stepper

  A collection of functions for controlling stepper motors

TFT

  Provides functions for drawing text, images, and shapes on the Arduino TFT screen

WiFi

  Supports the Arduino WiFi shield for wireless networking

Wire

  Supports the two-wire interface (TWI/I2C) for sending and receiving data over a network of devices or sensors

Esplora

  Provides functions to access the various actuators and sensors mounted on the Esplora board (used with Esplora only)

USB

  Used with the Leonardo, Micro, Due, and Esplora boards for serial I/O over the USB connection

Keyboard

  Sends keystrokes to an attached computer

Mouse

  Controls cursor movement on a connected computer

EEPROM

The EEPROM library supports reading and writing to "permanent" storage using an AVR's built-in EEPROM. The EEPROM is persistent, and it will retain whatever was written into it when power is removed from the board. Although the microcontroller's primary flash memory is also nonvolatile, the EEPROM is not disturbed when new executable code is uploaded to the board. It must be specifically accessed via software. The different types of AVR microcontrollers used on Arduino boards have different amounts of EEPROM storage available, ranging from 512 bytes for the ATmega168 to 4 KB for the ATmega1280 and ATmega2560. Refer to Chapter 3 for microcontroller-specific details.

The EEPROM class defines the functions used to read, write, and update the contents of the AVR on-board EEPROM. It must be instantiated before any of the EEPROM functions can be used.

The library include file, *EEPROM.h*, declares EEPROM as a static instance of the EEPROM class. If you want to use the class by instantiating an EEPROM object yourself you can do the following:

```
EEPROMClass eeprom;
```

Older versions of the EEPROM library have only two functions for accessing the built-in EEPROM of an AVR MCU:

read()

Reads a byte value from a specific address in the EEPROM. Uninitialized locations will contain a value of 255. This function returns the byte value read from the given address. Addresses start at zero (0).

```
uint8_t eeprom.read(int address);
```

write()

Writes a byte value to the microcontroller's EEPROM storage at a specific address. Addresses start at zero (0). About 3.3 ms is required to perform an EEPROM write operation, and the AVR EEPROM has a rated endurance of 100,000 write/erase cycles, so it should last a while. The function returns nothing.

```
void eeprom.write(int address, uint8_t value);
```

The latest version of the EEPROM library has four additional functions: update(), get(), put(), and EEPROM[]. The source code for the EEPROM library class is worth reading to see how the code deals with arbitrary data types.

update()

The update() function writes a byte of data to the EEPROM at a specific address, but only if the value currently at the address is different from the value supplied as an argument to the function. This function returns nothing.

```
void eeprom.update(int address, uint8_t value);
```

put()

The put() function writes any data type to the EEPROM, starting at the specified address. The data can be a primitive type (e.g., int or float), or it can be a structure. The function returns a reference to the data object passed in via the *data* argument. Note that this function uses update() to preform the write operation, so a write will not occur if the data at the specified address is the same as the data passed into the put() call.

```
data_ref eeprom.put(int address, data);
```

get()

The get() function reads and returns any data type or object from the EEPROM. The data read from the EEPROM is written to the address of the *data argument* in byte-wise fashion, with as many bytes as the size of the object pointed to by *data*. The function returns a reference to the data object passed in via the *data* argument.

```
data_ref eeprom.get(int address, data);
```

EEPROM[]

The EEPROM[] operator allows the EEPROM to be accessed like an array object. The actual EEPROM address is simply *EEPROM_base_address + int index*. The operator returns a reference to the EEPROM cell.

```
data_ref EEPROM[int index]
```

The following simple example is based loosely on what is provided by Arduino.cc in the EEPROM library documentation, but with a few twists. It writes to as well as reads from the EEPROM memory space, and it uses the remainder operator to determine if a value is odd or even:

```
#include <EEPROM.h>
// Instantiate our own copy of an EEPROMClass object rather than
// using the static declaration in EEPROM.h
EEPROMClass eeprom;
int a = 0;
int value;
void setup()
{
    Serial.begin(9600);
    // preload EEPROM with data pattern
    for (int i = 0; i < 512; i++) {
        if (i % 2)     // see if even or odd
            eeprom.write(i, 0);  // value is odd
        else
            eeprom.write(i, 1);  // value is even
    }
}
// An ATmega168 has 512 bytes of EEPROM, and an ATmega328 has 1024.
// Only 512 are used here, so it should be safe with any AVR that
// you might find in an Arduino.
void loop()
{
    value = eeprom.read(a);
    Serial.print(a);
    Serial.print("\t");
    Serial.print(value);
    Serial.println();
    // The variable a is declared outside of loop(), so it will
    // persist between calls from main().
```

```
        a++;
        if (a == 512)
            a = 0;
        delay(500);
    }
```

Ethernet

The Ethernet library provides the functionality necessary to interact with an Arduino Ethernet shield. As Arduino libraries go, it is rather complex, and provides both server and client functionality. It also supports four concurrent input or output connections, or a mix of either. The Ethernet shield uses the SPI interface to communicate with the host Arduino board.

The Ethernet library is comprised of a collection of five C++ classes. Most of the classes inherit from parent classes, but you don't need to bother with the details for most applications. However, if you do need to see the low-level class definitions, they can be found in the directory *hardware/arduino/avr/cores/arduino* in the Arduino source code. The following list shows the five classes of the Ethernet library and the public member functions of each class. We'll look at each of these in turn in the following sections:

- Ethernet class
  - begin()
  - localIP()
  - maintain()
- IPAddress class
- Server class
  - EthernetServer()
  - begin()
  - available()
  - write()
  - print()
  - println()
- Client class
  - EthernetClient()
  - connected()
  - connect()

— write()

— print()

— println()

— available()

— read()

— flush()

— stop()

- EthernetUDP class

— begin()

— read()

— write()

— beginPacket()

— endPacket()

— parsePacket()

— available()

— stop()

— remotePort()

Ethernet class: EthernetClass

The Ethernet class initializes the Ethernet library and network settings. The actual class name is EthernetClass, and an object named Ethernet is instantiated in the library file *Ethernet.cpp* and exported in *Ethernet.h*. You can create your own instance of EthernetClass if you wish, or just use what the library provides.

begin()

Used to initialize the Ethernet library and establish networking parameters. The begin() method (or, function, if you prefer) is overloaded so that there are five ways to invoke it.

All of the arguments are arrays of uint8_t bytes. The DHCP-only form of this method (Ethernet.begin(mac)) returns 1 if a DHCP lease was successfully obtained, or 0 if it failed. All other forms return nothing. The IPAddress class type is described in the next section.

```
int Ethernet.begin(uint8_t *mac);
void Ethernet.begin(uint8_t *mac, IPAddress ip);
void Ethernet.begin(uint8_t *mac, IPAddress ip, IPAddress dns);
void Ethernet.begin(uint8_t *mac, IPAddress ip, IPAddress dns,
```

```
                         IPAddress gateway);
      void Ethernet.begin(uint8_t *mac, IPAddress ip, IPAddress dns,
                         IPAddress gateway, IPAddress subnet);
```

localIP()

Obtains the IP address of the local host (i.e., the Ethernet shield). This is useful for determining the local IP address when DHCP is used. If the local Ethernet has been initialized successfully, `Ethernet.localIP()` will return an `IPAddress` object containing the assigned or specified IP address.

```
      IPAddress Ethernet.localIP();
```

maintain()

This method does not appear in older versions of the library. When a device is assigned an IP address by a DHCP server it is called a *lease*, and a DHCP lease is given for a specific period of time (it depends on how the DHCP server has been configured). The `Ethernet.maintain()` method is used to renew a DHCP lease.

`Ethernet.maintain()` will return 0 if nothing occurred, 1 if the lease renewal failed, 2 if the lease was successfully renewed, 3 if the DHCP rebind failed, and 4 if the rebind succeeded.

```
      int Ethernet.maintain();
```

IPAddress class

The `IPAddress` class defines a data object that is used to contain data for local and remote IP addressing. The class has four types of overloaded constructors. Each accepts a different form of IP address, as shown here:

```
IPAddress()
IPAddress(uint8_t first_octet,
          uint8_t second_octet,
          uint8_t third_octet,
          uint8_t fourth_octet)
IPAddress(uint32_t address)
IPAddress(const uint8_t *address)
```

An `IPAddress` object can hold a set of four IP address octets (the 192.168.1.100 format, for example, sans the periods), a 32-bit integer version of an IP address, or an array of unsigned bytes. The `IPAddress` class is used to create instances of address data types. For example:

```
IPAddress ip(192, 168, 0, 2);
IPAddress dnServer(192, 168, 0, 1);
IPAddress gateway(192, 168, 0, 1);
IPAddress subnet(255, 255, 255, 0);
```

ip, dnServer, gateway, and subnet are objects of type IPAddress. The Ethernet library knows to look for these names when initializing an Ethernet interface. Notice that they all use the multiple-octet form of initialization.

You can find the source files for IPAddress in the directory *Arduino/hardware/arduino/avr/cores/arduino* of the Arduino source.

Server class: EthernetServer

In Ethernet parlance, a server is a system (or host) that will accept a request for a connection from another system and establish a communications channel. The system requesting the connection is called a client. A server waits passively for clients to contact it; it doesn't initiate a connection. For a real-world example, consider a web server. The web server waits for browser clients to connect and request web pages. It returns the requested data to the client and then waits for the next request. Each time a link is selected, a button clicked, or text entered into a text field of a browser display, a request is created and sent to the web server.

The Server class is the base class for the EthernetServer class in the Ethernet library. It is not called directly. The other classes utilize it. As with IPAddress, the source for the Server class is located in the directory *Arduino/hardware/arduino/avr/cores/arduino* of the Arduino source.

EthernetServer()

Establishes the port to use when listening for a connection request from a client. The port is typically specified when an object of type EthernetServer is instantiated. It returns nothing (void).

```
EthernetServer server(int port);
```

Example:

```
EthernetServer newserv = EthernetServer(port);
```

The value of the *port* argument may be any number between 0 and 65535, but values between 0 and 1024 are typically reserved for system services such as FTP, SSH, and possibly a web server. Use high port values (greater than 9000, for instance) to avoid conflicts.

begin()

Commands the server to begin listening for connections from clients on the port set when the server object was created.

```
void newserv.begin();
```

available()

Returns a connection to a client that is connected and ready to communicate.

```
EthernetClient newclient = newserv.available();
```

```
write()
```
Writes data to a client connected to a server. The data is written to all connected clients. Accepts either a single char (or byte) value, or a pointer to an array of char values, and returns the number of bytes written.

```
int newserv.write(char data);
int newserv.write(char *data, int size);
```

```
print()
```
Prints data to all clients connected to a server as a sequence of ASCII characters.

```
println()
```
Similar to print(), but adds a newline character at the end of the output. Accepts data as char, byte (uint8_t), int, long, or string types. Returns the number of bytes written. With no arguments, the function simply sends a newline character.

```
int newserv.println();
int newserv.println(char *data);
int newserv.println(char *data, int BASE);
```

Client class: EthernetClient

The Client class creates client objects, which can connect to a server to send and receive data. A transaction, or exchange, of data between a server and a client is typically initiated by the client. A server listens and waits for a client to connect, and once connected the client can request data from the server, send data to the server, or request that the server perform some action on the client's behalf.

The Client class is the base class for the EthernetClient class in the Ethernet library. It is not called directly; the EthernetClient class inherits from it. As with Server, the source for the Client base class is located in the directory *Arduino/hardware/arduino/avr/cores/arduino* of the Arduino source.

```
EthernetClient()
```
Creates a client object that can connect to a server at a specific IP address using a specific port. The connect() method is used to define the server and the port to use. For example:

```
byte servaddr[] = {172, 120, 40, 10};
EthernetClient newclient;
newclient.connect(servaddr, 80);
```

```
connected()
```
Determines whether a client is connected or not. Returns true or false.

```
bool newclient.connected();
```

```
    if (newclient.connected()) {
        // do something
    }
```

connect()

Connects to a server at a specific IP address (an array of four bytes) and port. Instead of an IP address, a URL (web address) may be used.

```
int newclient.connect();
int newclient.connect(byte *servIP, int port);
int newclient.connect(char *URL, int port);
```

Returns an integer representing the connection status:

- SUCCESS = 1
- TIMED_OUT = -1
- INVALID_SERVER = -2
- TRUNCATED = -3
- INVALID_RESPONSE = -4

write()

Sends either a value or the contents of a buffer to the connected server. The data is sent as a series of bytes. The write() method returns the number of bytes written. This return value can be safely ignored.

```
int newclient.write(uint8_t value);
int newclient.write(uint8_t *buffer, int len);
```

print()

Prints data to the connected server as a sequence of ASCII characters. Accepts data as char, byte (uint8_t), int, long, or string types. Can also take a base specified. The valid base types are BIN (binary), DEC (base 10), OCT (base 8), and HEX (base 16). Returns the number of bytes written.

```
int newclient.print(data);
int newclient.print(data, base);
```

println()

Identical to the print() method except that a newline character is appended to the end of the output of ASCII characters. A println() with no parameters will send a single newline character to the connected server.

```
int newclient.println();
int newclient.println(data);
int newclient.println(data, base);
```

`available()`

Returns the number of characters available for reading from the connected server. Can be used to check for presence of incoming data.

```
int newclient.available();
```

`read()`

Reads the next available byte from the server. Use a loop to read multiple characters, or read and evaluate each character one at a time.

```
char newclient.read();
```

`flush()`

Discards any unread characters from the server that are in the receive buffer.

```
void newclient.flush();
```

`stop()`

Disconnects from the currently connected server. Once disconnected, the client may connect to another server (or to the same server again, of course).

```
void newclient.stop();
```

EthernetUDP class

Unlike TCP/IP, which is a stream protocol (i.e., it has no definite start and stop boundaries), UDP is a datagram protocol. Each item of data is a single packet, called a datagram, and the data must fit within the boundaries of the datagram packet. UDP does not have error detection, nor does it guarantee delivery of the data, but for short packets of noncritical data, or where the upper-level software can handle things like error detection and retries, it offers a fast and relatively simple way to move data around between hosts.

`begin()`

Initializes the UDP class to start listening for incoming data on a specific port.

```
byte UDP.begin(int port);
```

`read()`

Reads incoming data from the specified buffer. If no parameters are given, it will return one character from the current buffer. If the buffer and size are specified, it will return up to *maxsize* bytes from the buffer.

```
char UDP.read();
char *UDP.read(char *pkt_buffer, int maxsize)
```

Note that this function is intended to be used immediately after a call to `UDP.parsePacket()`.

write()

Sends data to a remote connection. The write() function must be placed between beginPacket() and endPacket() calls. beginPacket() initializes the data packet, and the endPacket() call actually sends the data to the remote host.

```
byte UDP.write(char *message);
byte UDP.write(char *buffer, int size)
```

beginPacket()

Opens a UDP connection to a remote host at a specific IP address and port. Returns 1 (true) if the connection succeeded or 0 on failure.

```
int UDP.beginPacket(byte *ip, int port);
```

endPacket()

Sends a UDP packet created by the write() function to the remote host specified by the beginPacket() function.

```
int UDP.endPacket();
```

parsePacket()

Checks an open UDP connection for the presence of a datagram packet and returns the size of the waiting data. parsePacket() must be called before the read() or available() function is used to retrieve the data (if any).

```
int UDP.parsePacket();
```

available()

Returns the number of bytes of received data currently in the receive buffer. Note that this should only be called after a call to parsePacket().

```
int UDP.available();
```

stop()

Disconnects from the remote UDP host and releases any resources used during the UDP session.

```
void USP.stop();
```

remoteIP()

Returns the IP address of the remote UDP connection as an array of 4 bytes. This function should only be called after a call to parsePacket().

```
byte *UDP.remoteIP();
```

remotePort()

Returns the UDP port of the remote UDP connection as an integer. This function should only be called after a call to parsePacket().

```
int UDP.remotePort();
```

Firmata

Firmata is an interesting library with a lot of potential applications. Firmata provides the means to use serial communications between an Arduino and an application on a host computer using a protocol similar to MIDI. It was developed with the intention of allowing as much of the functionality of an Arduino to be controlled from a host computer as possible—in other words, to use the Arduino as if it was an extension of the host's own I/O capabilities.

Before embarking on a Firmata project for the first time, you might want to try out the demonstration software. A suitable client can be downloaded from the old Firmata wiki (*http://firmata.org/wiki/Main_Page*), and the Arduino portion of the code is already included in the libraries distributed with the Arduino IDE.

This section provides only a summary of the functions available in the Firmata library. The following list shows the organization of the library components. Unfortunately, these components don't seem to be extensively documented, so some of what you might want to know in order to use them will need to be gleaned from the source code. For more details and usage examples, refer to the Firmata wiki (*http://firmata.org*) (now idle and no longer maintained) or check out the Firmata GitHub repository (*https://github.com/firmata*). You should pay particular attention to the protocol definition (*https://github.com/firmata/protocol*), as this is what the host application uses to communicate with a Firmata application running on an Arduino.

 The Firmata library code included with the Arduino IDE may not be the latest version. Check the GitHub (*https://github.com/firmata/arduino*) repository. The documentation presented here may refer to functions that your version does not have.

The Firmata library is organized as follows:

- Base methods
  - begin()
  - printVersion()
  - blinkVersion()
  - printFirmwareVersion()
  - setFirmwareVersion()
- Sending messages
  - sendAnalog()
  - sendDigitalPorts()

- — `sendDigitalPortPair()`
- — `sendSysex()`
- — `sendString()`
- — `sendString()`
- Receiving messages
  - — `available()`
  - — `processInput()`
- Callback functions
  - — `attach()`
  - — `detach()`
- Message types

Base methods

We'll look at each of these categories in turn in the following sections.

`begin()`

The basic form of `begin()` initializes the Firmata library and sets the serial data rate to a default of 57,600 baud. The second form accepts an argument of type long, which contains the desired baud rate for the communication between Firmata and a host system. The third form starts the library using a stream other than `Serial`. It is intended to work with any data stream that implements the Stream interface (Ethernet, WiFi, etc.). Refer to the issue discussions on Firmata's GitHub page (*https://github.com/firmata/arduino/issues*) for more information on the current status of this method.

```
void Firmata.begin();
void Firmata.begin(long);
void Firmata.begin(Stream &s);
```

`printVersion()`

Sends the library protocol version to the host computer.

```
void Firmata.printVersion();
```

`blinkVersion()`

Blinks the protocol version on pin 13 (the on-board LED on an Arduino).

```
void Firmata.blinkVersion();
```

`printFirmwareVersion()`

Sends the firmware name and version to the host computer.

```
void Firmata.printFirmwareVersion();
```

`setFirmwareVersion()`

Sets the firmware name and version using the sketch's filename, minus the extension.

```
void Firmata.setFirmwareVersion(const char *name,
                                byte vers_major, byte vers_minor);
```

Sending messages

`sendAnalog()`

Sends an analog data message.

```
void Firmata.sendAnalog(byte pin, int value);
```

`sendDigitalPort()`

Sends the state of digital ports as individual bytes.

```
void Firmata.sendDigitalPort(byte pin, int portData);
```

`sendString()`

Sends a string to the host computer.

```
void Firmata.sendString(const char* string);
```

`sendString()`

Sends a string to the host computer using a custom command type.

```
void Firmata.sendString(byte command, const char* string);
```

`sendSysex()`

Sends a command containing an arbitrary array of bytes.

```
void Firmata.sendSysex(byte command, byte bytec, byte* bytev);
```

Receiving messages

`available()`

Checks to see if there are any incoming messages in the input buffer.

```
int Firmata.available();
```

`processInput()`

Retreives and processes incoming messages from the input buffer and sends the data to registered callback functions.

```
void Firmata.processInput();
```

Callback functions

In order to attach a function to a specific message type, the function must match a callback function. There are three basic types of callback functions in Firmata. We'll look at each of these in turn in the following sections: generic, string, and sysex, and a fourth type to handle a system reset. The callback functions are:

attach()

Attaches a function to a specific incoming message type.

```
void attach(byte command, callbackFunction newFunction);
void attach(byte command, systemResetCallbackFunction newFunction);
void attach(byte command, stringCallbackFunction newFunction);
void attach(byte command, sysexCallbackFunction newFunction);
```

detach()

Detaches a function from a specific incoming message type.

```
void Firmata.detach(byte command);
```

Message types

A function may be attached to a specific message type. Firmata provides the following message types:

ANALOG_MESSAGE

The analog value for a single pin

DIGITAL_MESSAGE

Eight bits of digital pin data (one port)

REPORT_ANALOG

Enables/disables the reporting of an analog pin

REPORT_DIGITAL

Enables/disables the reporting of a digital port

SET_PIN_MODE

Changes the pin mode between INPUT/OUTPUT/PWM/etc.

FIRMATA_STRING

For C-style strings; uses `stringCallbackFunction` for the function type

SYSEX_START

For generic, arbitrary-length messages (via MIDI SysEx protocol); uses `sysex CallbackFunction` for the function type

SYSTEM_RESET

Resets firmware to its default state; uses `systemResetCallbackFunction` for the function type

GSM

The GSM library is used with the GSM shield to connect to a GSM/GPRS network. It is included with the 1.0.4 and later versions of the Arduino IDE. The GSM library supports most of the functions one would expect from a GSM phone, such as the ability to place and receive calls, send and receive SMS messages, and connect to the Internet via a GPRS network. GSM stands for global system for mobile communications, and GPRS is the acronym for General Packet Radio Service.

The GSM shield incorporates a modem to transfer data from a serial port to the GSM network. The modem utilizes AT-type commands to perform various functions. In normal usage each AT command is part of a longer series that performs a specific function. The GSM library relies on the SoftwareSerial library to support communication between the Arduino and the GSM modem.

The GSM library is a recent addition. If you have an older version of the IDE, then you may not have this library. Check the list of available libraries in the IDE to see if you do or do not have the GSM library available.

The suite of GSM library classes is complex, and a full description of all of the capabilities would be beyond the scope of this book. This section presents a summary of the functionality. For more detailed information, refer to the Arduino GSM (*https://www.arduino.cc/en/Reference/GSM*) library reference or check the built-in help in the Arduino IDE. Some vendors, such as Adafruit, also produce Arduino-compatible GSM shields, and they provide their own libraries for their products.

Ethernet library compatibility

The GSM library is largely compatible with the current Arduino Ethernet library, such that porting a program that uses the Arduino Ethernet or WiFi libraries to the GSM for use with the GSM shield should be relatively straightforward. Some minor library-specific modifications will be necessary, such as including the GSM- and GPRS-specific libraries and obtaining network settings from your cellular network provider.

Library structure

The GSM library is rather complex, and is comprised of 10 primary classes. The following list shows the functions in each of the GSM classes:

- GSM class
  - begin()
  - shutdown()

- GSMVoiceCall class
  - getVoiceCallStatus()
  - ready()
  - voiceCall()
  - answerCall()
  - hangCall()
  - retrieveCallingNumber()
- GSM_SMS class
  - beginSMS()
  - ready()
  - endSMS()
  - available()
  - remoteNumber()
  - read()
  - write()
  - print()
  - peek()
  - flush()
- GPRS class
  - attachGPRS()
- GSMClient class
  - ready()
  - connect()
  - beginWrite()
  - write()
  - endWrite()
  - connected()
  - read()
  - available()
  - peek()

- — flush()
- — stop()
- GSMServer class
  - — ready()
  - — beginWrite()
  - — write()
  - — endWrite()
  - — read()
  - — available()
  - — stop()
- GSMModem class
  - — begin()
  - — getIMEI()
- GSMScanner class
  - — begin()
  - — getCurrentCarrier()
  - — getSignalStrength()
  - — readNetworks()
- GSMPIN class
  - — begin()
  - — isPIN()
  - — checkPIN()
  - — checkPUK()
  - — changePIN()
  - — switchPIN()
  - — checkReg()
  - — getPINUsed()
  - — setPINUsed()
- GSMBand class
  - — begin()
  - — getBand()

— setBand()

GSM class

This class prepares the functions that will communicate with the modem. It manages the connectivity of the shield and performs the necessary system registration with the GSM infrastructure. All Arduino GSM/GPRS programs need to include an object of this class to handle the low-level communications functions.

This is the base class for all GSM-based functions. It should be instantiated as shown.

```
GSM gsmbase;
```

begin()

    Starts the GSM/GPRS modem and attaches to a GSM network. The full prototype for the begin() method looks like this:

```
begin(char* pin=0, bool restart=true, bool synchronous=true);
```

    The begin() method can be called four different ways because each argument has been assigned a default value. The first form takes no arguments, and it is assumed that the SIM has no configured pin.

```
gsmbase.begin();
gsmbase.begin(char *pin);
gsmbase.begin(char *pin, bool restart);
gsmbase.begin(char *pin, bool restart, bool sync);
```

shutdown()

    Shuts down the modem (power-off).

```
gsmbase.shutdown();
```

GSMVoiceCall class

The GSMVoiceCall class enables voice communication through the modem, provided that a microphone, a speaker, and a small amount of circuitry are connected to the GSM shield.

This is the base class for all GSM functions used to receive and make voice calls and should be instantiated as follows:

```
GSMVoiceCall gsmvc;
```

getVoiceCallStatus()

    Returns the status of a voice call as one of IDLE_CALL, CALLING, RECEIVINGCALL, or TALKING.

```
GSM3_voiceCall_st getVoiceCallStatus();

gsmvc.getVoiceCallStatus();
```

`ready()`

Returns the status of the last command: 1 if the last command was successful, 0 if the last command is still executing, and >1 if an error occurred.

```
int ready();

gsmvc.ready();
```

`voiceCall()`

Places a voice call in either asynchronous or synchronous mode. If asynchronous, `voiceCall()` returns while the number is ringing. In synchronous mode `voiceCall()` will not return until the call is either established or cancelled.

The first argument is a string containing the number to call. A country extension can be used or not. The buffer should not be released or used until `voiceCall()` is complete (the command is finished). The *timeout* argument is given in milliseconds, and is used only in synchronous mode. If set to 0, then `voiceCall()` will wait indefinitely for the other end to pick up.

```
int voiceCall(const char* to, unsigned long timeout=30000);

gsmvc.voiceCall(to);           // use default timeout
gsmvc.voiceCall(to, timeout);  // specify a timeout period
```

`answerCall()`

Accepts an incoming voice call. In asynchronous mode `answerCall()` returns 0 if the last command is still executing, 1 on success, and >1 if an error has occurred. In synchronous mode `answerCall()` returns 1 if the call is answered, and 0 if not.

```
gsmvc.answerCall();
```

`hangCall()`

Hangs up an established call or an incoming ring. In asynchronous mode `hangCall()` returns 0 if the last command is still executing, 1 on success, and >1 if an error has occurred. In synchronous mode `hangCall()` returns 1 if the call is answered, and 0 if not.

```
gsmvc.hangCall();
```

`retrieveCallingNumber()`

Retrieves the calling number and puts it into a buffer. The argument buffer is a pointer to a char buffer, and *bufsize* is the size of the buffer, in bytes. The buffer should be large enough to hold at least 10 characters.

In asynchronous mode `retrieveCallingNumber()` returns 0 if the last command is still executing, 1 on success, and >1 if an error has occurred. In synchronous mode retrieveCallingNumber() returns 1 if the number is correctly acquired and 0 if not.

```
int retrieveCallingNumber(char* buffer, int bufsize);

gsmvc.retrieveCallingNumber(buffer, bufsize);
```

GSM_SMS class

This class provides the capability to send and receive SMS (Short Message Service) messages.

beginSMS()

Defines the telephone number to receive an SMS message. The phone number is a char array. In asynchronous mode the function will return 0 if the last command is still active, 1 if it was successful, and a value >1 if an error occurred. In synchronous mode the function will return 1 if the previous command was successful and 0 if it failed.

```
int SMS.beginSMS(char *phone_number)
```

ready()

Returns the status of the last GSM SMS command. In asynchronous mode ready() will return 0 if the last command is still active, 1 if it was successful, and a value >1 if an error occurred. In synchronous mode the function will return 1 if the previous command was successful and 0 if it failed.

```
int SMS.ready()
```

endSMS()

Used to inform the modem that the message is complete and ready to send. In asynchronous mode the function returns 0 if it is still executing, 1 if successful, and >1 if an error has occurred. In synchronous mode it returns 1 if successful, and 0 otherwise.

```
int SMS.endSMS()
```

available()

If an SMS message is available to read, this function returns the number of characters in the message. If no message is available, it returns 0.

```
int SMS.available()
```

remoteNumber()

Extracts the remote phone number from an incoming SMS message and returns it in a char array. The size argument defines the maximum size of the array passed to remoteNumber(). In asynchronous mode the function returns 0 if still executing, 1 if successful, and >1 if an error has occurred. In synchronous mode the function returns 1 if successful, 0 otherwise.

```
int SMS.remoteNumber(char *remote_phone, int number_size)
```

read()

> Reads a byte (a character) from an SMS message. Returns the byte as an integer, or -1 if no data is available.

```
int SMS.read()
```

write()

> Writes a byte-sized character to an SMS message.

```
int write(int character)
```

print()

> Writes a character array to an SMS message. Returns the number of bytes successfully written.

```
int SMS.print(char *message)
```

peek()

> Returns the next available byte (a character) from an SMS message, without removing the character, or -1 if no data is available.

```
int SMS.peek()
```

flush()

> Clears the modem of any sent messages after all outbound characters have been sent.

```
void SMS.flush()
```

GPRS class

GPRS is the base class for all GPRS functions. This includes Internet client and server functions. This class is also responsible for including the files that are involved with TCP communication.

attachGPRS()

> Connects with a given access point name (APN) to initiate GPRS communications. Cellular providers have APNs the act as bridges between the cellular network and the Internet. Returns one of the following strings: ERROR, IDLE, CONNECTING, GSM_READY, GPRS_READY, TRANSPARENT_CONNECTED.

```
char *GPRS.attachGPRS(char *apn, char *user, char *password)
```

GSMClient class

This class creates clients that can connect to servers and send and receive data.

ready()

> Returns the status of the last command. In asynchronous mode the function returns 0 if still executing, 1 if successful, and >1 if an error has occurred. In synchronous mode the function returns 1 if successful, and 0 otherwise.

```
int GSMClient.ready()
```

connect(char *IP, int port)

> Connects to a specific port of a specified IP address. Returns true if the connection was successful, or false if not.

```
bool GSMClient.connect(char *hostip, int port)
```

beginWrite()

> Starts a write operation to the connected server.

```
void GSMClient.beginWrite()
```

write()

> Writes data to a connected server. Returns the number of bytes written.

```
int GSMClient.write(char data)
int GSMClient.write(char *data)
int GSMClient.write(char *data, int size)
```

endWrite()

> Stops writing data to a connected server.

```
void GSMClient.endWrite()
```

connected()

> Returns the connection status of a client. Note that a client is considered to be connected if the connection has been closed but there is still unread data in the buffer. Returns true if the client is connected, or false if not.

```
bool GSMClient.connected()
```

read()

> Reads the next available byte of data from the server the client is connected with. Returns the next byte, or -1 if no data is available.

```
int GSMClient.read()
```

available()

> Returns the number of bytes from the connected server that are waiting to be read.

```
int GSMClient.available()
```

peek()

Returns the next available byte of an incoming message without removing it from the incoming buffer. Successive calls to peek() will simply return the same byte.

```
int GSMClient.peek()
```

flush()

Discards any data currently waiting in the incoming buffer and resets the available data count to zero.

```
void GSMClient.flush()
```

stop()

Forces a disconnect from a server.

```
void GSMClient.stop()
```

GSMServer class

The GSMServer class creates servers that can send data to and receive data from connected clients. It implements network server functionality, similar to the Ethernet and WiFi libraries. Note that some network operators do not permit incoming network connections from outside their own network.

ready()

Returns the status of the last command. In asynchronous mode this function returns 0 if still executing, 1 if successful, and >1 if an error has occurred. In synchronous mode the function returns 1 if successful, and 0 otherwise.

```
int GSMServer.ready()
```

beginWrite()

Starts a write operation to all connected clients.

```
void GSMServer.beginWrite()
```

write()

Writes data to connected clients. Returns the number of bytes written.

```
int GSMServer.write(char data)
int GSMServer.write(char *data)
int GSMServer.write(char *data, int size)
```

endWrite()

Stops writing data to connected clients.

```
void GSMServer.endWrite()
```

read()

Reads the next available byte from a connected client. Returns the byte read, or -1 if no data is available.

```
int GSMServer.read()
```

available()

Listens for connection requests from clients. Returns the number of connected clients.

```
int GSMServer.available()
```

stop()

Stops the server from listening for client connection requests.

```
void GSMServer.stop()
```

GSMModem class

The GSMModem class provides diagnostic support functions for the internal GSM modem.

begin()

Checks the status of the modem and restarts it. This function must be called before a call to getIMEI(). Returns 1 if the modem is working correctly, or an error if it is not.

```
int GSMModen.begin()
```

getIMEI()

Queries the modem to retrieve its IMEI (International Mobile Equipment Identifier) number. The IMEI number is returned as a string. This function should only be called after a call to begin().

```
char *GSMModen.getIMEI()
```

GSMScanner class

The GSMScanner class provides functions to obtain diagnostic information about the network and carrier.

begin()

Resets the modem. Returns 1 if the modem is operating correctly, or an error code if it is not.

```
int GSMSScanner.begin()
```

getCurrentCarrier()

Returns the name of the current network server provider (the carrier) as a string.

```
char *GSMSScanner.getCurrentCarrier()
```

getSignalStrength()

Returns the relative signal strength of the network connection as a string with ASCII digits from 0 to 31 (31 indicates that the power is > 51 dBm), or 99 if no signal is detected.

```
char *GSMSScanner.getSignalStrength()
```

readNetworks()

Searches for available network carriers. Returns a string containing a list of the carriers detected.

```
char *GSMSScanner.readNetworks()
```

GSMPIN class

The GSMPIN class contains utility functions for communicating with the SIM card.

begin()

Resets the modem. Returns 1 if the modem is operating correctly, or an error code if it is not.

```
int GSMPIN.begin()
```

isPIN()

Examines the SIM card to determine if it is locked with a PIN or not. Returns 0 if the PIN lock is off, 1 if the lock is on, -1 if the PUK lock is on, and -2 if an error was encountered.

```
int GSMPIN.isPIN()
```

checkPIN()

Queries the SIM card with a PIN to determine whether it is valid or not. Returns 0 if the PIN is valid, and -1 if it is not.

```
int GSMPIN.checkPIN(char *PIN)
```

checkPUK()

Queries the SIM to determine if the PUK code is valid and establishes a new PIN code. Returns 0 if successful, and -1 if not.

```
int GSMPIN.checkPUK(char *PUK, char *PIN)
```

changePIN()

Changes the PIN code of a SIM card after verifying that the old PIN is valid.

```
void GSMPIN.changePIN(char *oldPIN, char *newPIN)
```

switchPIN()

Changes the PIN lock status.

```
void GSMPIN.switchPIN(char *PIN)
```

checkReg()

> Checks to determine if the modem is registered in a GSM/GPRS network. Returns 0 if the modem is registered, 1 if the modem is roaming, and -1 if an error was encountered.

```
int GSMPIN.checkReg()
```

getPINUsed()

> Checks to determine if the PIN lock is used. Returns true if locked, and false if not.

```
bool GSMPIN.getPINUsed()
```

setPINUsed()

> Sets the PIN lock status. If the argument is true, then the PIN is locked; if false, it is unlocked.

```
void GSMPIN.setPINUsed(bool used)
```

GSMBand class

The GSMBand class provides information about the frequency band the modem connects to. There are also methods for setting the band.

begin()

> Resets the modem. Returns 1 if the modem is operating correctly, or an error code if it is not.

```
int GSMBand.begin()
```

getBand()

> Returns the frequency band the modem is currently using for a connection.

```
char *GSMBand.getBand()
```

setBand()

> Sets the frequency band for the modem to use.

```
bool GSMBand.setBand(char *band)
```

LiquidCrystal

This class allows an Arduino board to control a liquid crystal display (LCD) module. It is specifically intended for LCDs that are based on the Hitachi HD44780 (or compatible) chipset, which is found on most text-based LCDs. The library supports either 4- or 8-bit interface mode, and also uses three of the Arduino pins for the RS (register select), clock enable, and R/W (read/write) control lines.

```
LiquidCrystal()
```
Creates an instance of a `LiquidCrystal` class object. The different forms of the class allow it to accommodate different LCD interface methods.

```
LiquidCrystal(uint8_t rs, uint8_t enable, uint8_t d0, uint8_t d1,
              uint8_t d2, uint8_t d3, uint8_t d4, uint8_t d5,
              uint8_t d6, uint8_t d7);

LiquidCrystal(uint8_t rs, uint8_t rw, uint8_t enable, uint8_t d0,
              uint8_t d1, uint8_t d2, uint8_t d3, uint8_t d4,
              uint8_t d5, uint8_t d6, uint8_t d7);

LiquidCrystal(uint8_t rs, uint8_t rw, uint8_t enable, uint8_t d0,
              uint8_t d1, uint8_t d2, uint8_t d3);

LiquidCrystal(uint8_t rs, uint8_t enable, uint8_t d0, uint8_t d1,
              uint8_t d2, uint8_t d3);
```
Where:

rs	The Arduino pin connected to the LCD's RS pin
rw	The Arduino pin connected to the LCD's RW pin
enable	The Arduino pin connected to the LCD's enable pin
d0 .. d7	The Arduino pins connected to the LCD's data pins

The use of the d4, d5, d6, and d7 signals is optional. If only four digital lines are used, these can be omitted.

Example:

```
LiquidCrystal lcd(12, 11, 10, 5, 4, 3, 2);
```

```
begin()
```
Initializes the interface to the LCD controller on the LCD module. The arguments specify the width and height of the LCD. The default character size is 5 × 8 pixels. This function must be called before any other LCD library functions can be used.

```
void lcd.begin(uint8_t cols, uint8_t rows,
               uint8_t charsize = LCD_5x8DOTS)
```

```
clear()
```
Clears the LCD screens and resets the cursor to the upper-left corner.

```
void lcd.clear()
```

```
home()
```
Positions the cursor at the upper-left position on the LCD. Does not clear the LCD; use the `clear()` function for that.

```
        void home()
```

setCursor()
Positions the cursor at the location specified by the *column* and *row* arguments.

```
        void setCursor(uint8_t column, uint8_t row)
```

write()
Writes a byte (char) of data to the LCD. Returns the number of bytes written.

```
        size_t write(uint8_t)
```

print()
This function is actually a part of the standard Arduino runtime AVR-GCC code, and it is overloaded to accept data of different types. The file *Print.h*, found at *hardware/arduino/avr/cores/arduino/Print.h*, defines the following forms of the print() function:

```
        size_t print(const __FlashStringHelper *);
        size_t print(const String &);
        size_t print(const char[]);
        size_t print(char);
        size_t print(unsigned char, int = DEC);
        size_t print(int, int = DEC);
        size_t print(unsigned int, int = DEC);
        size_t print(long, int = DEC);
        size_t print(unsigned long, int = DEC);
        size_t print(double, int = 2);
        size_t print(const Printable&);
```

cursor()
Enables the cursor, an underscored character, at the position where the next character will be written on the LCD screen.

```
        void cursor()
```

noCursor()
Disables the cursor, effectively hiding it. Does not affect the position where the next character will be displayed.

```
        void noCursor()
```

blink()
Displays a blinking cursor.

```
        void blink()
```

noBlink()
Turns off a blinking cursor.

```
        void noBlink()
```

display()

Enables the LCD, if it was initially disabled with the noDisplay() function. Restores the cursor and any text that was previously visible or which may have been added, deleted, or modified since the display was disabled.

```
void display()
```

noDisplay()

Disables the LCD display without altering any existing text on the screen.

```
void noDisplay()
```

scrollDisplayLeft()

Scrolls the text on the display one space to the left.

```
void scrollDisplayLeft()
```

scrollDisplayRight()

Scrolls the text on the display one space to the right.

```
void scrollDisplayRight()
```

autoscroll()

Enables automatic scrolling. As text is added to the display it moves the existing characters one space to either the left or the right, depending on the current text direction.

```
void autoscroll()
```

noAutoscroll()

Disables the autoscroll function of the LCD.

```
void noAutoscroll()
```

leftToRight()

Sets the direction the text will shift in when autoscroll is enabled, in this case from left to right.

```
void leftToRight()
```

rightToLeft()

Sets the direction the text will shift in when autoscroll is enabled, in this case from right to left.

```
void rightToLeft()
```

createChar()

Creates a custom 5 × 8-pixel character. The character is defined by an array of bytes, one per row. Only the five least significant bits of each byte are used.

```
void createChar(uint8_t, uint8_t[])
```

SD

The SD library provides support for reading and writing SD flash memory cards, both full-size and micro SD types (they're identical in terms of interface and functions, just different sizes). The library is based on *sdfatlib* by William Greiman.

This library treats an SD card as a small disk with either a FAT16 or a FAT32 filesystem. It uses short filenames (8.3 format). Filenames passed to the SD library functions may include a path, with directory names separated by forward slashes (like on Linux, not the backslashes used by MS-DOS or Windows).

The SPI interface is used to communicate with the SD card. This uses the digital pins 11, 12, and 13 on a standard Arduino board. One additional pin, usually pin 10, is used as the select pin, or another pin can be assigned to this role. Note that even if another pin is used for the select, the SS pin (pin 10) must remain as an output for the library to work.

SD class

The SD class provides functions for accessing the SD card and manipulating files and directories.

begin()

> Initializes the SD library and the interface with the SD card. The optional argument *csPin* defines the pin to use as the select. The default is to use pin 10 (*SD_CHIP_SELECT_PIN*). This function must be called before any other of the SD functions are used. Returns true if successful, or false if not.
>
> ```
> bool SD.begin(uint8_t csPin = SD_CHIP_SELECT_PIN);
> ```

exists()

> Tests for the presence of a file or directory on the SD card. The string *filepath* may be a fully qualified path name (FQPN). Returns true if the file or directory exists, or false if not.
>
> ```
> bool SD.exists(char *filepath);
> ```

mkdir()

> Creates a directory on the SD card. It will also create any necessary intermediate directories. Returns true if the directory was created successfully, or false if not.
>
> ```
> bool SD.mkdir(char *filepath);
> ```

open()

> Opens a file on an SD card for reading or writing. If the *mode* argument is not provided, the default is to open the file for reading. Returns a File object that

can be tested as a Boolean value. If the file could not be opened, then `File` will evaluate to `false`. The available modes are `FILE_READ` and `FILE_WRITE`.

```
File SD.open(const char *filename, uint8_t mode = FILE_READ);
```

`remove()`

Deletes (removes) a file from the SD card. *filepath* is an FQPN. Returns `true` if the removal succeeded, or `false` if not.

```
bool SD.remove(char *filepath);
```

`rmdir()`

Removes an empty directory from an SD card. Returns `true` if the directory was successfully deleted, or `false` if an error occurred (such as the directory not empty).

```
bool SD.rmdir(char *filepath);
```

File class

The `File` class provides functions for reading and writing individual files on an SD card. Objects of type `File` are created by the `SD.open()` function:

```
fname = SD.open("data.txt", FILE_WRITE);
```

There are a number of methods in a `File` object to manipulate the contents of a file:

`available()`

Returns the number of available bytes to read from a file.

```
int fname.available()
```

`close()`

Closes a file, ensuring that any remaining data is written to the file beforehand.

```
void fname.close()
```

`flush()`

Writes any remaining data in the file buffer to the file. Does not close the file.

```
void fname.flush()
```

`peek()`

Reads a byte from a file without advancing the internal data pointer. Successive calls to `peek()` will return the same byte.

```
int fname.peek()
```

`position()`

Returns the current position in the file that the next byte will be read from or written to.

```
uint32_t fname.position()
```

print()

Writes data to a file that has been opened for writing. Accepts data as char, byte (uint8_t), int, long, or string types. Can also take a base specified. The valid base types are BIN (binary), DEC (base 10), OCT (base 8), and HEX (base 16). Returns the number of bytes written.

```
int fname.print(data)
int fname.print(char *data, int BASE)
```

Note: string data is shown in this example.

println()

Writes data to a file that has been opened for writing followed by a carriage return and newline character pair. Accepts data as char, byte (uint8_t), int, long, or string types. Can also take a base specified. The valid base types are BIN (binary), DEC (base 10), OCT (base 8), and HEX (base 16). Returns the number of bytes written.

```
int fname.println()
int fname.println(data)
int fname.println(data, int BASE)
```

seek()

Moves the internal pointer to a new position in the file. The position must be between 0 and the size of the file, inclusive. Returns true if successful, or false if an error occurs (seeking beyond the end of the file, for example).

```
bool fname.seek(uint32_t pos)
```

size()

Returns the size of the file, in bytes.

```
uint32_t fname.size()
```

read()

Reads the next byte from the file, or returns a value of -1 if no data is available.

```
int fname.read(void *buf, uint16_t nbyte)
```

write()

Writes data to a file. Accepts either a single byte, or a data object that may be a byte, character, or string. The size argument defines the amount of data to write to the SD card. Returns the number of bytes written.

```
size_t fname.write(uint8_t)
size_t fname.write(const uint8_t *buf, size_t size)
```

isDirectory()

Returns true if the fname object refers to a directory, or false otherwise.

```
bool fname.isDirectory()
```

openNextFile()

Opens the next file folder in a directory and returns a new instance of a File object.

```
File openNextFile(uint8_t mode = O_RDONLY)
```

rewindDirectory()

Used with openNextFile(), this function returns to the first file or subdirectory in a directory.

```
void fname.rewindDirectory()
```

Servo

The Servo library is a collection of functions for controlling servo motors, such as the ones used with RC aircraft. Once an instance of the Servo class has been created, the attach() function is used to pass in the pin number to use with the servo. The pulses that control a servo are generated in the background. The class is instantiated as follows:

```
Servo servo;
```

attach()

Attaches a servo motor to an I/O pin. The second form allows the caller to specify the minimum and maximum write time values in microseconds. Returns the channel number, or 0 if the function fails.

```
uint8_t servo.attach(int pin)
uint8_t servo.attach(int pin, int min, int max)
```

write()

Sets the servo angle in degrees. If the value is > 200 it is treated as a pulse width in microseconds.

```
void servo.write(int value)
```

read()

Returns the last written servo pulse width as an angle between 0 and 180 degrees.

```
int servo.read()
```

writeMicroseconds()

Sets the servo pulse width in microseconds.

```
void servo.writeMicroseconds(int value)
```

readMicroseconds()
> Returns the current pulse width in microseconds for this servo.

```
int servo.readMicroseconds()
```

attached()
> Returns true if the servo object has been attached to a physical servo.

```
bool servo.attached()
```

detach()
> Stops an attached servo object from generating pulses on its assigned I/O pin.

```
void servo.detach()
```

SPI

The SPI library supports the use of the Serial Peripheral Interface (SPI) bus for communication with SPI-compatible peripherals, typically chips with a built-in SPI interface. It can also be used for communications between two microcontrollers.

The SPISettings class is used to configure the SPI port. The arguments are combined into a single SPISettings object, which is passed to SPI.beginTransaction().

```
SPISettings(uint32_t clock, uint8_t bitOrder, uint8_t dataMode)
```

Example:

```
SPISettings spiset(uint32_t clock,
                   uint8_t bitOrder,
                   uint8_t dataMode)
```

beginTransaction()
> Initializes the SPI interface using the settings defined in an SPISettings object.

```
void SPI.beginTransaction(SPISettings)
```

endTransaction()
> Stops communication with the SPI interface. Typically called after the select pin is de-asserted to allow other libraries to use the SPI interface.

```
void SPI.endTransaction()
```

usingInterrupt()
> Used when SPI communications will occur within the context of an interrupt.

```
void SPI.usingInterrupt(uint8_t interruptNumber)
```

begin()
> Starts the SPI library and initializes the SPI interface. Sets the SCK, MOSI, and SS pins to output mode, and pulls SCK and MOSI low while setting SS high.

```
void SPI.begin()
```

end()

Disables the SPI interface but leaves the pin modes (in or out) unchanged.

```
void SPI.end()
```

transfer()

Transfers one byte over an SPI interface, either sending or receiving.

```
uint8_t SPI.transfer(uint8_t data)
```

setBitOrder()

Sets the order of the bit shifted out to the SPI interface. The two choices are LSBFIRST (least significant bit first) and MSBFIRST (most significant bit first). This function should not be used with new projects. Use the beginTransaction() function to configure the SPI interface.

```
void SPI.setBitOrder(uint8_t bitOrder)
```

setClockDivider()

Sets the SPI clock divider relative to the system clock. For AVR-based Arduino boards the valid divisors are 2, 4, 8, 16, 32, 64, or 128. This function should not be used with new projects. Use the beginTransaction() function to configure the SPI interface.

```
void SPI.setClockDivider(uint8_t clockDiv)
```

setDataMode()

Sets the clock polarity and phase of the SPI interface. This function should not be used with new projects. Use the beginTransaction() function to configure the SPI interface.

```
void SPI.setDataMode(uint8_t dataMode)
```

SoftwareSerial

The SoftwareSerial library implements software-based serial communication on the digital I/O pins of an Arduino. In other words, it is a "bit-banger" that emulates a conventional serial interface. It is useful when more than one serial interface is required, but the built-in USART in the AVR microcontroller is assigned to some other function (such as a USB interface).

The SoftwareSerial library supports multiple serial interfaces, each with a speed of up to 115,200 bits per second. When using multiple instances of SoftwareSerial only one can receive data at a time. The I/O pins used must support pin change interrupts. SoftwareSerial provides a 64-byte receive buffer for each instance of a serial interface.

An object of type `SoftwareSerial` is created to use with other serial I/O operations. The class constructor is passed the digital pins to use for input (rx) and output (tx).

```
SoftwareSerial(uint8_t rxPin, uint8_t txPin, bool inv_logic = false)
```

Example:

```
SoftwareSerial serial = SoftwareSerial(rxPin, txPin)
```

available()

Returns the number of bytes in the serial buffer that are available for reading.

```
int serial.available()
```

begin()

Sets the baud rate (speed) of the serial interface. Valid baud rates are 300, 600, 1200, 2400, 4800, 9600, 14400, 19200, 28800, 31250, 57600, and 115200.

```
void serial.begin(long speed)
```

isListening()

Tests the serial interface to see if it is listening for input. Returns `true` if the interface is actively waiting for input, `false` otherwise.

```
bool serial.isListening()
```

overflow()

If the input exceeds the 64-byte size of the receive buffer in the `SoftwareSerial` object, a flag is set. Calling the `overflow()` function will return the flag. A return value of `true` indicates that an overflow has occurred. Calling the `overflow()` function clears the flag.

```
bool serial.overflow()
```

peek()

Returns the oldest character from the serial input buffer, but does not remove the character. Subsequent calls will return the same character. If there are no bytes in the buffer, then `peek()` will return -1.

```
int serial.peek()
```

read()

Returns a character from the receive buffer and removes the character. The `read()` function is typically used in a loop. Returns -1 if no data is available. Note that only one instance of `SoftwareSerial` can receive incoming data at any given time. The `listen()` function is used to select the active interface.

```
int serial.read()
```

print()

The print() function behaves the same as the Serial.print() function. Accepts any data type that the Serial.print() will accept, which includes char, byte (uint8_t), int, long, or string types, and returns the number of bytes written.

```
int serial.print(data)
```

println()

Identical to the serial.print() function, except that a carriage return/line feed (CR/LF) pair is appended to the output. Returns the number of bytes written. If no data is provided, it will simply emit a CR/LF.

```
int serial.println(data)
```

listen()

Enables listening (data receive) for an instance of SoftwareSerial. Only one instance of a SoftwareSerial object can receive data at any given time, and the object whose listen() function is called becomes the active listener. Data that arrives on other interface instances will be discarded.

```
bool serial.listen()
```

write()

Transmits data from the serial interface as raw bytes. Behaves the same as the Serial.write() function. Returns the number of bytes written.

```
size_t serial.write(uint8_t byte)
```

Stepper

The Stepper library can be used to control both unipolar and bipolar stepper motors with the appropriate hardware to handle the required current.

The Stepper library has two forms of constructors, one for unipolar motors and one for bipolar motor types. Each creates a new instance of the Stepper class. It is called at the start of a sketch, before the setup() and loop() functions. The *steps* argument defines the number of steps in a full revolution of the motor's output shaft. The *pin1*, *pin2*, *pin3*, and *pin4* arguments specify the digital pins to use.

```
Stepper(int steps, int pin1, int pin2);
Stepper(int steps, int pin1, int pin2, int pin3, int pin4);
```

Example:

```
Stepper stepdrv = Stepper(100, 3, 4);
```

setSpeed()

Sets the speed (step rate) in terms of RPM. Does not cause the motor to turn; it just sets the speed to use when the step() function is called.

```
void stepdrv.setSpeed(long speed);
```

step()

Commands the motor to move a specific number of steps. A positive count turns the motor one direction, and a negative count causes it to turn in the opposite direction.

```
void stepdrv.step(int stepcount);
```

TFT

The TFT (thin-film transistor) display library provides functions for drawing text, images, and shapes on a TFT display. It is included with versions 1.0.5 and later of the Arduino IDE. This library simplifies the process for displaying graphics on a display. It is based on the Adafruit ST7735H library, which can be found on GitHub (*http://bit.ly/ada-st7735*). The ST7735H library is based on the Adafruit GFX library, also available on GitHub (*http://bit.ly/ada-gfx*).

The TFT library is designed to work with interfaces that use the SPI communications capabilities of an AVR microcontroller. If the TFT shield includes an SD card slot, then the SD library can be used to read and write data by using a separate select signal from an Arduino. The TFT library relies on the SPI library for communication with the screen and SD card, and it also needs to be included in all sketches that use the TFT library.

TFT class

The TFT class constructor is available in two forms. One is used when the standard Arduino SPI pins are used (the hardware SPI), and the second form allows you to specify which pins to use:

```
TFT(uint8_t cs, uint8_t dc, uint8_t rst)
TFT(uint8_t cs, uint8_t dc, uint8_t mosi, uint8_t sclk, uint8_t rst)
```

Where:

cs   Chip select pin

dc   Data or command mode select

rst   Reset pin

mosi Pin used for MOSI if not using hardware SPI

sclk Pin used for clock if not using hardware SPI

Example:

```
#define cs  10
#define dc  9
```

```
#define rst 8
TFT disp = TFT(cs, ds, rst);
```

The Esplora version of the TFT library uses predefined pins. All that is necessary is to instantiate the TFT object:

```
EsploraTFT disp = EsploraTFT;
```

begin()

Called to initialize the TFT library components. Must be called before any other functions are used. Typically called in the setup() function of a sketch.

```
void disp.begin()
```

background()

Overwrites the entire display screen with a solid color. May be used to clear the display. Note that the screen cannot actually display 256 unique levels per color, but instead uses 5 bits for the blue and red colors, and 6 bits for green.

```
void disp.background(uint8_t red, uint8_t green, uint8_t blue)
```

stroke()

Called before drawing on the screen, and sets the color of lines and borders. Like the background() function, stroke() uses 5 bits for the blue and red colors, and 6 bits for green.

```
void disp.stroke(uint8_t red, uint8_t green, uint8_t blue)
```

noStroke()

Removes all outline stroke color.

```
void disp.noStroke()
```

fill()

Sets the fill color of objects and text on the screen. Like the stroke() function, fill() uses 5 bits for the blue and red colors, and 6 bits for green.

```
void disp.fill(uint8_t red, uint8_t green, uint8_t blue)
```

noFill()

Disables color fills for objects and text.

```
void disp.noFill()
```

setTextSize()

Sets the size of the text written by a call to the text() function. The default text size is 1, or 10 pixels. Each increase in the text size increases the height of the text on the screen by 10 pixels.

```
void disp.setTextSize(uint8_t size)
```

text()

Writes text to the display at the specified coordinates. The text color is set by calling the fill() function before calling text().

```
void disp.text(const char * text, int16_t x, int16_t y),
```

point()

Draws a point at a specific location on the screen. The point color will be what was specified by a preceding fill() function call.

```
void disp.point(int16_t x, int16_t y)
```

line()

Draws a line between start and end coordinates using the color set by the stroke() function.

```
void disp.line(int16_t x1, int16_t y1, int16_t x2, int16_t y2)
```

rect()

Draws a rectangle starting at an upper-left point (*x*, *y*) with a specified width and height.

```
void disp.rect(int16_t x, int16_t y, int16_t width, int16_t height)
```

width()

Reports the width of the TFT screen in pixels.

```
int disp.width()
```

height()

Reports the height of the TFT screen in pixels.

```
int disp.height()
```

circle()

Draws a circle on the display with a center point of (*x*, *y*), and radius of *r*.

```
int disp.circle(int16_t x, int16_t y, int16_t r)
```

image()

Draws an image loaded from an SD card onto the screen at a specified position.

```
void image(PImage image, int xpos, int yPos)
```

loadImage()

Creates an instance of a PImage object using the image file name provided. The image file must be a 24-bit BMP type, and it must reside on the root directory of the SD card. Uses the PImage function loadImage().

```
PImage disp.loadImage(char *imgname)
```

PImage

The `PImage` class contains functions to encapsulate and draw a bitmap image on a TFT display.

`PImage.height()`

Once an image has been encapsulated in a `PImage` object it may be queried to obtain its height. This function returns the height as an int value.

`PImage.width()`

Returns the width of an encapsulated image object as an int value.

`PImage.isValid()`

Returns a Boolean `true` if the image object contains a valid bitmap file, or `false` if it does not.

WiFi

The WiFi library gives an Arduino the ability to connect to a wireless network. The descriptions here don't define all the available functions in detail, since many of them are similar or identical to those found in the Ethernet library. The built-in help in early versions of the Arduino IDE (which, unfortunately, seems to be all that some Linux distributions have available at the time of writing) do not have the WiFi library reference pages, but later versions do. The library source code does seem to be installed with the older versions of the IDE, or at least it is on my Kubuntu development system.

The WiFi library is used with the SPI library to communicate with the WiFi module and an optional SD memory card. A baseline-type Arduino (see Chapter 4) communicates with the WiFi shield using the SPI pins 10, 11, 12, and 13. The Mega-type boards use pins 10, 50, 51, and 52. Also, on the Arduino WiFi shield pin 7 is used as a handshake signal between the Arduino and the WiFi shield, so it should not be used for anything else. Other WiFi shields may have similar restrictions.

The Arduino WiFi shield can act either as a server for accepting incoming connections, or as a client to make a connection with an existing server. The library provides WEP and WPA2 Personal encryption modes, but it does not support WPA2 Enterprise encryption. Also, if a server node does not broadcast its SSID (Service Set Identifier), the WiFi shield will not be able to make a connection.

Like the Ethernet library, the WiFi library is comprised of a collection of five C++ classes. Most of the classes inherit from parent classes, but you don't need to bother with the details for most applications. However, if you need to see the low-level class definitions for `Client`, `Server`, `UDP`, and others, they can be found in the directory *libraries/WiFi* in the Arduino source code. The following list shows the five classes of the WiFi library and the public member functions of each class:

- WiFi class
  - begin()
  - disconnect()
  - config()
  - setDNS()
  - SSID()
  - BSSID()
  - RSSI()
  - encryptionType()
  - scanNetworks()
  - getSocket()
  - macAddress()
- IPAddress class
  - localIP()
  - subnetMask()
  - gatewayIP()
- Server class
  - WiFiServer()
  - begin()
  - available()
  - write()
  - print()
  - println()
- Client class
  - WiFiClient()
  - connected()
  - connect()
  - write()
  - print()

- — println()
- — available()
- — read()
- — flush()
- — stop()
- UDP class
  - — begin()
  - — available()
  - — beginPacket()
  - — endPacket()
  - — write()
  - — parsePacket()
  - — peek()
  - — read()
  - — flush()
  - — stop()
  - — remoteIP()
  - — remotePort()

 The Arduino WiFi shield is based on the HDG204 802.11b/g chip. Be aware that other WiFi shields, such as Adafruit's WiFi shield based on the TI CC3000 WiFi chip, may use a different library specifically for a particular WiFi chip. Much of the functionality should be similar to what is listed here, but there will still be some differences to take into consideration. The Adafruit library is available on GitHub (*http://bit.ly/ada-cc3000*). Refer to the Adafruit website (*http://bit.ly/ada-cc3000-wifi*) for details.

WiFi class

The following is a quick summary of the WiFi classes. For function descriptions, refer to the Ethernet library.

The WiFi class contains functions to initialize the library and the network settings. The class definition cna be found in the include file *WiFi.h*.

```
WiFiClass()

int begin(char* ssid)
int begin(char* ssid, uint8_t key_idx, const char* key)
int begin(char* ssid, const char *passphrase)

int disconnect()

void config(IPAddress local_ip)
void config(IPAddress local_ip, IPAddress dns_server)
void config(IPAddress local_ip, IPAddress dns_server, IPAddress gateway)
void config(IPAddress local_ip, IPAddress dns_server,
            IPAddress gateway, IPAddress subnet)

void setDNS(IPAddress dns_server1)
void setDNS(IPAddress dns_server1, IPAddress dns_server2)

char* SSID()
char* SSID(uint8_t networkItem)
uint8_t* BSSID(uint8_t* bssid)
int32_t RSSI()
int32_t RSSI(uint8_t networkItem)

uint8_t encryptionType()
uint8_t encryptionType(uint8_t networkItem)
int8_t scanNetworks()

static uint8_t getSocket()
uint8_t* macAddress(uint8_t* mac)
static char* firmwareVersion()
uint8_t status()
int hostByName(const char* aHostname, IPAddress& aResult)
```

IPAddress class

Like the IPAddress in the Ethernet library, the IPAddress class in the WiFi library
provides a container for information about the network configuration.

```
IPAddress localIP()
IPAddress subnetMask()
IPAddress gatewayIP()
```

Server class

The Server class creates servers that can accept connections from clients to exchange
data. A client may be another Arduino with a WiFi shield, a desktop PC, notebook
computer, or just about any device with compatible WiFi capability. Refer to "Server
class: EthernetServer" on page 165 for descriptions of print() and println().

```
WiFiServer(uint16_t)
WiFiClient available(uint8_t* status = NULL)

void begin()
```

```
int print(data)
int print(data, base)

int println()
int println(data)
int println(data, base)

size_t write(uint8_t)
size_t write(const uint8_t *buf, size_t size)

uint8_t status()
```

Client class

The Client class creates WiFi clients that can connect to servers in order to send and receive data. A server may be another Arduino with a WiFi shield, a desktop PC, and notebook computer, or just about any device with compatible WiFi server capability. Refer to "Server class: EthernetServer" on page 165 for descriptions of print() and println().

```
WiFiClient()
WiFiClient(uint8_t sock)

uint8_t connected()

int connect(IPAddress ip, uint16_t port)
int connect(const char *host, uint16_t port)

size_t write(uint8_t)
size_t write(const uint8_t *buf, size_t size)

int print(data)
int print(data, base)

int println()
int println(data)
int println(data, base)

int available()
int read()
int read(uint8_t *buf, size_t size)
int peek()

void flush()
void stop()
```

UDP class

The UDP class enables short messages to be sent and received using the UDP protocol. Unlike TCP/IP, which is a stream protocol (i.e., it has no definite start and stop boundaries), UDP is a datagram protocol. In this case, each item of data is a single packet, called a datagram, and the data must fit within the boundaries of the datagram packet. UDP does not have error detection, nor does it guarantee delivery of the data, but for short packets of noncritical data, or where the upper-level software can handle things like error detection and retries, it offers a fast and relatively simple way to move data around between hosts.

```
WiFiUDP()

uint8_t begin(uint16_t)
void stop()

int beginPacket(IPAddress ip, uint16_t port)
int beginPacket(const char *host, uint16_t port)
int endPacket()

size_t write(uint8_t)
size_t write(const uint8_t *buffer, size_t size)

int parsePacket()
int available()
int read()
int read(unsigned char* buffer, size_t len)
int peek()
void flush()

IPAddress remoteIP()
uint16_t remotePort()
```

Wire

The Wire library is used to communicate with TWI- or I2C-type devices. Refer to Chapters 2 and 3 for more information about the TWI capabilities of the AVR microcontrollers. Chapter 8 describes some shields that use I2C for communications with the Arduino.

The following table defines where the TWI pins are located on different types of Arduino boards. Refer to Chapter 4 for board pinout diagrams.

Board	SDA	SCL
Uno, Ethernet	A4	A5
Mega2560	20	21
Leonardo	2	3

The core of the Wire library is the TwoWire class.

Example:

```
TwoWire twi = TwoWire()
```

begin()

Initializes the TWI library and activates the I2C interface in either master or servant mode. If the address is not specified, the I2C interface defaults to master mode.

```
void twi.begin()
void twi.begin(uint8_t addr)
void twi.begin(int addr)
```

requestFrom()

Used by the interface master to request data from a servant device. The data bytes are retrieved with the `available()` and `read()` functions. Returns the number of bytes of data read from the addressed device.

```
uint8_t twi.requestFrom(uint8_t addr, uint8_t quantity)
uint8_t twi.requestFrom(uint8_t addr, uint8_t quantity, uint8_t stop)
uint8_t twi.requestFrom(int addr, int quantity)
uint8_t twi.requestFrom(int addr, int quantity, int stop)
```

beginTransmission()

Begins a data transmission to an I2C servant device at the spefifiend address. Data is queued for transmission using the `write()` function and then actually transmitted using the `endTransmission()` function.

```
void twi.beginTransmission(uint8_t addr)
void twi.beginTransmission(int addr)
```

endTransmission()

Transmits the bytes that were queued by `write()` to a servant device and then ends a transmission that was initiated by `beginTransmission()`.

```
uint8_t twi.endTransmission()
uint8_t twi.endTransmission(uint8_t stop)
```

write()

Writes the supplied data to a queue for transmission from a master to a servant device, or from a servant device to a master in response to a data request. Returns the number of bytes written into the queue.

```
size_t twi.write(uint8_t data)
size_t twi.write(const uint8_t *data)
size_t twi.write(const uint8_t *data, size_t len)
```

available()

Returns the number of bytes available to the `read()` function. Called by a master device after a call to `requestFrom()`, and on a servant device after a data receive event.

```
int twi.available()
```

read()

Reads a byte that was transferred from master to servant, or vice versa.

```
int twi.read()
```

onReceive()

Registers the function to call (a handler function) when a servant device receives data from a master.

```
void twi.onReceive(void (*)(int))
```

onRequest()

Registers the function to call when a master requests data from a servant device.

```
void twi.onRequest(void (*)(void))
```

Esplora

The Arduino Esplora library provides a set of functions for easily interfacing with the sensors and actuators on the Esplora board via the Esplora class. For pinout information refer to Chapter 4.

The sensors available on the board are:

- Two-axis analog joystick
- Center pushbutton of the joystick
- Four pushbuttons
- Microphone
- Light sensor
- Temperature sensor
- Three-axis accelerometer
- Two TinkerKit input connectors

The actuators available on the board are:

- Bright RGB (Red-Green-Blue) LED
- Piezo buzzer
- 2 TinkerKit output connectors

Esplora()

Creates an instance of an Esplora object.

```
Esplora esp = Esplora()
```

readSlider()

Returns an integer value corresponding to the current position of the slider control. The value can range from 0 to 1023.

```
unsigned int esp.readSlider()
```

readLightSensor()

Returns an integer value corresponding to the amount of light impinging on the light sensor on an Esplora board.

```
unsigned int esp.readLightSensor()
```

readTemperature()

Returns a signed integer with the current ambient temperature in either Fahrenheit or Celsius. The *scale* argument takes either *DEGREES_C* or *DEGREES_F*. The temperature ranges are –40C to 150C, and –40F to 302F.

```
int esp.readTemperature(const byte scale);
```

readMicrophone()

Returns an integer value corresponding to the amount of ambient noise detected by the microphone. The returned value can range from 0 to 1023.

```
unsigned int esp.readMicrophone()
```

readJoystickSwitch()

Reads the joystick button and returns either 0 or 1023. An alternative is the read JoystickButton() function.

```
unsigned int esp.readJoystickSwitch()
```

readJoystickButton()

Reads the joystick's button and returns either LOW or HIGH (pressed or not pressed). This function performs the same function as readJoystickSwitch(), but it returns a value that is consistent with the readButton() function.

```
bool readJoystickButton()
```

readJoystickX()

Returns the x-axis position of the joystick as a value between –512 and 512.

```
int esp.readJoystickX()
```

readJoystickY()

Returns the y-axis position of the joystick as a value between –512 and 512.

```
int esp.readJoystickY()
```

readAccelerometer()

Returns the current value for a selected axis, with the possible values for the *axis* argument being X_AXIS, Y_AXIS, and Z_AXIS. A return value of 0 indicates that the accelerometer is perpendicular to the direction of gravity, and positive or negative values indicate the direction and rate of acceleration.

```
int esp.readAccelerometer(const byte axis)
```

readButton()

Reads the current state of a particular button on an Esplora. Returns a low (false) value if the button is pressed, or a high (true) value if not.

```
bool esp.readButton(byte channel)
```

writeRGB()

Writes a set of values defining the brightness levels of the red, green, and blue elements in the Esplora's RGB LED.

```
void esp.writeRGB(byte red, byte green, byte blue)
```

writeRed()

Accepts an argument that defines the brightness of the red LED with a range of 0 to 255.

```
void esp.writeRed(byte red)
```

writeGreen()

Accepts an argument that defines the brightness of the green LED with a range of 0 to 255.

```
void esp.writeGreen(byte green)
```

writeBlue()

Accepts an argument that defines the brightness of the blue LED with a range of 0 to 255.

```
void esp.writeBlue(byte blue)
```

readRed()

Returns the value last used to set the brightness of the red LED.

```
byte esp.readRed()
```

readGreen()

Returns the value last used to set the brightness of the green LED.

```
byte esp.readGreen()
```

readBlue()

Returns the value last used to set the brightness of the blue LED.

```
byte esp.readBlue()
```

tone()

Emits a tone from the Esplora's on-board annunciator at a given frequency. If no *duration* argument is supplied the tone will continue until the noTone() function is called. Only one frequency at a time can be used. Note that using the tone() function will interfere with controlling the level of the red LED.

```
void esp.tone(unsigned int freq)
void esp.tone(unsigned int freq, unsigned long duration)
```

noTone()

Terminates the output of the square wave signal to the annunciator.

```
void esp.noTone()
```

USB libraries

The core USB libraries allow an Arduino Leonardo or Micro to appear as a mouse and/or keyboard device to a host computer.

 If the Mouse or Keyboard library is constantly running, it will be difficult to program the Arduino. Functions such as Mouse.move() and Keyboard.print() should only be called when the host is ready to handle them. One way to deal with this is to use a control system or a physical switch to control when the Arduino will emit mouse or keyboard messages.

Mouse

The mouse functions allow a Leonardo or Micro to control cursor movement on a host computer. The reported cursor position is always relative to the cursor's previous location; it is not absolute.

```
Mouse.begin()
Mouse.click()
Mouse.end()
Mouse.move()
Mouse.press()
Mouse.release()
Mouse.isPressed()
```

Keyboard

The keyboard functions allow a Leonardo or Micro to send keystrokes to an attached host computer. While not every possible ASCII character, particularly the nonprinting ones, can be sent with the Keyboard library, the library does support the use of modifier keys.

```
Keyboard.begin()
Keyboard.end()
Keyboard.press()
Keyboard.print()
Keyboard.println()
Keyboard.release()
Keyboard.releaseAll()
Keyboard.write()
```

Modifier keys change the behavior of another key when pressed simultaneously. Table 7-1 lists the modifier keys supported by the Leonardo.

Table 7-1. USB keyboard modifier keys

Key	Hex value	Decimal value	Key	Hex value	Decimal value
KEY_LEFT_CTRL	0x80	128	KEY_PAGE_UP	0xD3	211
KEY_LEFT_SHIFT	0x81	129	KEY_PAGE_DOWN	0xD6	214
KEY_LEFT_ALT	0x82	130	KEY_HOME	0xD2	210
KEY_LEFT_GUI	0x83	131	KEY_END	0xD5	213
KEY_RIGHT_CTRL	0x84	132	KEY_CAPS_LOCK	0xC1	193
KEY_RIGHT_SHIFT	0x85	133	KEY_F1	0xC2	194
KEY_RIGHT_ALT	0x86	134	KEY_F2	0xC3	195
KEY_RIGHT_GUI	0x87	135	KEY_F3	0xC4	196
KEY_UP_ARROW	0xDA	218	KEY_F4	0xC5	197
KEY_DOWN_ARROW	0xD9	217	KEY_F5	0xC6	198
KEY_LEFT_ARROW	0xD8	216	KEY_F6	0xC7	199
KEY_RIGHT_ARROW	0xD7	215	KEY_F7	0xC8	200
KEY_BACKSPACE	0xB2	178	KEY_F8	0xC9	201
KEY_TAB	0xB3	179	KEY_F9	0xCA	202
KEY_RETURN	0xB0	176	KEY_F10	0xCB	203
KEY_ESC	0xB1	177	KEY_F11	0xCC	204
KEY_INSERT	0xD1	209	KEY_F12	0xCD	205

Contributed Libraries

There are many contributed libraries available for the Arduino boards. Some have been created by individuals, others by firms that sell and support Arduino hardware and accessories. In addition, some vendors also provide libraries or support software for their products, and a search of the appropriate website or examination of an eBay listing will often uncover this code.

Tables 7-2 through 7-8 list a selection of these libraries, broken down by category. The descriptions are necessarily brief; there is just no way they could all be described sufficiently to do them justice and still have this be a compact-format book. For links for further details, see *http://www.arduino.cc/en/Reference/Libraries*.

Table 7-2. Communication (networking and protocols)

Library	Description
Messenger	For processing text-based messages from the computer
NewSoftSerial	An improved version of the SoftwareSerial library
OneWire	For controlling devices (from Dallas Semiconductor) that use the One Wire protocol
PS2Keyboard	For reading characters from a PS2 keyboard
Simple Message System	For sending messages between the Arduino and the computer
SSerial2Mobile	For sending text messages or emails using a cell phone (via AT commands over SoftwareSerial)
Webduino	An extensible web server library (for use with the Arduino Ethernet shield)
X10	For sending X10 signals over AC power lines
XBee	For communicating with XBees in API mode
SerialControl	For remote control of other Arduinos over a serial connection

Table 7-3. Sensing

Library	Description
Capacitive Sensing	For turning two or more pins into capacitive sensors
Debounce	For reading noisy digital inputs (e.g., from buttons)

Table 7-4. Displays and LEDs

Library	Description
GFX	Base class with standard graphics routines (by Adafruit Industries)
GLCD	Graphics routines for LCDs based on the KS0108 or equivalent chipset
LedControl	For controlling LED matrixes or 7-segment displays with a MAX7221 or MAX7219
LedControl	An alternative to the Matrix library for driving multiple LEDs with Maxim chips
LedDisplay	For controlling an HCMS-29xx scrolling LED display
Matrix	Basic LED matrix display manipulation library
PCD8544	For the LCD controller on Nokia 55100-like displays (by Adafruit Industries)
Sprite	Basic image sprite manipulation library for use in animations with an LED matrix
ST7735	For the LCD controller on a 1.8″, 128 × 160-pixel TFT screen (by Adafruit Industries)

Table 7-5. Audio and waveforms

Library	Description
FFT	For frequency analysis of audio or other analog signals
Tone	For generating audio frequency square waves in the background on any microcontroller pin

Table 7-6. Motors and PWM

Library	Description
TLC5940	For controlling the TLC5940 IC, a 16-channel, 12-bit PWM unit

Table 7-7. Timing

Library	Description
DateTime	A library for keeping track of the current date and time in software
Metro	Helps you time actions at regular intervals
MsTimer2	Uses the timer 2 interrupt to trigger an action every N milliseconds

Table 7-8. Utilities

Library	Description
PString	A lightweight class for printing to buffers
Streaming	A library that simplifies print statements

CHAPTER 8

Shields

An Arduino shield is an add-on circuit board designed to work with the connectors on a standard Arduino board like an Uno, Duemilanove, Leonardo, or Mega. A shield has pins that interface with the Arduino so that things like DC power, digital I/O, analog I/O, and so on are available to the shield. This chapter covers some of the Arduino-compatible shields that are available, and Chapter 10 describes the process of creating a custom shield.

Shields are available for a variety of applications, ranging from minimal boards for prototyping to motor controllers, Ethernet interfaces, SD flash memory, and displays. Many shields have the ability to be stacked, allowing a base Arduino board to interface with two or more shields at once.

 This chapter references many different vendors and manufacturers, but it is not intended to be a specific endorsement of any of them. The shields shown here are representative of what is available, and for any given shield type you can likely find different vendors selling the same, or an equivalent, product. Shop around.

This chapter is by no means a complete list of all the various types of shields that are available. There is a quiet cottage industry that specializes in creating new variations on existing shields and new shields that have never been seen before. The selection of shields described here is broadly representative of what is available, and links are provided if you want to learn more, or perhaps buy a shield or two. In some cases I've included more detailed information to supplement what the vendor provides (or doesn't, in some cases), but that doesn't mean I'm especially fond of any particular shield. I'm just fond of documentation, so I can get the information I need to move on to what I want to do. You may also find yourself in a position of having a really useful-looking shield for which there's little or no documentation, or where what there is happens to be in Chinese (or some other language you might not know). Hopefully this chapter will help to fill some of the gaps, or at least give you some ideas on where to look.

 Some of the shields shown here are no longer available from the original vendor, but can be purchased from other sources. Most vendors provide active links to documentation, so if you find a look-alike shield (the hardware is open source, after all) you can often still get the technical information you need.

One thing to remember when looking for a shield is that some people seem to be unclear on what a shield is. It is not a module with a row of pins down one side (these are discussed in Chapter 9). A shield is a board that meets the physical characteristics described in "Physical Characteristics of Shields" on page 217. Anything else can be considered to be a module, and modules may or may not plug into an Arduino directly (they usually don't simply plug in, but need wiring of some sort to route power and signals to the appropriate pins on an Arduino board).

Electrical Characteristics of Shields

If you compare the pinout diagrams of the various shields in this chapter, you may notice a pattern emerging: shields that employ a two-wire interface (the TWI or I2C interface) always use pins A4 and A5 of the Arduino. On a baseline-type board (Diecimila, Duemilanove, and Uno R2 or SMD) these will be found on the analog I/O connector, and on the newer extended layout boards (Uno R3, Ethernet, Leonardo) the signals also appear on the extended pin header (in the upper corner by the USB or RJ45 connector). A shield that uses I2C can utilize either set of pins, and you can assume that A4 and A5 will not be available for other uses without some clever programming.

By the same token, shields that use the SPI interface will typically use the D10, D11, D12, and D13 pins (SS, MOSI, MISO, and SCK, respectively). The use of D10 as the select is optional, and some shields use a different digital pin for that purpose. Remember that with SPI each "slave" device must be explicitly selected with a select signal before it will accept incoming data from the master device. Shields with more than one attached SPI device may also use more than one digital I/O pin as a select signal.

Shields that use the UART (or USART, as Atmel calls it) will typically use pins D0 and D1 (Rx and Tx, respectively, or RxD0 and TxD0 on a Mega-type Arduino). Some Bluetooth shields use the UART interface to communicate with a Bluetooth module, as do RS-232 and RS-485 interface shields.

Then there are shields that use almost every single Arduino pin. The DIY Multi-function shield described in "Miscellaneous Shields" on page 268 is like this, but it does have pins to connect to the signals that are not specifically used on the board (three digital and one analog, in this case). In general, you can safely assume that I/O extension shields will use most or all of the available Arduino pins, and shields that support something like a display will generally not have any connection points for accessing unused signals. For this reason these types of shields are usually nonstacking, and should be placed at the top of a stack of shields on an Arduino.

Most shields do not have extremely complicated circuits but are relatively simple things based on existing ICs or components of some type. Like the Arduino boards, they are essentially carriers for various types of ICs, relays, and so on. The electrical characteristics of a shield are those of the chip or components it is designed around. This simplicity is what helps to keep the cost of a shield low, and the capabilities of the ICs or components it uses are what make a shield useful.

Lastly, some shields may buffer the signals to and from an Arduino, using active circuits or devices such as optical isolators or relays, but most of them simply serve as places to mount connectors or components, such as the extension shields shown in "I/O Extension Shields" on page 222 that utilize multipin connectors. The connections, be they sockets or pin headers, are just extensions of the Arduino's own pins, and there is nothing to protect against connecting 12 volts to a 5V (or even 3.3V) digital input and converting the AVR microcontroller on an Arduino into charcoal. Always observe the voltage and current limitations of the AVR microcontroller.

Physical Characteristics of Shields

Physically a typical shield is as wide as a baseline Arduino board (see Chapter 4 for dimensions), and it can be the same length as an Arduino, or it might be longer or shorter. It is possible to make a shield that is wider than a baseline Arduino board, as the only real constraint is that the pins on the shield line up with the pin sockets on the underlying Arduino board (see Chapter 4 for locations and dimensions). Installing a shield is just a matter of connecting the shield as shown in Figure 8-1.

Newer Arduino boards that use the R3 pin layout will have two pin sockets at the end of each row that are not used by the shield. This is unimportant, as these extra sockets are either duplicates of existing pins or not connected. In the case of the Mega boards the shield will mount as shown in Figure 4-20 in Chapter 4, with most of the pins on the Mega board not connected to the shield, and some others made inaccessible by the overlying shield PCB. All of the baseline pins and signals are available to the shield.

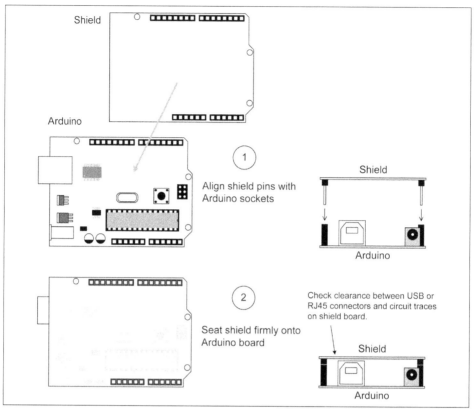

Figure 8-1. Shield mounting on host Arduino board

 Always check the clearance between components on the base Arduino board and the shield PCB. In some cases a USB connector or an RJ45 jack can interfere with the shield and potentially cause a short circuit.

Sometimes you might have a problem with parts on the Arduino board colliding with the circuit traces or pads on the underside of a shield board. A small piece of electrical tape or even some heavy card stock paper can be used as an insulating shim, but the better way to deal with this is to use spacers or standoffs to physically separate the two boards. These can be short metal or nylon tubes (7/16 to 1/2 inch, or about 11 to 12 mm in length) with a center hole sufficient for a machine screw (which would typically be a 2-56 SAE type, or a suitable metric size). The difference between a standoff and a spacer is simple: standoffs have internal threads, spacers do not. A spacer or standoff serves to raise the upper board enough to prevent shorts, as shown in Figure 9-3 in Chapter 9. It also results in firm mechanical coupling between two or more boards.

When stacking two or more shields on an Arduino you can elect to use long machine screws with spacers, or you might want to consider a type of standoff that has a threaded screw-like projection at one end and a threaded hole at the other end. Also known as "jack screws," these are common in PCs, and if you've ever assembled your own computer from scratch you've already encountered them. Figure 8-2 shows some examples of the types of spacers and standoffs that are available. These parts can be made from nylon, aluminum, stainless steel, brass, and plastic.

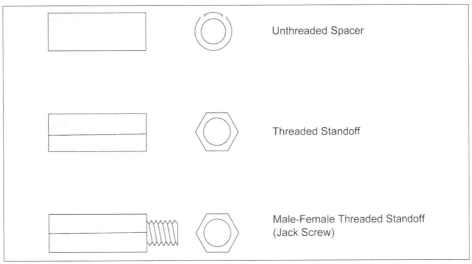

Figure 8-2. Spacers and standoff types

Spacers can also be made from strips cut from a prototyping PCB and slipped over the pins of the upper board. This technique won't do much for physically coupling the boards, but it will add enough space to prevent collisions.

You can purchase spacers and standoffs from various sources, including Amazon (*http://amazon.com*), McMaster-Carr (*http://www.mcmaster.com*), Mouser (*http://www.mouser.com*), and Digi-Key (*http://www.digikey.com*). If you happen to have a fabrication hardware supplier in your city, you can also purchase these and other useful things (like 1-72 or 2-56 machine screws and nuts) locally.

Stacking Shields

Sometimes it doesn't make sense to use extended-lead pin sockets to make a shield fit into a stack. If, for example, a shield has a large number of connectors that require vertical access, then there simply may not be enough room for another shield to mount on top of it. But you will sometimes encounter a shield that could stack but doesn't have the correct connector types.

One of two techniques is used to allow a shield to be stacked on another shield: extended socket pins or offset pins and sockets. The extended pin approach allows the stacked shield to stay in vertical alignment. The offset pin and socket design results in stacked shields that are shifted over by the amount of offset between rows of pins and sockets. If the shields all shift the same way, the result can look like stairs, and the holes in the boards for mounting screws will not line up correctly.

The use of extended pins requires pin sockets with long pins for mounting on a PCB. The protruding pins on the back, or solder, side of the shield board will plug into the pin sockets on an Arduino, and another shield can be mounted on the top of the stackable shield. The offset technique uses separate pin strips to connect the shield to the Arduino and separate socket strips to accept another shield. These are mounted side-by-side, usually as closely as possible to minimize the offset between shields.

Common Arduino Shields

This section reviews some of the readily available shields. This is by no means a comprehensive list, as new shields appear continually, and older shields are discontinued or may otherwise no longer be available. In this age of rapid prototyping, fast-turnaround production, and low-cost manufacturing, a shield may appear and then vanish a few months later. To stay abreast of what is available, you might want to check out the many Chinese vendors on eBay, and the listings on Amazon.com, Adafruit, SparkFun, SainSmart, and other websites. See Appendix C for known sources of Arduino shields. Table 8-2 at the end of this chapter lists the vendors and manufacturers covered here.

The following list provides a quick reference of the shields discussed here, broken down by category:

- Input/output
  - I/O extension shields
  - I/O expansion shields
  - Relay shields
  - Signal routing shields

- Memory
  - SD and microSD card flash memory
- Communication
  - Serial I/O
  - MIDI
  - Ethernet

- — Bluetooth
- — USB
- — ZigBee
- — CAN
- Prototyping
- Motion control
  - — DC and stepper motor control
  - — PWM and servo control
- Displays
  - — LED arrays
  - — 7-Segment LED displays
  - — LCD displays
  - — Color TFT displays
- Instrumentation shields
  - — Data logging
  - — Logic analyzer
  - — 24-bit ADC
  - — 12-bit DAC
- Adapter shields
  - — Nano adapters
  - — Terminal block adapters
- Miscellaneous Shields
  - — Terminal block/prototyping
  - — Multifunction shield

In this section we will also take a quick look at some uncommon shields designed for specific applications: a CNC engraver control interface, a RepRap control interface, and an FPGA game controller.

Input/Output

Input/output (I/O) shields are available that bring out the various I/O pins of the Arduino to connectors that are more robust than the pins on the Arduino circuit board (or, in the case of the Arduino Nano, the pins below the board are connected to terminal block–type connectors on a carrier for the Nano PCB).

I/O shields can be broadly classified as either extension shields or expansion shields, although the term "expansion shield" is often applied to both types. An extension shield brings out the I/O pins from an Arduino without altering the signals—it just uses different types of connectors. A true expansion shield, on the other hand, employs active electronics to increase the number of discrete digital I/O channels. These types of shields use either SPI or I2C to communicate with the host Arduino board.

I/O Extension Shields

This category of shield is used to route the input/output signals from the AVR chip to connectors that are more robust than the pin sockets used on an Arduino board. Some I/O extension shields may offer active buffering of some type, but most simply bring out the signals from the Arduino board. These are sometimes referred to as expansion shields, but that is not really correct. They simply transfer the existing signals from one connector on the Arduino to another on the shield.

SainSmart Sensor Shield (http://bit.ly/sainsmart-sensor-shield)
This is a stackable shield (note the offset pin and socket strips in Figure 8-3). The AVR I/O is brought out as latching multipin sockets, pin header blocks, and positions for two 10-pin headers suitable for use with ribbon cable IDC-type connectors. A reset switch is also provided. This shield can be used with any Arduino board with the baseline pin configuration, including the Mega boards.

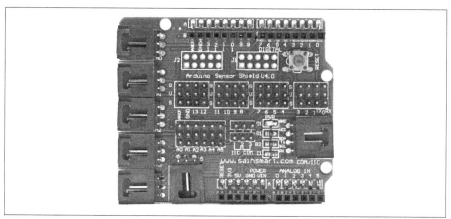

Figure 8-3. SainSmart I/O expansion (sensor interface) shield

Figure 8-4 explains the large modular connectors along the edges of the shield PCB. These are multipin connectors, sometimes referred to as "buckled" connectors, that mate with corresponding three- and four-pin plugs. The multiconductor cables are common and can be obtained from multiple vendors. One source (other than SainSmart) is TrossenRobotics (*http://bit.ly/trossen-robotgeek*).

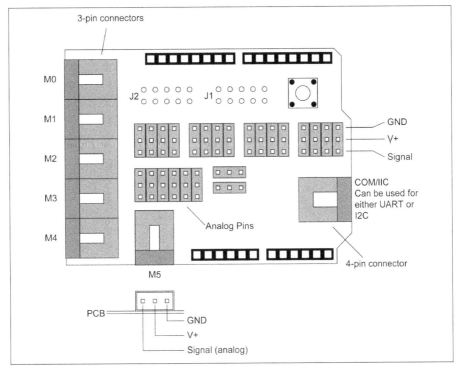

Figure 8-4. SainSmart I/O expansion shield pin and connector layout

TinkerKit Sensor Shield (http://bit.ly/tinkerkit-sensor)

This stackable shield (Figure 8-5) uses long-lead pin sockets, 12 three-pin connectors, and 2 four-pin connectors. A reset switch is located between the four-pin connectors.

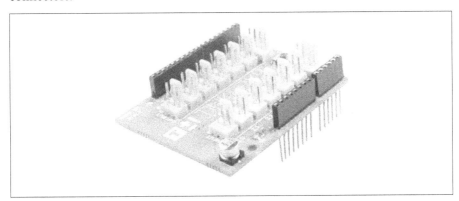

Figure 8-5. TinkerKit I/O expansion shield

The TinkerKit sensor shield was originally designed to work with the various sensor and motor modules produced by TinkerKit, but it can be used like any other I/O extension shield. Figure 8-6 shows the layout of the connectors on the PCB. These employ a three-wire scheme like that used with the SainSmart board shown previously. For more about the TinkerKit modules designed for use with this shield, refer to Chapter 9.

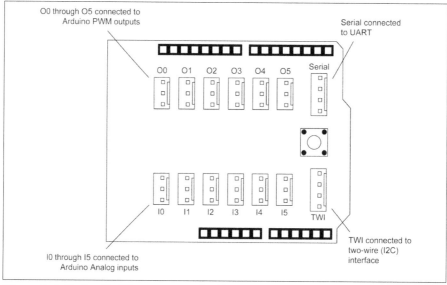

Figure 8-6. TinkerKit I/O expansion shield connectors

Although the status of TinkerKit is currently in limbo, the products are still available from Mouser (*http://www.mouser.com*) and other sources. The software libraries are available on GitHub (*https://github.com/TinkerKit*).

TinkerKit Mega Sensor Shield (http://bit.ly/tinkerkitmega-sensor)
The TinkerKit Mega Sensor Shield (Figure 8-7) is designed to bring out the additional I/O pins of an Arduino Mega, Mega2650, or Mega ADK board. It utilizes long-lead pin sockets for stackability, and a reset switch is provided on the PCB. It is essentially a larger version of the TinkerKit shield described previously.

Grove Base Shield
A module and shield system gaining popularity among Arduino users are the Grove components sold by Seeed Studio. These are a large number of modules to select from, and a base board is available that features an integrated Arduino-compatible ATMEGA328p MCU. The board, designed by Linaro.com (96Boards.org) is shown in Figure 8-8.

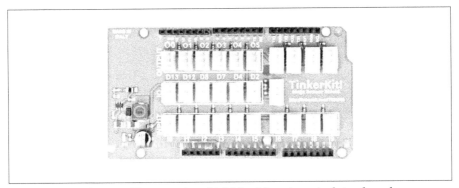

Figure 8-7. TinkerKit I/O expansion shield for Mega-type Arduino boards

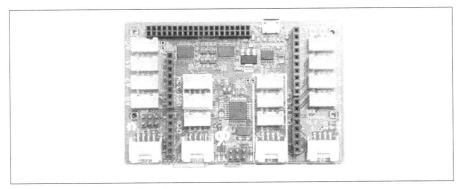

Figure 8-8. Grove Base Shield

Seed Studio also sold a passive (that is, no onboard MCU) expansion shield for the Grove module system, but while you may still be able to find some, they have been discontinued. The Passive Seeed Studio Grove Base Shield is shown in Figure 8-9.

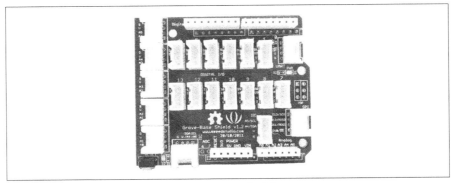

Figure 8-9. Seeed Studio Passive Grove Base Shield

For more information about the Grove series of modules and compatible interface shields, visit the Seeed Studio wiki (*http://bit.ly/seeed-grove*).

CuteDigi Sensor Expansion Shield (http://bit.ly/cutedigi)
Technically an I/O extension shield rather than a true expansion shield, this shield from CuteDigi (Figure 8-10) uses pin headers instead of connectors to bring out the signals from an Arduino Mega-type board. It is not stackable, but given the vertical arrangement of the pins on the PCB it wouldn't make sense to stack something on top of this board. The labels on the PCB are clear and the functions obvious.

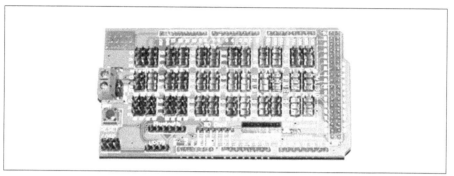

Figure 8-10. CuteDigi Mega extension shield with SD flash and bluetooth connections

This shield is interesting in that it also includes a header with right-angle pins for connecting to an SD-type flash card carrier, and there is a pin header for use with a Bluetooth module as well. The connectors employ the same S-V-G (signal, V+, ground) scheme seen on other I/O extension shields.

I/O Expansion Shields

Unlike the I/O extension shields listed in the previous section, an I/O *expansion* shield provides additional I/O capabilities, usually in the form of discrete digital I/O (although some shields do have analog capabilities). Because these shields have active circuitry in addition to various connectors, they are more expensive than extender shields. Their big advantage lies in providing multiple I/O channels using only an I2C or SPI connection to the underlying Arduino board. This leaves the remaining pins on the Arduino available for other applications.

Macetech Centipede Shield (http://bit.ly/macetech-centipede)
The Macetech Centipede Shield (Figure 8-11) uses the Arduino I2C interface to provide 64 general-purpose discrete digital I/O pins. The pins are arranged as 4 groups of 16 pins, with each group controlled by an I2C I/O expander chip.

Figure 8-11. Macetech Centipede I/O expansion shield

Figure 8-12 shows the layout of the I/O pins used on the Macetech shield. Each of the MUX (multiplexer) ICs controls 16 pins, or one block of I/O pins. Notice how the pin numbering is arranged on each block, with the numbering "wrapping" around the block.

LinkSprite I/O Expander Shield (http://bit.ly/linksprite)

This is a stackable shield that uses an MCP23017 I2C I/O expander chip to provide an additional 16 discrete digital I/O pins (Figure 8-13). Note that this shield is designed for use with R3-style Arduino boards and uses the last two pins (9 and 10, SDA and SCL) on the extended connector found on Uno R3 and Leonardo boards.

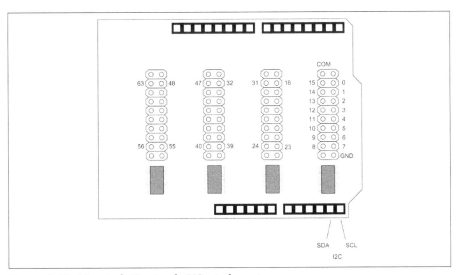

Figure 8-12. Macetech Centipede I/O pin layout

The expanded I/O pins are arranged as two sets of eight discrete channels, designated GPIOA and GPIOB. As shown in Figure 8-14, these are positioned next to the digital I/O pin socket strips. This might make it awkward to use these pins if another shield is stacked on top of this one, so be aware of little "gotchas" like that. But if it's the only shield, or the top shield in a stack, then it shouldn't be a problem.

Numato Digital and Analog IO Expander Shield (http://bit.ly/numato-digital-analog)
The digital and analog expander shield from Numato (Figure 8-15) provides 28 additional discrete digital I/O channels and 16 analog inputs using two MCP23017 I2C digital I/O chips and an NXP 74HC4067 analog multiplexer IC for analog signals.

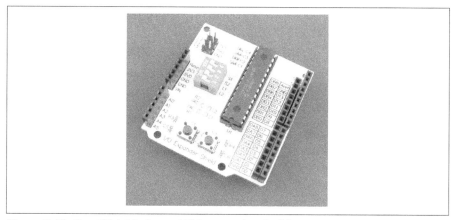

Figure 8-13. CuteDigi 16-channel I/O expander shield

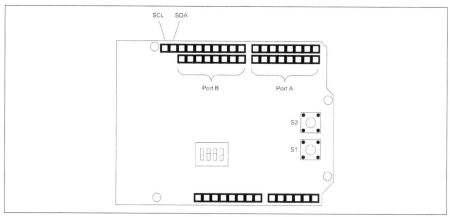

Figure 8-14. CuteDigi 16-channel I/O expander shield pin layout

As shown in Figure 8-16, the primary interface to an Arduino is via the I2C interface on pins A4 and A5. The interrupt pins on the MCP23017 chips are also brought out on the digital I/O pin blocks. The six pin headers in the middle of the board are used with jumper blocks to select the I2C addresses of the two MCP23017 chips.

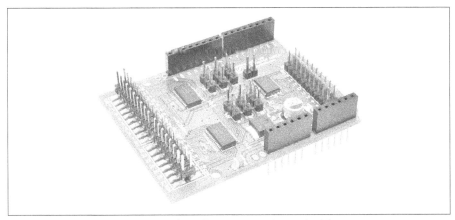

Figure 8-15. Numato digital and analog I/O expansion shield

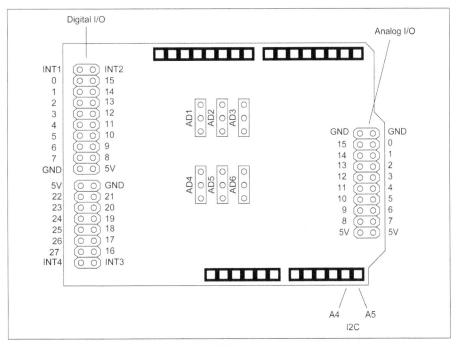

Figure 8-16. Numato I/O expansion shield pin layout

Relay Shields

Relay shields are available with one or more relays. The relays used in these shields may be 5- or 10-ampere types like the ones shown on the shields listed here, as well as reed relays in DIP packages.

 When evaluating relay boards take care to note what the vendor gives as maximum ratings for the shield. The modular relays used on a shield may have contacts rated for 10Λ at 120V AC, but the connectors and traces on the shield PCB may not be rated for that level of current. Also note that not all vendors will derate the current capacity to account for PCB or connector constraints. It is prudent to take a moment and look up the relay specification from the part number shown in the vendor's photos or schematics. You might want to think twice about a shield where the part numbers have been removed or otherwise obliterated (and that goes for any shield, not just relay shields).

DFRobot Relay Shield (http://bit.ly/dfrobot-relay)
  This shield shows how the form factor of a shield can be "tweaked" to accommodate larger parts. In the case of the DFRobot relay shield (Figure 8-17), the four relays are mounted in an expanded part of the shield PCB. The relay contacts are rated at 3A nominal at 24V DC or 120V AC, with 5A maximum current capacity. This is a stackable shield, although the vertical I/O pins may be difficult to use with a shield on top of this board.

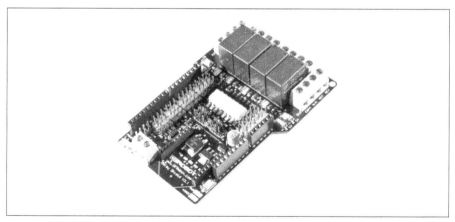

Figure 8-17. DFRobot relay shield

The board uses a set of jumpers to route the digital signals for the relay drivers and an XBee module's pins, shown in Figure 8-18. All of the Arduino's digital and analog pins are brought out in blocks of pin headers.

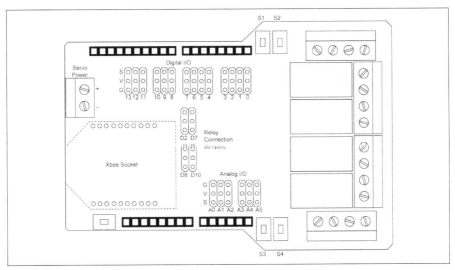

Figure 8-18. DFRobot relay shield board layout

Numato Relay Shield (http://bit.ly/numato-relay)

This shield (Figure 8-19) uses two low-power modular relays with contacts rated for 1A at 120V AC and 2A at 24V DC. The Numato shield uses the Arduino's digital pins 2 and 3. Miniature terminal blocks bring out the relay terminals. This is a stackable shield.

Figure 8-19. Numato relay shield

Seeed Studio Relay Shield (http://bit.ly/seeedstudio-relay)

The Seeed Studio relay shield (Figure 8-20) uses four relays with contacts rated at 10A at 120V AC. It uses the Arduino's digital output pins 4 through 7, one per relay. Each relay has an LED to indicate activity. This is a stackable shield.

Figure 8-20. Seeed Studio relay shield

Signal Routing Shields

There aren't a lot of signal routing shields available. What shields are available typically come in one of two styles: passive routing or active MUX (multiplexer) routing.

Adafruit Patch Shield (http://www.adafruit.com/products/256)

The passive patch shield available from Adafruit, shown in Figure 8-21, allows you to route signals between four RJ45-type connectors (also known as 8P8C connectors) and the underlying Arduino board using short patch wires inserted into blocks of pin sockets.

 This is a kit, not an assembled shield. The photo shows what the assembled shield should look like.

Figure 8-21. Adafruit patch shield

The main idea behind the patch shield is to route specific signals to and from an Arduino through conventional Ethernet cables to remote connection points, as shown in Figure 8-22. The kit includes four satellite PCBs with 8P8C (RJ45) jacks and pins to connect to a solderless breadboard, a sensor module, or another Arduino. There are no active components on this shield or the satellite boards; it just routes signals.

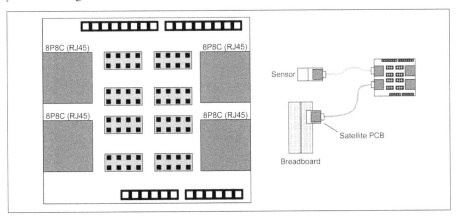

Figure 8-22. Adafruit patch shield layout and usage

Mayhew Labs Go-Between Shield (http://bit.ly/mayhew-gobtw)

The Go-Between shield (Figure 8-23) employs a matrix of solder jumper locations to route signals from the base Arduino or a lower shield to an upper shield. It could be handy if you want to stack two shields that use the same pins for I/O functions. If the pins of the upper shield could be moved to different pins on the lower shield without any shield hacking, that would solve the problem. A minor change to the software would make it all work.

Figure 8-23. Mayhew Labs Go-Between shield

Mayhew Labs Mux Shield II (http://bit.ly/mayhew-mux)

The Mux Shield II (Figure 8-24) is an active shield that supports up to 48 inputs or outputs using three Texas Instruments CD74HC4067 multiplexer chips and three output shift register circuits. The board uses four of the Arduino digital pins to control the MUX chips and shift registers. The default digital pins are 2, 4, 6, and 7. Pins A0, A1, and A2 of the Arduino are used as inputs from the MUX chips.

The 3 × 16 array of pads (Figure 8-25) can be used with pin headers or pin sockets. Each channel is bidirectional, routing signals to a common output or from a common input. Each MUX chip is similar to an array of switches, each with a slight resistance while closed, and a very high impedance when open.

Figure 8-24. Mayhew Labs active signal multiplexer shield (Mux II)

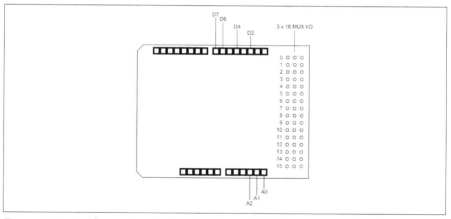

Figure 8-25. Mayhew Labs signal multiplexer shield pins

Mayhew Labs Mux Shield (http://mayhewlabs.com/arduino-mux-shield)

Similar to the shield described previously, this active multiplexer shield provides 48 programmable I/O lines using three TI CD75HC4067 MUX chips (see Figure 8-26). It also provides two large blocks of pin socket headers to access the signals. The Arduino's digital pins 2 through 5 are used to address the MUX chips, and pins A0, A1, and A2 are the analog inputs from the MUX chips. This is a stackable shield.

Memory

Without a doubt, the SD and microSD flash memory formats are the most popular way to add some file-like memory to an Arduino. External flash memory is accessed through the SPI interface, and an SD or microSD socket is often found as an added feature on shields that are using the SPI for the primary function (Ethernet, WiFi, USB host, etc.). Removable flash memory is a convenient way to log data from a standalone Arduino, and then later load the data into your PC and do whatever it is you want to do with it.

These descriptions don't have accompanying diagrams, mainly because the SD or microSD interface to an Arduino is just an SPI interface. One shield has the select signal on an unusual pin. That might create a conflict with existing software.

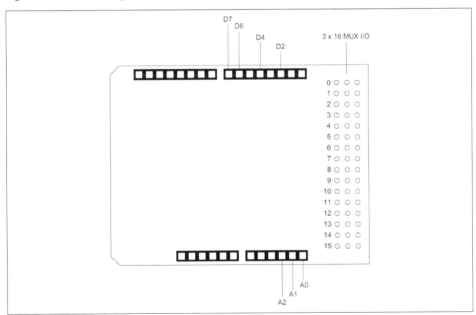

Figure 8-26. Mayhew Labs active signal multiplexer shield (Mux)

Seeed Studio SD Card Shield (http://bit.ly/seeedstudio-sd-card)

Designed for full-size SD flash memory cards, this shield (shown in Figure 8-27) can easily be used with microSD cards with an adapter. Arduino digital pins D4, D11, D12, and D13 are used for the SPI interface. The shield also brings out the ICSP, I2C, and UART pins to connectors on the PCB. This is a stackable shield.

Figure 8-27. Seeed Studio SD memory card shield

SHD-SD SD Card Shield (http://bit.ly/shd-sd-sdcard)

This SD shield, pictured in Figure 8-28, features a small prototyping area for adding your own circuitry. It can accept microSD cards with an adapter. The shield uses the D10, D11, D12, and D13 pins on an Arduino for the SPI interface to the SD memory. It also uses the 3.3V DC from the Arduino. The shield is shorter than a conventional shield, with a baseline Arduino pin arrangement. It is a stackable shield.

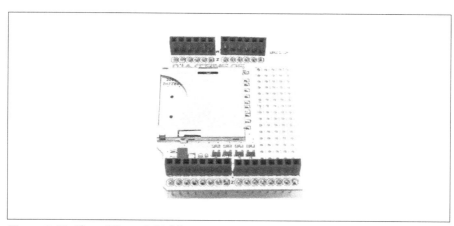

Figure 8-28. Short SD card shield

SparkFun microSD Shield (http://bit.ly/sparkfun-microsd)

The SparkFun microSD shield (Figure 8-29) accepts only microSD cards. It includes a large 12 by 13 prototyping area. It does not come with pin sockets or headers, but these can be ordered separately. It uses the D10, D11, D12, and D14 pins on an Arduino.

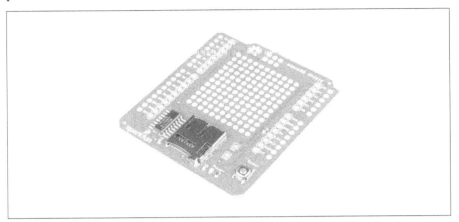

Figure 8-29. SparkFun microSD shield

Communication

Although an Arduino might have a USB interface (most do) with the ability to act as a serial port from the host system's perspective, or an Ethernet jack as found with the Arduino Ethernet, a basic Arduino like an Uno or Leonardo doesn't really have much in the way of plug-and-play data communications interfaces. It is possible to attach level-shifting chips and use the built-in UART or write a so-called "bit-banger" to send serial data, but it is sometimes more convenient to use something with an SPI interface to the Arduino and let it do the serial sending and receiving. For other forms of data communication the necessary external hardware can get rather involved, so it's definitely easier to use a ready-made shield.

Serial I/O and MIDI

Although even Ethernet can be considered a form of serial data communication, serial I/O here refers to the old standards of RS-232 and RS-485. While these are old, they are ubiquitous. Older PCs have RS-232 connectors (and quite a few newer models also have at least one), and RS-485 is common in instrumentation, testing, and distributed measurement systems.

CuteDigi RS232 Shield (http://bit.ly/cutedigi-rs232)

This RS-232 shield (Figure 8-30) employs a MAX232 IC to perform the signal-level shifting necessary to send and receive RS-232–compatible signals. A bank of jumpers is used to configure the serial interface, and any two digital pins from D0 to D7 on an Arduino may be used for the serial interface. With the supplied pin sockets installed it becomes a stackable shield.

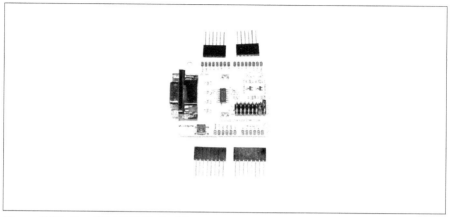

Figure 8-30. CuteDigi RS232 shield

CuteDigi RS-485 Shield (http://bit.ly/cutedigi-rs485)

CuteDigi's RS485 shield (Figure 8-31) uses a MAX481CSA chip to provide the RS485 electrical interface using the Arduino's Rx and Tx ports (D2 and D3, respectively). A mounting position is provided in the shield PCB for an optional DB-9 connector. This is a stackable shield.

Figure 8-31. CuteDigi RS485 shield

SparkFun MIDI Shield (http://bit.ly/sparkfun-midi)

MIDI is a venerable serial protocol that has been around for over 30 years. It is used to control musical synthesizers, sequencers, drum machines, and mixers, among other things. The SparkFun MIDI shield (Figure 8-32) uses the Arduino's USART pins to send and receive MIDI event messages.

Figure 8-32. SparkFun MIDI shield

The MIDI shield uses the D0 and D1 pins (Rx and Tx, respectively) for MIDI serial I/O. It also has pushbuttons on D2, D3, and D4; LEDs connected to D6 and D7; and potentiometers connected to A0 and A1.

Ethernet

Ethernet shields are popular, and the Arduino IDE comes with a fairly comprehensive Ethernet library suite (see Chapter 7 for details). Be aware that the communication between the AVR MCU on the Arduino board and the Ethernet controller on the shield uses the SPI interface. The AVR does not have DMA (direct memory access) capability, and it has no external memory to directly access in any case.

With Ethernet shields that use SPI as the interface with an Arduino there is an inherent limit on how fast data can move between the AVR MCU and the Ethernet I/O chip, and consequently on how fast data can move over the Ethernet connection. It is simply not possible to get 100 Mb/s (100Base-T) data rates with a processor running at 20 MHz using an SPI interface, and 10 Mb/s (10BASE-T) is an unlikely stretch. 5 Mb/s is a more realistic expectation. The data is still sent out over the physical layer (the actual Ethernet) at 10 Mb/s, just in byte-sized dribbles rather than as a continuous stream. It all depends on how quickly the software running on the AVR can assemble outbound data and send it to the Ethernet chip. So, while it is possible to create a web server that can fit into a tiny enclosure like an old mint tin, it isn't going to be very fast, and it won't handle a lot of connections at the same time.

Where the Ethernet interface really shines is when it is used as the end node of a remote sensing or control system. You can attach it to the Internet, implement some password protection, and use it to retrieve data from some remote location. It can be used to report data back to a central controller in an industrial setting such as a factory, or it can be used to sense temperature, humidity, and other parameters for a distributed HVAC (heating, ventilation, and air conditioning) system controller like the one described in Chapter 12.

Vetco Ethernet Shield with microSD Card Reader (http://bit.ly/vetco-ethernet)
This Ethernet shield from Vetco (Figure 8-33) also includes a microSD card carrier and a reset button. The Arduino's digital pins D10, D11, D12, and D13 are used for the SPI interface used by the WIZnet W5100 Ethernet chip and the microSD card socket. It is a stackable shield.

Figure 8-33. Vetco Ethernet shield

Arduino Ethernet Shield R3 with microSD Connector (http://bit.ly/ethernet-r3)
The official Ethernet shield from Arduino comes with a microSD card reader, a reset button, and all the necessary electronics to implement the Ethernet interface (see Figure 8-34). It uses digital pins D10, D11, D12, and D13 for the SPI interface. This is a stackable shield.

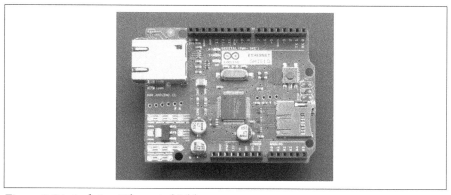

Figure 8-34. Arduino Ethernet shield

Bluetooth

Bluetooth is a low-power, short-range wireless communication technology originally intended to replace the cables strung between a computer and external devices such as printers, keyboards, mice, and so on. While it is still used for these applications, it has found use in many other types of communications applications. There are a number of Bluetooth shields available.

Bluetooth Shield (http://bit.ly/dx-bluetooth)
This shield comes assembled with a Bluetooth module already mounted on the PCB. Note that it is a stackable shield that is shorter than a typical shield. The antenna is the gold pattern emerging from the end of the Bluetooth module (see Figure 8-35). It uses the Rx and Tx pins (D2 and D3, respectively).

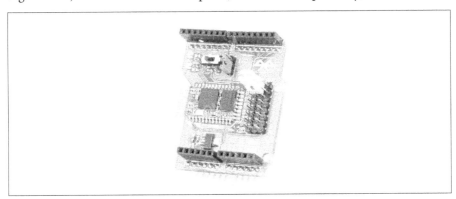

Figure 8-35. DealeXtreme Bluetooth shield

Seeed Studio Bluetooth Shield (http://bit.ly/seeed-bluetooth)
This compact, stackable Bluetooth shield (Figure 8-36) uses an Arduino's Rx and Tx pins. It also brings out the pins for analog and digital signals to interface with sensor modules.

ITEAD Bluetooth Wireless BT Module Shield Kit (http://bit.ly/itead-bluetooth)
This shield (Figure 8-37) uses a standard Bluetooth module, and comes with a prototyping area on the PCB. It is a "short" shield, in that it doesn't cover the whole length of an Arduino board. It is stackable.

DFRobot Gravity:IO Expansion Shield (http://bit.ly/dfrobot-gravityio)
This multifunction board from DFRobot (Figure 8-38) combines I/O extension capabilities with a set of pin sockets for a Bluetooth or ZigBee wireless module. It is not a stackable board.

Figure 8-36. Compact Bluetooth shield from Seeed Studio

Figure 8-37. ITEAD Bluetooth shield with prototyping area

Figure 8-38. DFRobot multifunction shield with Bluetooth

USB

One thing an 8-bit Arduino can't do is act as a USB host to other USB devices. A USB host shield allows you connect USB devices such as keyboards, printers, some test instruments, and various toys to an Arduino.

ITEAD USB Host Shield (http://bit.ly/itead-usb)
This is a stackable shield with USB host functionality. It also provides pins for the Arduino digital and analog signals, and two 10-pin positions on the PCB for either connectors or pin headers (see Figure 8-39). It is based on a MAX3421E chip with an SPI interface to an Arduino.

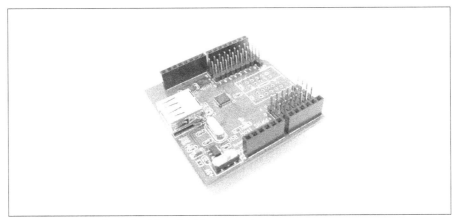

Figure 8-39. USB host shield with I/O connections

Circuits@Home USB Host Shield (http://bit.ly/circuitsathome-usb)
This shield (Figure 8-40) supports USB 2.0 full-speed operation and uses an SPI interface to an Arduino. The digital and analog signals from the Arduino are available on the shield's PCB, and with the right pin sockets it could be stackable. It employs a MAX3421E chip with an SPI interface to an Arduino using pins D10, D11, D12, and D13. It does not come with pin sockets or headers.

Arduino USB Host Shield (http://bit.ly/arduino-usb)
Like other USB host shields, this board uses the MAX3421E chip with an SPI interface to the Arduino using pins D10, D11, D12, and D13. Three- and four-pin connectors bring out input and output ports that will work directly with TinkerKit modules (the TinkerKit modules are described in Chapter 9). This is a stackable shield (see Figure 8-41).

Figure 8-40. Circuits@Home USB host shield

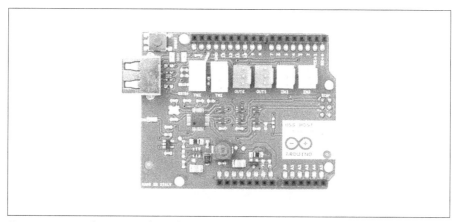

Figure 8-41. Arduino USB host shield with I/O connectors

ZigBee

ZigBee is a popular low-power wireless protocol. Many of the available ZigBee Arduino shields use readily available XBee modules, but most shields will accommodate any RF module with the correct pinout. Some are available with an XBee module, and some are available without. A 1 mW XBee module costs around $25.

Arduino Wireless SD Shield (http://bit.ly/arduino-wireless)
   On this ZigBee shield (Figure 8-42), two inline pin sockets are provided for a Digi XBee module, or any module with a compatible pin arrangement. This shield uses an Arduino's pin D4 as the select, and pins D11, D12, and D13 for SPI communication. A microSD carrier on the shield also uses the SPI interface.

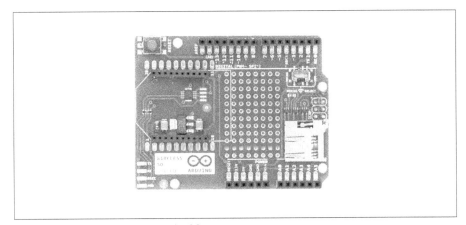

Figure 8-42. Arduino ZigBee shield

SainSmart XBee Shield (http://bit.ly/sainsm-xbee)
This shield comes without an XBee module, but the pin layout will accept a standard XBee module, or any module with a compatible pin arrangement. Note that there is no microSD carrier. This is an offset stacking shield (note the locations of pin headers and pin sockets in Figure 8-43).

Figure 8-43. SainSmart compact ZigBee shield

Seeed Studio XBee Shield (http://bit.ly/seeedst-xbee)
The XBee shield from Seeed Studio (Figure 8-44) has the expected mounting position for a common XBee module, and it also provides a prototyping area. This shield uses the Rx and Tx pins from an Arduino, and a block of pins for jumpers is used to route the Rx and Tx signals from the Arduino to the wireless module.

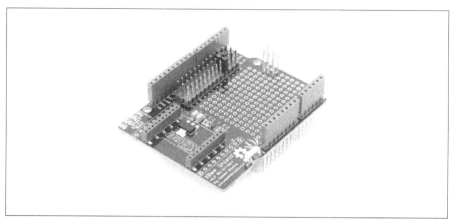
Figure 8-44. SainSmart ZigBee shield with prototyping area

CAN

The Controller Area Network (CAN, also known as *CAN bus*), is a differential signalling relative of RS-485 found in vehicles, industrial settings, and some military equipment. It is relatively fast (up to 1 Mb/s), incorporates signal collision detection and error detection, and supports multiple nodes. It is used with the OBD-II on-board diagnostics found in late-model automobiles, in electric vehicles, and with distributed sensors in scientific instruments, and has even been integrated into some high-end bicycles.

Seeed Studio CAN-BUS Shield (http://bit.ly/seeed-canbus)
The CAN interface shield from Seeed Studio (Figure 8-45) utilizes an MCP2515 CAN bus controller with an SPI interface and an MCP2551 CAN transceiver chip. Both a terminal block and a DB-9 connector are provided for the CAN bus signals. The shield also brings out the I2C and UART communications from the Arduino. The pin layout is shown in Figure 8-46.

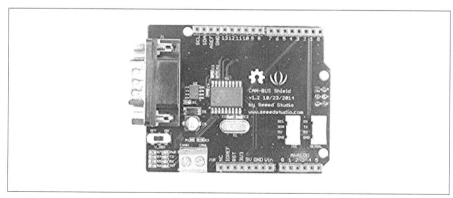
Figure 8-45. Seeed Studio CAN shield with auxiliary I/O connectors

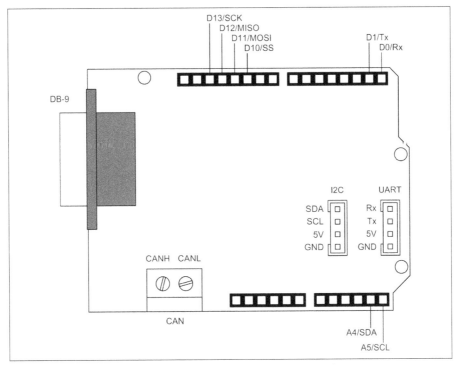

Figure 8-46. Seeed Studio CAN shield with auxiliary I/O pin layout

SparkFun CAN-BUS Shield (http://bit.ly/sparkfun-canbus)

This CAN shield from SparkFun (Figure 8-47) incorporates many of the features one might want to use when creating an OBD-II readout and data capture device. It has a DB-9 connector for CAN bus signals, and a 4-pin header also provides the signals. Connection points are provided for an external LCD display and an EM406 GPS module.

An interesting feature is a four-position binary joystick, and it incorporates a microSD flash card carrier. The joystick is connected to the Arduino analog inputs. The CAN chip and the SD flash are separately selected via digital pins D9 and D10. Pins D3, D4, D5, and A0 are not used by the shield. The pin layout is shown in Figure 8-48.

LinkSprite CAN-BUS Shield (http://bit.ly/linksprite-can-bus)

The LinkSprite CAN shield (Figure 8-49) is physically similar to the Seeed shield. It has both a DB-9 connector and a two-position terminal block for the CAN signals. Pins D10, D11, D12, and D13 are used for the SPI interface to an MCP2515 CAN chip.

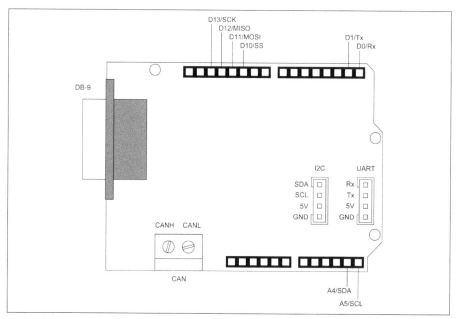

Figure 8-47. SparkFun CAN shield with microSD carrier and digital joystick

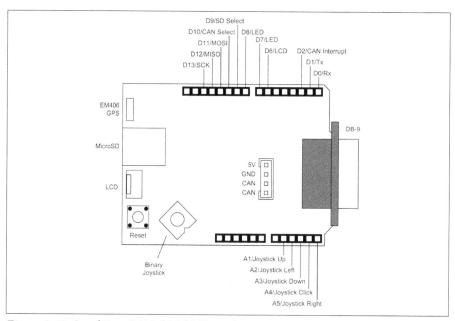

Figure 8-48. SparkFun CAN shield I/O pin layout

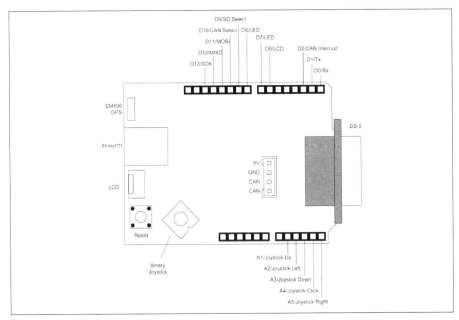

Figure 8-49. LinkSprite CAN shield

Prototyping

If you want to create your own shield you can build a prototype (or even a permanent shield) using a prototyping shield board. This is not the same as the process of creating a shield described in Chapter 10, which involves laying out a PCB for (possible) mass production. Figure 8-50 shows a prototype shield similar to those described in this section with a temperature sensor and a relay mounted on it. A potentiometer is connected to the +5V, ground, and A0 (analog input 0) pins passed through from the underlying Arduino board. The pot controls the temperature set point.

This prototype was used to control an ancient (and very dangerous) portable electric heater that used a bimetallic thermostat that couldn't seem to hold the temperature to better than +/– 15 degrees. Since the relay is only rated to 10 amperes at 120 VAC and the heating elements were rated for 15 amps, it was used along with a 24 VAC transformer to control a 20-amp contactor. It worked pretty well, and kept my office relatively comfortable during the winter. I plan to add a tilt sensor, output temperature sensor, and fan motion detector to it. Just to be safe.

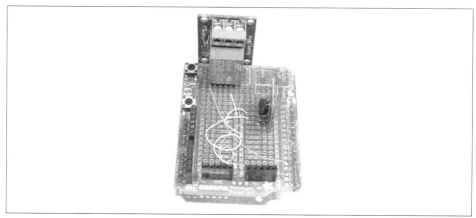

Figure 8-50. Prototype temperature sensor/controller shield

The shields in this section are representative of what is available, and none are particularly complicated. What the Arduino pins are used for is entirely up to you (it is a prototype, after all), so there isn't much need for diagrams.

Adafruit Stackable R3 Proto Shield (http://bit.ly/stackable-r3)
   This shield (Figure 8-51) comes as a kit, which means a bare PCB and a bag of parts. It's not hard to assemble, but some soldering skill is essential.

Figure 8-51. Adafruit stackable prototype shield kit

Adafruit Mega Proto Shield (http://bit.ly/ada-mega-proto)

Another kit from Adafruit, this stackable prototyping shield comes with all the bits you can see in Figure 8-52. Note the double rows of solder pads for the connections along the sides of the PCB. This allows you to solder in short-lead pin sockets to gain easy access to the signals from an underlying Mega-type Arduino.

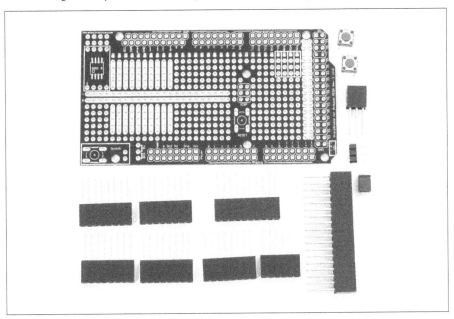

Figure 8-52. Adafruit Mega prototype shield kit

CuteDigi Assembled Protoshield for Arduino (http://bit.ly/cutedigi-assembled)

This prototyping shield (Figure 8-53) comes fully assembled. It also has pin sockets on the PCB for access to the Arduino signals. This is not a stacking shield.

CuteDigi Assembled Protoshield for Arduino MEGA (http://bit.ly/assembled-mega)

Designed to work with a Mega-type Arduino, this shield (Figure 8-54) also features a small-outline (SOIC) mounting location for an IC. This is not a stacking shield.

CuteDigi Protoshield for Arduino with Mini Breadboard (http://bit.ly/proto-bboard)

This shield, pictured in Figure 8-55, includes a small solderless breadboard for your own circuit creations.

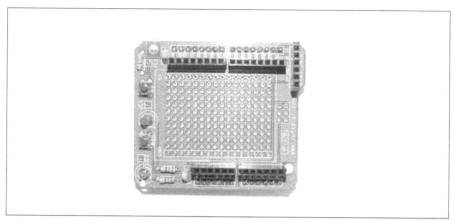

Figure 8-53. CuteDigi prototyping shield

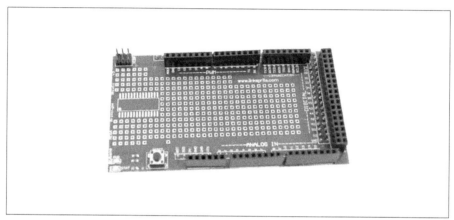

Figure 8-54. CuteDigi Mega prototyping shield

Figure 8-55. CuteDigi prototyping shield with breadboard

Creating a Custom Prototype Shield

You can whip up a workable shield using nothing more than a prototyping PCB and some pin and socket connectors. The size of the PCB doesn't really matter, so long as the pins line up with the sockets on an Arduino board.

Adafruit DIY Shield Kit (http://bit.ly/ada-diy)

If there was award for the simplest shield kit, this would definitely be at the top of the list of contenders. Consisting of a prototyping PCB and four long-lead pin socket connectors (Figure 8-56), this shield lets you put anything you like on it. This is handy for prototyping a new shield design, or just quickly throwing something together, and it's great for using an existing module of some type that was never intended to be connected to an Arduino.

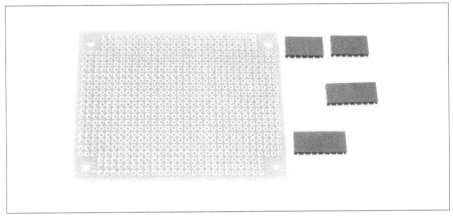

Figure 8-56. Adafruit DIY shield kit

Unfortunately the product has been discontinued by Adafruit, but all you need is a prototyping PCB (these are readily available from multiple sources) and the pin sockets, which Adafruit and other vendors carry. Since the pin pads on an Arduino use industry-standard 0.1 inch (2.54 mm) spacing, it's easy to make something with basic prototyping supplies that can serve as a shield.

Motion Control

Motion control is a big area of interest in the Arduino domain. From programmable mobile robots to CNC engravers, 3D printers to laser scanners, and even automated sun-following tracking controllers for solar panels and kinetic sculptures, Arduinos have been used to control DC motors, servos, and stepper motors from the outset. As you might expect, a number of shields are available for each type of motor, and this is just a small sampling of what's available.

DC and Stepper Motor Control

Motor controller shields based on an H-bridge (a type of solid-state current routing switch) can usually be used to control either brush-type DC motors or stepper motors. Basically, these types of shields can be used to control any inductive DC load, including solenoids and relays.

Rugged Motor Driver shield (http://bit.ly/rugged_motor)
> The Rugged Motor Driver Shield from Rugged Circuits (Figure 8-57) can drive either two brush-type DC motors, or one bipolar stepper motor. It is rated for up to 30V at 2.8A peak current. The shield uses the D3, D11, D12, and D13 pins for enable and direction control inputs. The enable inputs can be driven with a PWM signal for smooth control of a DC motor. Check the website for more details regarding current handling and software.

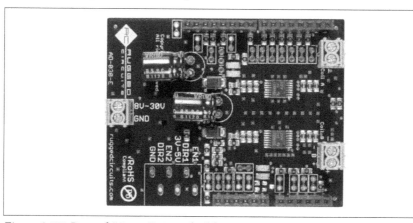

Figure 8-57. Rugged Motor Driver shield

SainSmart Motor Drive Shield (http://bit.ly/sainsm-motor)
> The SainSmart motor shield (Figure 8-58) is based on an L293D four-channel driver IC. It can drive four brush-type DC motors or two stepper motors at up to 10V. It features terminals for an external power supply. Check the website for more details regarding current handling and software.

Arduino Motor Shield (http://bit.ly/arduino-motor)
> The motor shield from Arduino.cc (Figure 8-59) is based on an L298 dual-driver IC. It can be used with relays, solenoids, DC motors, and stepper motors. An interesting feature is the ability to measure current consumption, which can be handy for detecting a stalled motor. Also note that it has modular connectors that are compatible with various TinkerKit modules (described in Chapter 9).

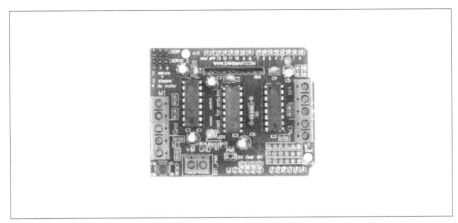

Figure 8-58. SainSmart L239D Motor Drive shield

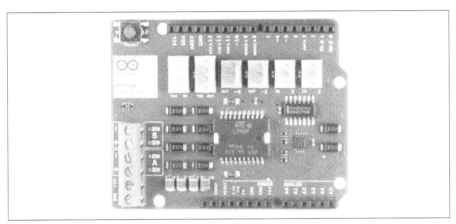

Figure 8-59. Arduino motor shield

PWM and Servo Control

The small servos used in RC models and small-scale robotics work by positioning an armature relative to a series of control pulses of varying width but at a steady frequency. The width (the "on" time) of the pulses determines the rotation angle of the servo. A PWM/servo shield can also be used to drive a DC motor, precisely control the brightness of one or more LEDs, or operate a linear actuator.

16-Channel 12-bit PWM/Servo Shield (http://bit.ly/ada-16-pwm)
This shield (Figure 8-60) utilizes a PCA9685 16-channel PWM controller IC with an I2C interface. It has the ability to generate a unique programmable PWM signal on each output, and it doesn't require the constant attention of the Arduino.

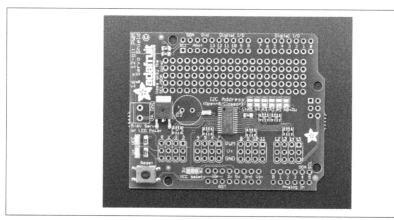

Figure 8-60. Adafruit PWM/servo shield with I2C interface

LinkSprite 27-Channel PWM Servo Shield (http://bit.ly/linksprite-27-pwm)
This shield from LinkSprite (Figure 8-61) uses an STM32F103C8T6 microcontroller IC to generate up to 27 unique PWM outputs. It is worth noting that the STM32F103C8T6 is an ARM Cortex-M3 32-bit RISC device with up to 128 KB of flash memory and 20 KB of SRAM. The microcontroller on this shield is actually more computationally powerful than the AVR on the Arduino it is mounted on. It communicates with an Arduino using the SPI interface.

Figure 8-61. LinkSprite servo shield

SparkFun PWM Shield (http://bit.ly/sparkfun-pwm)
The PWM shield from SparkFun (Figure 8-62) utilizes a TLC5940 IC and is capable of producing 16 PWM outputs. The TLC5940 is capable of driving LEDs or servo motors. It uses an SPI-type clocked serial interface, but only receives data. More information and software libraries are available from SparkFun.

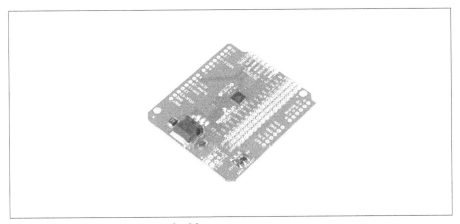

Figure 8-62. SparkFun PWM shield

Displays

Display shields for Arduino boards might contain LED (light-emitting diode) read-outs, LED arrays, an LCD (liquid-crystal display), or a color graphical display. Some of the shields utilize multiple digital outputs from an Arduino; others use the SPI or TWI (I2C) interfaces. Whatever it is you want to display, chances are there's a display shield that will do the job.

LED arrays

By itself a single LED array is fun, but when they're set side-by-side it is possible to create marquee displays in a variety of colors. For more information about the shields listed here, refer to the websites for each:

- Adafruit LoL Shield (*http://bit.ly/ada-lol*) (Figure 8-63)

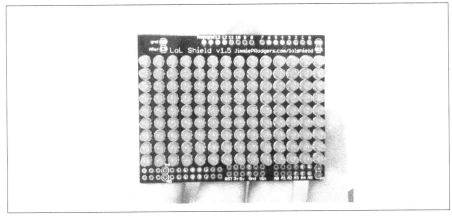

Figure 8-63. Adafruit 9 × 14 LED array shield

- Solarbotics SMD LoL Shield (*http://bit.ly/solarbio-smd-lol*) (Figure 8-64)

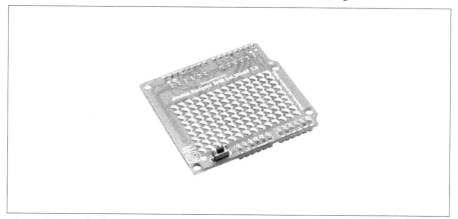

Figure 8-64. Solarbotics 9 × 14 LED array shield

- Adafruit NeoPixel Shield (*http://bit.ly/ada-neo*) (Figure 8-65)

Figure 8-65. Adafruit 40 RGB LED pixel matrix

7-segment LED displays

The 7-segment LED display has been around for almost as long as there have been LEDs. While now considered rather quaint, the 7-segment display still has a role to play when you need big, bright digits you can easily see from across the room. Also check out the multifunction shield listed in "Miscellaneous Shields" on page 268, which features a 4-digit numeric LED display.

- 4x 7-Segment Arduino Compatible Digit Shield (*http://bit.ly/arduino-7seg*) (Figure 8-66)

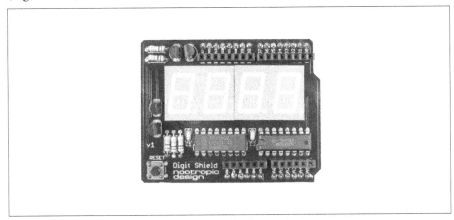

Figure 8-66. Nootropic Design 7-segment display shield

LCD displays

Many low-cost character-based LCD shields utilize 16 × 2 (16 characters in 2 rows) displays with white letters on a blue screen, red letters on a black screen, or black letters on a green screen. Other combinations are also available, including 16 × 4 and 20 × 4 configurations. Most of these types of shields are based on the Hitachi HD44780 LCD controller, or something similar.

There are also pixel-addressable and bitmap-capable LCD displays available. Some of these, like the popular Nokia 5110, are available from various vendors and are easy to interface to an Arduino. You can also find displays with resolutions of 128 × 64 and 160 × 128 pixels without looking too hard, but not many of these are available in the form of an Arduino-compatible shield. See Chapter 9 for more information about bare (nonshield) display components.

SainSmart LCD Keypad Shield (http://bit.ly/sainsmart-keypad)
   This is a common LCD shield design, shown in Figure 8-67, that uses a 16 × 2 LCD display module and the Hitachi HD44780 LCD controller (the display module is available separately, and one is used in the signal generator in Chapter 11 and in the thermostat in Chapter 12).

   This LCD shield uses a voltage divider for the five pushbutton switches, so each button press results in a different voltage. Figure 8-68 shows how this works. The advantage of this approach is that five switch inputs are routed through one analog input.

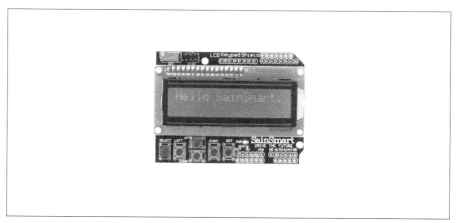

Figure 8-67. SainSmart 16 × 2 LCD keypad shield

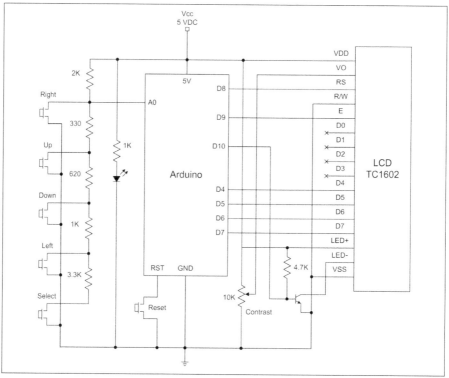

Figure 8-68. SainSmart LCD shield schematic

DFRobot LCD Keypad Shield (http://bit.ly/dfrobot-keypad)
    This is similar to the SainSmart LCD shield, except with analog pin connections to the underlying Arduino (see Figure 8-69). According to the vendor's docu-

mentation the additional pins are mainly intended for interfacing to an APC220 radio module or a Bluetooth module.

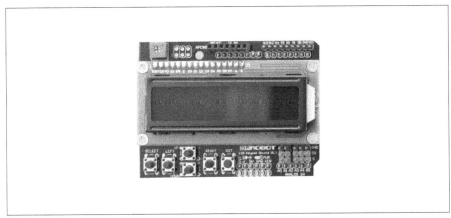

Figure 8-69. DFRobot 16 × 2 LCD keypad shield with analog pins

Adafruit LCD Shield Kit (http://bit.ly/adafruit-lcd)

The 16 × 2 LCD shield kit from Adafruit, shown in Figure 8-70, uses the I2C interface and an MCP232017 I/O expander IC (also used in Chapter 10) to control the LCD display. This results in the shield only using two of the Arduino's pins, A4 and A5, for the I2C interface. The LCD and the pushbuttons are all connected to the MCP23017 IC, and it does not use a resistor divider. Note that this is a kit, but it's not too hard to assemble.

Figure 8-70. Adafruit LCD shield kit with 16 × 2 character display

Nokia LCD5110 Module with SD (http://bit.ly/nokia_lcd)

This shield, pictured in Figure 8-71, combines a Nokia 5110 LCD display with an SD flash card socket. It uses pins D3, D4, D5, D5, and D7 for the display, and pins D10, D11, D12, and D13 for the SD card. Pins D0, D1, and D2 are available for other applications.

Figure 8-71. Nokia 5110 LCD shield from Elechouse

Unlike the 16 × 2 character-based displays, the 5110 is a graphics-capable LCD with a 48 × 84 display area. Originally manufactured for cell phones, all the currently available units are surplus. Some may have scratches or other slight blemishes. Also bear in mind that when they are gone, that's it. So, it's not a good idea to design a new product using these, but they are fun to play with and they are relatively inexpensive.

TFT displays

The TFT LCD (thin-film transistor liquid crystal display), or just TFT for short, is a common display type found in computer monitors, cash register displays, cell phones, tablets, and just about anything else with a color graphical display. A color TFT shield for an Arduino can display thousands of colors at resolutions such as 240 × 320 pixels. Larger displays are available, but these generally don't fit on a shield. TFT shields are generally inexpensive; most use an SPI interface, and some have a parallel digital interface for high-speed image generation.

ITEAD 2.4" TFT LCD Touch Shield (http://bit.ly/ITEAD-tft)

This shield, pictured in Figure 8-72, uses a parallel digital interface with an Arduino. An 8-bit interface is used with an S6D1121 TFT controller on the shield, and the touchscreen functions are handled by the TSC2046 chip. For additional detailed information, refer to the vendor's website.

Figure 8-72. ITEAD 2.4 inch color TFT shield with touchscreen

Adafruit 2.8" TFT Touch Shield (http://bit.ly/ada-tft)

The 2.8 inch TFT shield from Adafruit (Figure 8-73) uses a high-speed SPI interface for both an ILI9341 display controller with a built-in video RAM buffer and
an STMPE610 touchscreen controller. It also incorporates a microSD flash card
carrier. The shield uses the Arduino digital pins D8 through D13, the
touchscreen controller uses pin D8, and the microSD carrier select is on pin D4.

Figure 8-73. Adafruit 2.8 inch TFT shield with resistive touchscreen

Instrumentation Shields

Although not as plentiful as some other shield types, there are instrumentation-type
shields available. These include data logging shields, logic analyzers, and precision
analog-to-digital (A/D) converters. Instrumentation, in this case, refers to the ability
to sense and capture data from the physical world, or generate an analog signal.

There aren't many shields available with on-board data capture and conversion capabilities, mainly because the AVR MCU on an Arduino already has most of these functions in the chip itself. The built-in A/D converter (ADC) in the AVR MCUs used with Arduino boards has 10-bit resolution, which gives a conversion resolution of 1/1,024 per DN, or digital number. If you need better resolution (12, 16, or even 24 bits, for example), then you will need to consider some type of add-on module or a shield.

Adafruit Data Logging Shield (http://bit.ly/1Tjlu5V)
   The Adafruit data logging shield, shown in Figure 8-74, includes an SD flash card carrier and a real-time clock (RTC) chip. The RTC can be powered by a battery when the main power from the base Arduino is off. A small prototyping grid is supplied for custom circuitry. This is not a stacking shield.

Figure 8-74. Adafruit assembled data logging shield

Adafruit Ultimate GPS Logger Shield (http://bit.ly/ada-gps)
   This shield, shown in Figure 8-75, incorporates a GPS receiver as well as an SD flash carrier and an RTC chip. The output of the GPS can be automatically logged to the flash memory card. This is not a stacking shield.

HobbyLab Logic Analyzer and Signal Generator Shield (http://www.arduinolab.us)
   This is actually a standalone logic analyzer on a shield. It monitors the Arduino signals without interfering with them, which is handy to see what's happening on the I/O pins. In addition to the logic analyzer capability, it also includes an SPI decoder, a UART decoder, and a one-wire monitor. It does not communicate directly with the Arduino, but uses a USB interface to interact with a host computer. This is a stacking shield (see Figure 8-76).

Figure 8-75. Adafruit GPS data logging shield

Figure 8-76. HobbyLab logic analyzer and signal generator

Iowa Scaled Engineering 16-Channel 24-Bit ADC Shield (http://bit.ly/ard-ltc2499)
This shield incorporates a 24-bit A/D converter and a precision voltage reference to obtain readings from multiple single-ended or differential inputs. It also has on-board EEPROM for storing and reading calibration and configuration data. It communicates with an Arduino via the I2C (TWI) interface. This is a stacking shield (see Figure 8-77).

Visgence Power DAC Shield (http://bit.ly/power-dac)
One function the AVR MCU, and by extension an Arduino, lacks is a built-in digital-to-analog converter (DAC). The AVR's internal ADC incorporates a 10-bit DAC, but the output is not available externally. Although audio output shields are readily available, there don't seem to be a lot of pure DC output DAC shields available. The Visgence Power DAC Shield (Figure 8-78) provides three channels of analog output with the ability to source up to 250 mA of current.

Figure 8-77. Iowa Scaled Engineering 24-bit ADC data acquisition shield

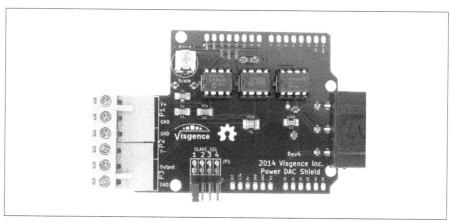

Figure 8-78. Visgence 12-bit Power DAC Shield

Adapter Shields

An adapter shield is used as a physical interface between what would otherwise be two physically incompatible modules. The primary difference between an adapter shield and a signal routing shield (see "Signal Routing Shields" on page 232), at least with regard to how the shields are organized in this chapter, is that an adapter is intended as a physical interface. A signal routing shield doesn't deal with physical differences, just signals.

Tronixlabs Australia Expansion Shield for Arduino Nano (http://bit.ly/exp-nano)
A Nano is every bit as capable as a larger baseline Arduino, but it won't work with conventional shields. This board, shown in Figure 8-79, addresses that by bringing out the pins from a Nano to pin headers, and optionally to standard pin socket connectors.

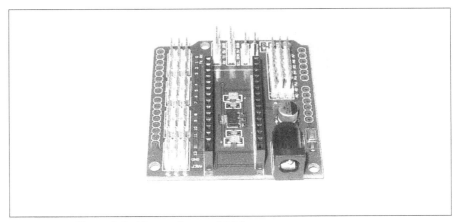

Figure 8-79. Tronixlabs Nano adapter shield

Arduino Nano I/O Expansion Board (eBay) (http://bit.ly/exp-nano-ebay)
Another example of a Nano adapter (Figure 8-80). This was found on eBay, but there are others like it available.

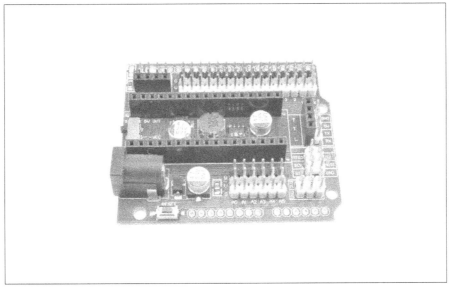

Figure 8-80. Expansion board for Arduino Nano

Screw Terminal Shield (http://bit.ly/screw-term)
Although not technically a shield, per se, these handy block adapters allow you to connect up to 18-gauge insulated wire to an Arduino. They are readily available from a variety of sources. Notice that these parts have the ability to stack (see Figure 8-81).

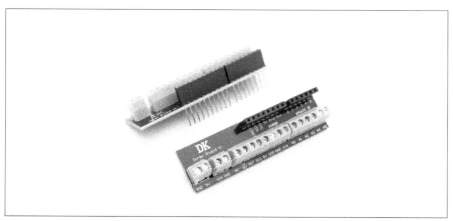

Figure 8-81. Terminal block adapters

Miscellaneous Shields

This section describes some useful shields that don't really fit into a neat category. I like the so-called "wing shields" because they allow for neater wiring, and the multi-function shield shown here has a multitude of uses.

Adafruit Proto-ScrewShield (Wingshield) (http://bit.ly/ada-proto-screw)
The Wingshield, also known as a Proto-ScrewShield, is a passive shield with two sets of miniature terminal blocks (see Figure 8-82). If you plan to incorporate an Arduino into a commercial product or a laboratory setup, then you may want to consider a shield like this. The screws in the terminal blocks provide for a much more secure and reliable connection than pin jumpers. The prototyping area in the middle of the shield can be used to mount sensor modules, or it can hold a custom circuit. The shield includes a reset switch and an LED, and it's also a stackable shield. Be aware that this is a kit, so it comes as a bare PCB and a bag of parts. Assembly isn't hard, but it does require some soldering skill.

DFRobot Screw ProtoShield (http://bit.ly/screw-protoshield)
Like the Adafruit Wingshield this shield (Figure 8-83) provides screw terminals for each of the signals from an Arduino, but it comes fully assembled. It is also a stackable shield, so you can put it under other shields and still be able to access the terminal blocks.

DealeXtreme DIY Multifunction Shield (http://bit.ly/diy-multi)
This is an interesting shield: it has a 4-digit numeric LED readout; an interface point for an APC220 Bluetooth module; several pushbutton switches connected to Arduino pins A1, A2, and A3; a reset switch; and a potentiometer connected to the A0 input (see Figure 8-84). Four LEDs are connected to digital pins D10, D11, D12, and D13, and a 3 × 4-pin header brings out pins D5, D6, D9, and A5,

along with +5V and ground. Unfortunately this shield is also rather poorly documented, and it can take some digging to get useful information. You can find additional information on the HobbyComponents.com (*http://bit.ly/dx-diy-hc*) forum. A collection of example sketches and a schematic can be downloaded from *http://bit.ly/dx-diy-sketch*. Note that the directory names in the ZIP archive are all in Chinese, but most of the sketches have comments in English.

Figure 8-82. Adafruit Wingshield

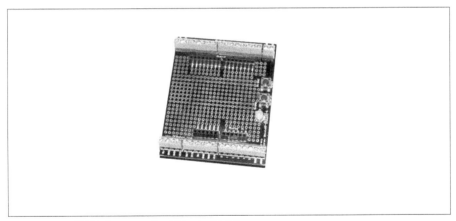

Figure 8-83. DFRobot terminal block shield

The arrangement of the various components of the multifunction shield is shown in Figure 8-85. Note that this is not a stacking shield, which makes sense, because stacking another shield on top would make the LED display useless.

Figure 8-84. DX DIY multifunction shield expansion board

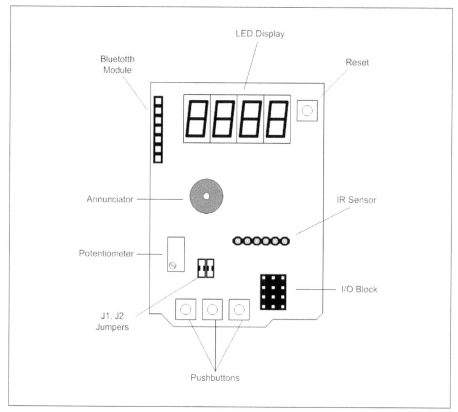

Figure 8-85. Multifunction shield main functional features

The schematic provided for this shield is somewhat cryptic, so I've created an alternate version, shown in Figure 8-86. Notice that all the Arduino pins are used by this shield. Table 8-1 lists the Arduino pins and what the multifunction shield uses them for.

This shield is a good example of what one will often encounter with a new shield board. The documentation may be minimal, and much of it might be in a language you don't understand (Chinese, in this case, but English can be just as difficult for other people). The schematic is correct, but might be difficult to understand at just a glance, and there is no detailed pinout description. (Well, actually, there is now.)

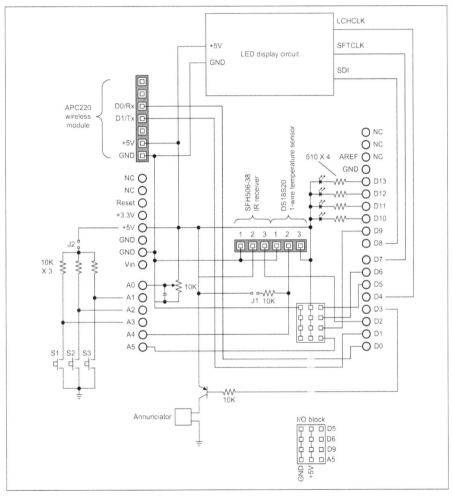

Figure 8-86. Multifunction shield schematic

Table 8-1. Multifunction shield Arduino pin functions

Pin	Use	Pin	Use
D0	Rx from wireless module	D10	LED
D1	Tx to wireless module	D11	LED
D2	ID received input	D12	LED
D3	Annunciator control	D13	LED
D4	LED display latch	A0	Potentiometer wiper
D5	To D5 on I/O block	A1	Switch S3
D6	To D6 on I/O block	A2	Switch S2
D7	LED display clock	A3	Switch S1
D8	LED display serial data input	A4	Temperature sensor input
D9	To D9 on I/O block	A5	A5 on I/O block

Uncommon Arduino Shields

In addition to a large collection of shields for everything from RS-232 I/O to PWM servo control, there are also shields created for specific applications. Some open source 3D printers utilize an Arduino as the primary controller, and the control interfaces, in the form of a shield, can be found easily. Some of these shields are intended for Mega-type Arduinos and are sized accordingly.

Other uncommon shields don't really fit into any of the categories in this chapter, but are interesting nonetheless. The Gameduino is an example of this type of shield. It's essentially a carrier for an FPGA (field-programmable gate array) chip, and has potential for other applications besides playing video games.

Qunqi CNC Shield for Arduino V3 Engraver (http://bit.ly/qunqi-cnc)
This shield, pictured in Figure 8-87, does not come with the motor driver modules, but these are readily available. A typical driver module uses Allegro's A4988 DMOS microstepping driver chip.

SainSmart RepRap Arduino Mega Pololu Shield (http://bit.ly/sainsmart-reprap)
The RepRap 3D fabricator (*http://reprap.org*), billed as "humanity's first general-purpose self-replicating manufacturing machine," is a compact fabricator that can create parts for other RepRap machines, along with many other things. This shield, shown in Figure 8-88, is designed to replace the electronics on a RepRap-type device and uses a Mega-type Arduino as its processor.

excamera Gameduino (http://bit.ly/excam-gameduino)
This shield uses a Xilinx FPGA (Figure 8-89) to control the graphics and sound for a homemade games console. An Arduino is used to interface with the controls and direct the game play. The Gameduino is open source, so all the techni-

cal details are available, and one could conceivably repurpose the FPGA for something other than games.

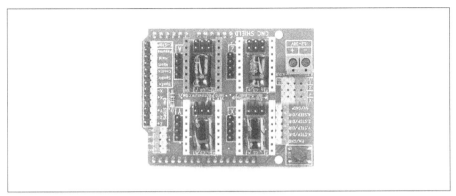

Figure 8-87. Qunqi A4988 driver CNC shield expansion board

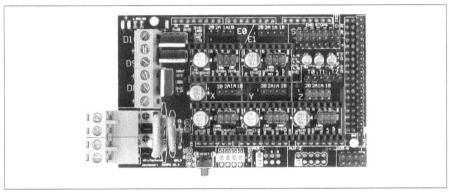

Figure 8-88. SainSmart RAMPS 1.4 RepRap shield for 3D printer

Figure 8-89. excamera Gameduino game controller shield

Sources

Table 8-2 lists the vendors and manufacturers referenced in this chapter. There are, of course, many others not listed here that are also good places to seek useful or novel shields. Entering "Arduino shield" in the Google search bar returns around 400,000 results, so there's no shortage of places to look for products and information. Just because a vendor or manufacturer is not here doesn't mean they aren't worth considering; it's just that trying to be all-inclusive with a market as volatile as this, with so many different products available, would be a Sisyphean task.

Table 8-2. List of shield vendors and manufacturers

Name	URL	Name	URL
Adafruit	www.adafruit.com	Macetech	www.macetech.com/store/
Arduino	store.arduino.cc	Mayhew Labs	www.mayhewlabs.com
Arduino Lab	www.arduinolab.us	Nootropic Design	www.nootropicdesign.com
Circuits@Home	www.circuitsathome.com	Numato	www.numato.com
CuteDigi	store.cutedigi.com	RobotShop	www.robotshop.com
DFRobot	www.dfrobot.com	Rugged Circuits	www.ruggedcircuits.com
DealeXtreme (DX)	www.dx.com	SainSmart	www.sainsmart.com
Elecfreaks	www.elecfreaks.com	Seeed Studio	www.seeedstudio.com
Elechouse	www.elechouse.com	SparkFun	www.sparkfun.com
excamera	www.excamera.com	Tindie	www.tindie.com
Iowa Scaled Engineering	www.iascaled.com	Tronixlabs	www.tronixlabs.com
iMall	imall.itead.cc	Vetco	www.vetco.net

CHAPTER 9

Modules and I/O Components

While many of the shields available for the Arduino have a lot of interesting and useful functions already built in, they don't have everything. Nor should they, given that there are a multitude of different types of sensors, controls, and actuator interfaces available that can be used with an Arduino. Many vendors offer single-function add-on sensor components and small PCB modules for the Arduino. These include temperature and humidity sensors, vibration detectors, photo detectors, keypads, joysticks, and even solid-state lasers.

Almost any sensor, control, or actuator device that can be used with a microcontroller can be used with an Arduino. There are some limitations in terms of DC supply voltage, depending on the type of microcontroller in the Arduino itself (3.3V versus 5V), but for the most part this is a relatively minor detail that can be resolved with simple interface electronics and an appropriate power supply.

This chapter looks at both I/O modules and individual components. I/O modules are small PCBs that perform a specific function and use only a few active components, if any at all. They are small, about the size of a postage stamp or less, and they use pins for the connections. They work well with female-to-male or female-to-female jumpers, and in some cases special multiwire cables can be used to connect modules to a shield made for just that purpose. The products from KEYES, SainSmart, and Tinker-Kit are featured here, primarily because they are good representatives of modules in general. Other modules worth considering are the Grove series of modules and interface shields available from Seeed Studio, and the modules from TinyCircuits.

Individual I/O components cover the spectrum from LEDs to graphical displays, and from mechanical sensors like switches and reed relays to self-contained temperature and humidity sensors. The individual sensors are intentionally covered after the discussion of modules because many of the modules use the components described there. Cross-references are provided between the sections as appropriate.

Most people appreciate neatness and reliability. Unfortunately, using the ubiquitous jumper wires to connect modules and other components can quickly become anything but neat, and the push-on crimp connectors used at each end of the jumper wire have a tendency to work loose from a module's pins.

Rather than resorting to soldering directly to the pins of a module, or covering the jumper connectors with a blob of silicon adhesive to hold them in place, you can use modular connectors. These can be custom-made for your specific application using simple hand tools, or you can opt to use a system like TinkerKit, Grove, or TinyCircuits. This chapter wraps up with an overview of methods for connecting modules to an Arduino that don't involve a tangle of jumper wires.

Modules

Sensor and I/O modules are certainly the most convenient way to connect a sensor, switch, relay, or microphone to an Arduino and experiment with it. Figure 9-1 shows some different module types.

Figure 9-1. Different sensor module types and sizes

Over time it is easy to end up with a large collection of modules, some of which are more useful than others. I would suggest starting off with the largest kit of modules you can afford, and then figuring out which ones you use the most. Keep several of those on hand, and save the other less-used modules for a future project.

The descriptions found online for how various modules work are not always correct. It might be a translation issue, but sometimes an online description will state that a module will generate a high output when active, when in reality it will generate a low output. Always check the operation of a module with a digital multimeter (DMM) before committing it to a circuit. Most of the modules listed in this section have been tested to determine how they really work, and the descriptions here reflect those findings. That being said, I can't claim that every module that may look like one of the modules listed in this chapter will behave the same or have the same pin functions. In this corner of the universe, standardization has yet to take hold.

As for schematics for the modules listed here, well, there really aren't any official schematics that I have been able to locate. Several brave souls across the Internet have taken it upon themselves to trace out some of the module PCBs. I've attempted to gather what I could find and combine it with my own efforts at reverse engineering. In some cases the result is a complete schematic, and in others I just wanted to verify that the pins really did what the available (and rather minimal) documentation stated.

Physical Form Factors

Modules can vary in size from 1.8 by 1.5 cm (approx. 3/4 by 9/16 inch) to 4.2 by 1.7 cm (approx. 1 11/16 by 5/8 inch), with some as large as 3.4 by 2.5 cm (approx. 1 5/16 by 1 inch) for a 5V modular relay. Figure 9-2 shows a variety of module dimensions. Note that these are approximate dimensions. The actual PCB dimensions may vary by about +/- 1 mm (0.04 inches), depending on where the modules were produced.

Many modules have mounting holes in the PCB large enough for a #2 machine screw. The metric equivalent is typically an M1.6 or M1.8 size. Figure 9-3 shows a stack of two modules made using 2-56 machine screws and nylon spacers. This happens to be a one-wire temperature sensor with a mercury tilt switch module.

Unfortunately, not all modules will stack nicely. Sometimes the mounting holes don't have the same spacing, or they might be in the wrong places on the PCB to allow modules to stack. Some modules don't have mounting holes at all, so always check before assuming that you will be able to create a stack of modules.

Interfaces

The pinouts used with the various PCB modules can vary from one type to another, so other than the TinkerKit series there really isn't a lot of standardization. Figure 9-4 shows some of the variations you can expect to encounter.

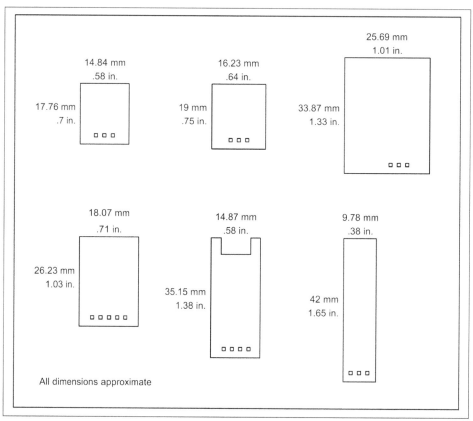

Figure 9-2. Examples of commonly encountered module dimensions

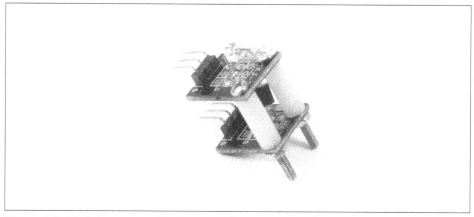

Figure 9-3. Modules mounted using #2 machine screws and spacers

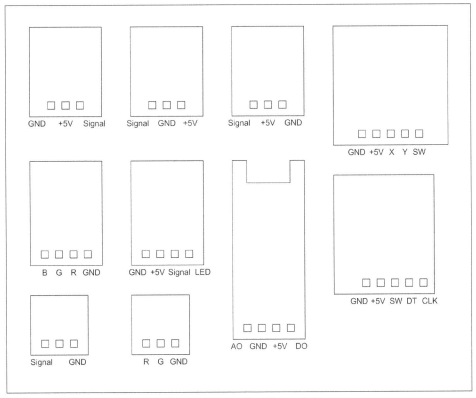

Figure 9-4. Typical pinout configurations used with modules

 Although you might assume that modules with three or four pins would be compatible with I/O extension shields that feature modular connectors or I/O pins arranged in blocks (such as those described in Chapter 8), that is not always the case. Pin-to-pin compatibility between modules and shields is only guaranteed when connecting a family of modules to an interface shield designed to work with those components. TinkerKit is one example of this, but only when connecting TinkerKit modules to a TinkerKit interface extender shield. Always check the module pins to verify the voltage and signal positions before connecting a module.

Because the AVR is a rather robust device it is possible to connect many sensors directly to the inputs of an Arduino, and some output devices as well. Power-hungry output devices, like a solid-state laser or an RGB LED, really need some type of driver to boost the current beyond what the AVR chip on an Arduino can supply directly (recall the current source and current sink specifications given in Chapter 3).

Some output modules have a high current interface built in, but some do not. A simple circuit, like the one shown in Figure 9-5, can be used to safely boost the current supplied to something like a relay or laser LED module. The value of R in the right-hand circuit in Figure 9-5 would be determined by the LED and the amount of current it needs to operate. So long as the current doesn't exceed the rating of the transistor it should work fine.

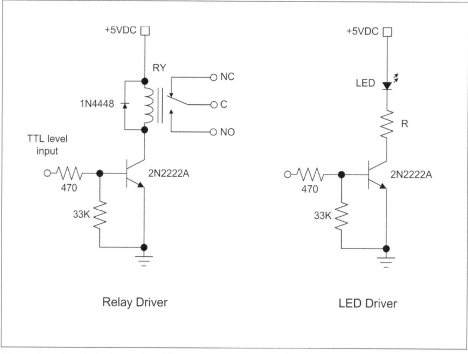

Figure 9-5. Output circuit for driving high-current devices

Another option is a special-purpose IC like the MAX4896, shown in Figure 9-6. It uses an SPI interface, and an Arduino can interface directly to the IC. While intended to drive small relays, this IC can handle large LEDs just as easily.

Module Sources

Single-PCB modules and kits of sensors with 24, 36, or more modules are available from the sources listed in Table 9-1. They can also be found on eBay and Amazon. Figure 9-7 shows a plastic storage container with a full suite of modules (37 different types; the last bin contains two different modules).

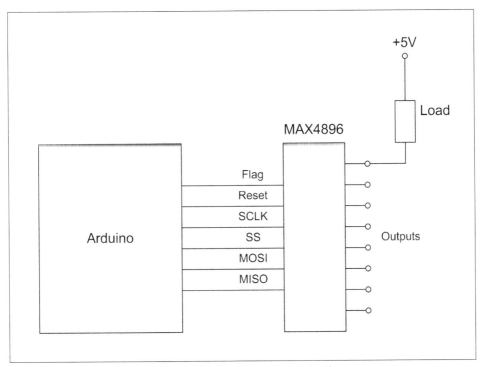

Figure 9-6. Output driver IC for multiple relays or other loads

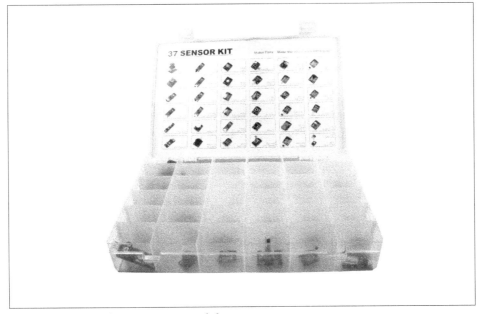

Figure 9-7. A set of input/output modules

Most of the suppliers listed in Table 9-1 also sell individual modules, as well as jumper wires, interconnection cables, and "bare" input and output components. You can buy most any module you might need in small quantities, although not every module in a kit may be available as a single item.

Table 9-1. Partial list of sensor and output module vendors and manufacturers

Name	URL
Adafruit	*www.adafruit.com*
CuteDigi	*store.cutedigi.com*
DealeXtreme (DX)	*www.dx.com*
KEYES	*en.keyes-robot.com*
SainSmart	*www.sainsmart.com*
Seeed Studio	*www.seeedstudio.com*
TinyCircuits	*www.tiny-circuits.com*
Trossen Robotics	*www.trossenrobotics.com*
Vetco	*www.vetco.net*

Module Descriptions

This section lists modules from three sources: KEYES, SainSmart, and TinkerKit. The descriptions are, by necessity, terse. The main emphasis here is on the physical form and the electrical connections of the modules listed in the tables. "Sensors" on page 312 describes the sensor components in more detail, and the modules are cross-referenced to the detailed descriptions where appropriate.

After working with various modules for a time you may notice that many of the modules described here use the same basic circuit. This typically consists of an LM393 comparator and some type of sensor. In each case a potentiometer sets the threshold comparison voltage, and the output of the LM393 is wired to the signal pin of the module. The KEYES modules with similar circuits are listed in Table 9-2, and the SainSmart modules with similar circuits are listed in Table 9-3.

Table 9-2. KEYES modules with similar circuits (Figure 9-8)

Part no.	Name
KY-025	Reed Switch Module
KY-026	Flame Sensor
KY-036	Conductive Contact Sensor
KY-037	Sensitive Microphone Sensor
KY-038	Microphone Sensor

Table 9-3. SainSmart modules with similar circuits (Figure 9-8)

Part no.	Name
20-011-981	Photosensitive Sensor
20-011-982	Vibration/Shock Sensor
20-011-983	Hall Effect Sensor
20-011-984	Flame Sensor

Figure 9-8 shows a generic schematic representation of the LM393 comparator circuit used in the modules listed in Table 9-2 and Table 9-3. The actual component values may vary somewhat, but this same basic circuit is used in multiple modules. The sensor (IR flame, microphone, LDR, etc.) is the major difference between the modules.

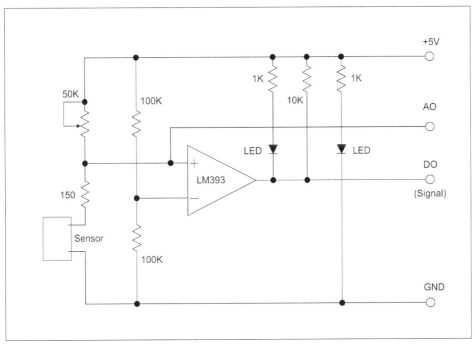

Figure 9-8. Generic module circuit with comparator IC

The DO (digital output) terminal comes directly from the comparator. It is high when the noninverting (+) input is greater than the inverting (–) input. The output of the comparator will go low if the noninverting input is lower than the inverting input. Some modules may be designed with the inputs to the comparator IC arranged opposite to what it shown in Figure 9-8, but the operating principle is the same.

The voltage on the noninverting input is set by the potentiometer, or pot. It is one-half of a voltage divider, with the sensor and perhaps a current-limiting resistor making up the other half. A second voltage divider is used to apply about one-half of the

+5V VCC to the inverting input, or about 2.5V. This is the reference input. The pot is used to set the voltage at which the comparator will change its output as the sensor changes its resistance in response to some type of input.

On the output end of the circuit a 10K resistor is used as a pull-up on the output pin of the IC. The AO (analog output) is the "raw" value from the sensor at the input of the LM393 IC (in this case). On those modules with just a single output, only the output of the comparator is brought out to a terminal pin as the DO signal. A module may also have an LED to indicate when power is present, and some have an additional LED to indicate when the comparator output is low (it becomes a current sink for the LED).

How all this relates to a sensor depends on how the sensor responds to input, and you may need to experiment with it a bit to get a feel for how it behaves. In many cases an active sensor will exhibit decreased resistance, which will cause the noninverting input to go below the reference voltage on the inverting input. When this happens the output of the comparator circuit shown in Figure 9-8 will go low, and the LED on the output of the IC will be active (the IC serves as a current sink for the LED).

Comparator circuits that exhibit a low output when the input is in an active state or below some threshold are called *active-low* circuits. In these circuits, a low voltage on the output is equivalent to a true condition. Circuits that exhibit a high output level when the sensor is active or the input is above some threshold level are called *active-high* circuits. In an active-high circuit a high output is equivalent to a true condition. True and false in this sense just mean that the sensor is either receiving input or not, respectively.

 Spend a few moments with a DMM (digital multimeter) and a magnifying glass or loupe and carefully examine a module before connecting it to an Arduino. This is particularly important with the bargain modules, where I have discovered missing parts, solder bridges (shorts) between pads, and connection pins that go nowhere. One module I examined had a factory-installed piece of wire soldered between traces connected to the +5V and ground pins! That would have caused some problems. Just because a module has pins for ground, +5V, and signal doesn't mean that all the pins are actually used, or even that the pins do what the assigned names imply. Once any issues are addressed, the modules tend to work just fine (they only have a small number of parts on them, after all). Keep notes and save yourself from headaches later on.

Many common modules designed by the Chinese company Shenzhen KEYES DIY Robot Co. Ltd. (also known simply as KEYES) can be found as single units or bundled into kits of modules, usually with 36 or so modules per kit. You may also encounter modules with the letters "HXJ" on them. They are functionally identical to the KEYES modules, but the PCB layout may be slightly different.

Table 9-4 is a summary list of the KEYES modules covered in this section, and Table 9-7 contains images, notes, and pinout diagrams, from KY-001 to KY-040. Some of the models shown are KEYES, and some are from other vendors but are otherwise identical (the HXJ units, for example). They all share the same pinout and functions, and often have the same name (KY-002, KY-027, and so on). Notice that there are no KY-007, KY-014, KY-029, or KY-030 modules. I have no idea why

Table 9-4. Commonly available KEYES I/O modules

Part no.	Name	Part no.	Name
KY-001	Temperature Sensor	KY-021	Mini Reed Switch
KY-002	Vibration Sensor	KY-022	Infrared Sensor/Receiver
KY-003	Hall Effect Magnetic Field Sensor	KY-023	2 Axis Joystick
KY-004	Pushbutton Switch	KY-024	Linear Hall Effect Sensor
KY-005	Infrared Emitter	KY-025	Reed Switch Module
KY-006	Passive Buzzer	KY-026	Flame Sensor
KY-008	Laser LED	KY-027	Magic Light Cup Module
KY-009	RGB Color LED	KY-028	Temperature Sensor
KY-010	Optical Interrupter	KY-031	Shock Sensor
KY-011	2 Color LED	KY-032	IR Proximity Sensor
KY-012	Active Buzzer	KY-033	IR Line Following Sensor
KY-013	Analog Temperature Sensor	KY-034	Automatic Flashing Color LED
KY-015	Temperature and Humidity Sensor	KY-035	Hall Effect Magnetic Sensor
KY-016	3 Color RGB LED	KY-036	Touch Sensor
KY-017	Mercury Tilt Switch	KY-037	Sensitive Microphone Sensor
KY-018	LDR Module	KY-038	Microphone Sensor
KY-019	5V Relay	KY-039	LED Heartbeat Sensor
KY-020	Tilt Switch	KY-040	Rotary Encoder

SainSmart is another manufacturer of sensor modules. Table 9-5 is a summary of what is available in a SainSmart module kit, and Table 9-8 contains images, notes, and pinout diagrams. This is a representative list, as the contents of the kits can vary. The modules with a SainSmart Part no. are also available individually, and some of them are unique.

Table 9-5. SainSmart modules available as a kit

Part no.	Name
Part no.	Name
N/A	Relay Module
20-011-985	Touch Sensor
20-019-100	Ultrasonic Distance Sensor HC-SR04
20-011-984	Flame Sensor
20-011-986	Temperature & Relative Humidity Sensor
N/A	Active Buzzer
20-011-982	Vibration/Shock Sensor
N/A	Passive Buzzer
20-011-987	Tracking Sensor
20-011-983	Hall Effect Sensor
20-011-981	Photosensitive Sensor
N/A	Infrared Receiver
20-011-944	Joystick Module
20-011-946	Water Sensor

Lastly, there are the TinkerKit modules. Table 9-6 is a summary listing, and Table 9-9 has details on each of the modules found in the "Pro" kit. The TinkerKit series of modules are designed to work with a TinkerKit interface shield (described in Chapter 8). Although the status of TinkerKit as a company is currently in limbo, the products are still available from Mouser (*http://www.mouser.com*), Newark/Element14 (*http://www.newark.com*), and other sources. Software libraries are available on GitHub (*https://github.com/TinkerKit*). A set of basic datasheets for the modules are available from Mouser (*http://bit.ly/mouser-tinkerkit*).

Table 9-9 doesn't show pinout diagrams because the TinkerKit modules all use a standard pinout. The only difference between modules is between the discrete digital (on/off) and analog modules. Both types use the same basic connector pinout shown in Figure 9-9.

Circuitry on the TinkerKit modules handles the electrical interface, which might be level sensing, amplification, and so on. So bear in mind that with many of the Tinker-Kit modules the Arduino isn't communicating directly with the sensor, LED, input control, or output device, but rather with an interface circuit. Flip a module over and check the back side to see what circuitry is installed. Even the LED modules have a driver transistor installed. The objective with the TinkerKit products was to create something that was easy to connect to an Arduino and relatively insensitive to minor errors, so some of the low-level interface interaction capability was given up to achieve that.

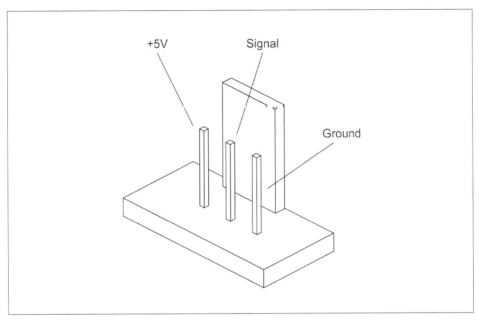

Figure 9-9. TinkerKit module common connector

Table 9-6. TinkerKit modules summary listing

Part no.	Name	Part no.	Name
T000020	Accelerometer Module	T010020	Mosfet Module
T000030	Joystick Module	T010110	High Power LED Module
T000070	Hall Effect Sensor	T010111	5 mm Blue LED Module
T000090	LDR Module	T010112	5 mm Green LED Module
T000140	Rotary Potentiometer Module	T010113	5 mm Yellow LED Module
T000150	Linear Potentiometer Module	T010114	5 mm Red LED Module
T000180	Pushbutton Module	T010115	10 mm Blue LED Module
T000190	Tilt Module	T010116	10 mm Green LED Module
T000200	Thermistor Module	T010117	10 mm Yellow LED Module
T000220	Touch Sensor Module	T010118	10 mm Red LED Module

KEYES modules

Table 9-7. KEYES-type sensor and output modules

Part no.	Name and description	Image	Pinout
KY-001	**Temperature Sensor** Uses a DS18B20 one-wire temperature sensor IC in a TO-92 package.		GND +5V Signal
KY-002	**Vibration Sensor** A sealed shock sensor closes the circuit between the GND and signal pins. The +5V doesn't seem to be connected, but there is a position for a resistor on the bottom side of the PCB. Very sensitive.		Signal GND +5V
KY-003	**Hall Effect Magnetic Field Sensor** Detects the presence of a magnetic field. The output is the open collector of an NPN transistor in the A3144 Hall effect device. The output is pulled to ground when the sensor is active. This is not a linear sensor, whereas the KY-024 is.		GND +5V Signal

Part no.	Name and description	Image	Pinout
KY-004	**Pushbutton Switch** A simple pushbutton switch. The output is pulled low when the switch is pressed.		Signal +5V GND
KY-005	**Infrared Emitter** An IR LED suitable for use with a KY-022. Note that an external current-limiting resistor must be used. The GND pin was not connected on the module tested.		Signal +5V GND
KY-006	**Passive Buzzer** This is a small speaker with a metal diaphragm. The +5V pin is not used.		GND +5V Signal

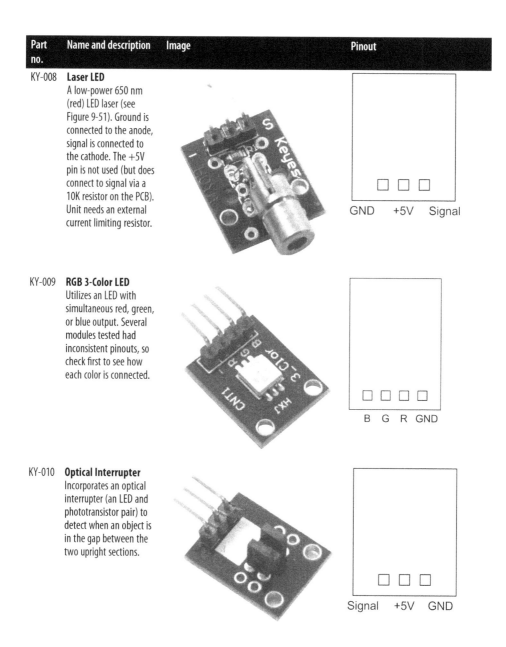

KY-008 · **Laser LED**
A low-power 650 nm (red) LED laser (see Figure 9-51). Ground is connected to the anode, signal is connected to the cathode. The +5V pin is not used (but does connect to signal via a 10K resistor on the PCB). Unit needs an external current limiting resistor.

Pinout: GND   +5V   Signal

KY-009 · **RGB 3-Color LED**
Utilizes an LED with simultaneous red, green, or blue output. Several modules tested had inconsistent pinouts, so check first to see how each color is connected.

Pinout: B   G   R   GND

KY-010 · **Optical Interrupter**
Incorporates an optical interrupter (an LED and phototransistor pair) to detect when an object is in the gap between the two upright sections.

Pinout: Signal   +5V   GND

Part no.	Name and description	Image	Pinout
KY-011	**2 Color LED** Utilizes an LED capable of green or red output, or both at once. This is a three-lead part, which means that the internal LEDs share a common connection and each can be operated independently.	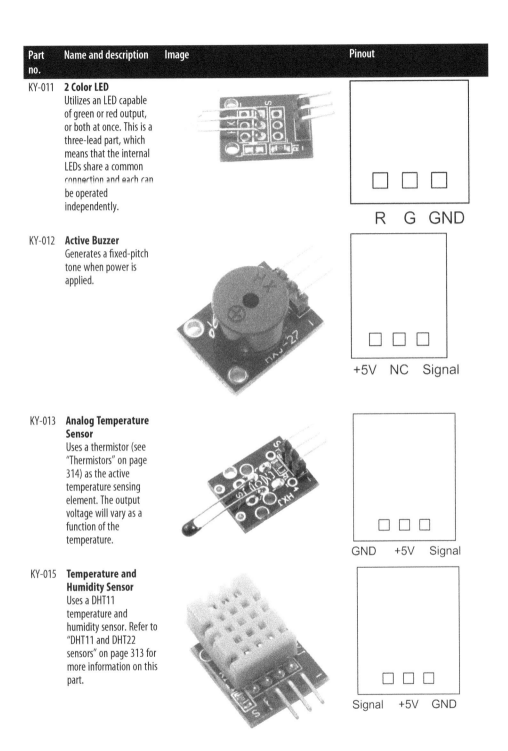	R G GND
KY-012	**Active Buzzer** Generates a fixed-pitch tone when power is applied.		+5V NC Signal
KY-013	**Analog Temperature Sensor** Uses a thermistor (see "Thermistors" on page 314) as the active temperature sensing element. The output voltage will vary as a function of the temperature.		GND +5V Signal
KY-015	**Temperature and Humidity Sensor** Uses a DHT11 temperature and humidity sensor. Refer to "DHT11 and DHT22 sensors" on page 313 for more information on this part.		Signal +5V GND

Part no.	Name and description	Image	Pinout

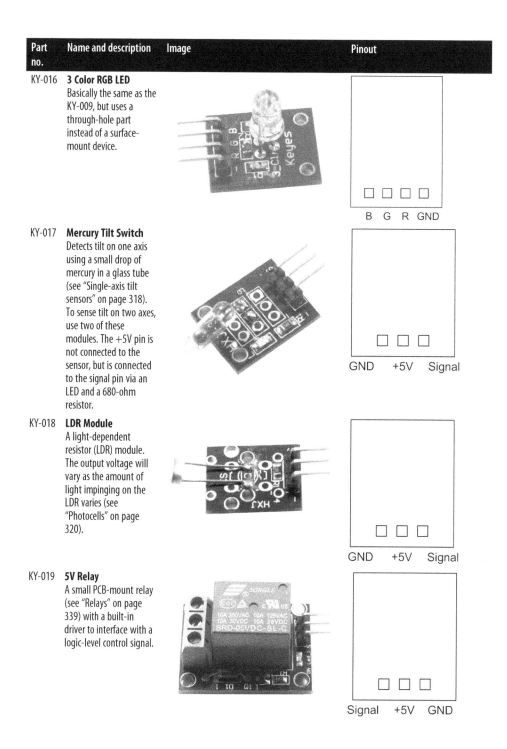

KY-016 — **3 Color RGB LED**
Basically the same as the KY-009, but uses a through-hole part instead of a surface-mount device.

B   G   R   GND

KY-017 — **Mercury Tilt Switch**
Detects tilt on one axis using a small drop of mercury in a glass tube (see "Single-axis tilt sensors" on page 318). To sense tilt on two axes, use two of these modules. The +5V pin is not connected to the sensor, but is connected to the signal pin via an LED and a 680-ohm resistor.

GND   +5V   Signal

KY-018 — **LDR Module**
A light-dependent resistor (LDR) module. The output voltage will vary as the amount of light impinging on the LDR varies (see "Photocells" on page 320).

GND   +5V   Signal

KY-019 — **5V Relay**
A small PCB-mount relay (see "Relays" on page 339) with a built-in driver to interface with a logic-level control signal.

Signal   +5V   GND

Part no.	Name and description	Image	Pinout
KY-020	**Tilt Switch** Similar to the KY-017, but utilizes a metal ball inside a small enclosure.		GND +5V Signal
KY-021	**Mini Reed Switch** A reed switch encapsulated in a small glass tube that closes when in the presence of a magnetic field.		GND +5V Signal
KY-022	**Infrared Sensor/ Receiver** An 1838 IR sensor like those used for remote control of televisions and other appliances. Operates at a carrier frequency of 37.9 KHz. The presence of the IR carrier causes the output to go high.		GND +5V Signal
KY-023	**2 Axis Joystick** Contains two potentiometers mounted at a right angle to one another to detect the x–y motion of the center shaft.		GND +5V X Y SW

Part no.	Name and description	Image	Pinout
KY-024	**Linear Hall Effect Sensor** Utilizes an SS49E linear Hall effect sensor. An LM393 voltage comparator is used with a potentiometer to adjust the sensitivity of the circuit.	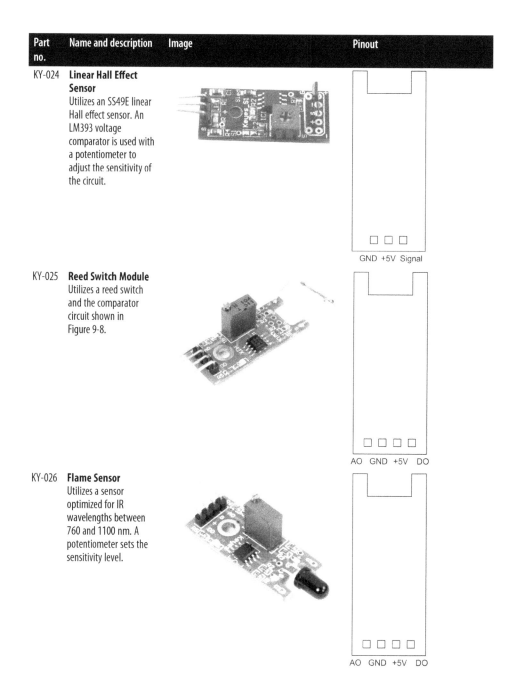	GND +5V Signal
KY-025	**Reed Switch Module** Utilizes a reed switch and the comparator circuit shown in Figure 9-8.		AO GND +5V DO
KY-026	**Flame Sensor** Utilizes a sensor optimized for IR wavelengths between 760 and 1100 nm. A potentiometer sets the sensitivity level.		AO GND +5V DO

Part no.	Name and description	Image	Pinout
KY-027	**Magic Light Cup Module** To be honest, I'm not sure what these are intended to be used for. The KY-027 is basically a KY-017 with an LED on a separate circuit.		GND +5V Signal LED
KY-028	**Temperature Sensor** Uses a thermistor and an comparator IC to detect a threshold (see "Thermistors" on page 314). The potentiometer sets the threshold point.		AO GND +5V DO
KY-031	**Shock (Impact) Sensor** Detects a sharp impact and generates an output. Does not detect tilt like the KY-002, KY-017, and KY-020.		Signal +5V GND

Part no.	Name and description	Image	Pinout
KY-032	**IR Proximity Sensor** Uses the reflectance of an IR light source to detect a nearby surface or obstacle. The potentiometer sets the sensitivity. Note that there are at least two variants of this module. Check the pinout before applying power.		G +5V Signal
KY-033	**IR Line Following Sensor** Uses the reflection of IR light from a surface to detect the difference between light and dark. Typically used to create a line-following robot that will track a dark line on a white surface. The potentiometer sets the sensitivity. Basically the same as the KY-032 but with the LED and IR sensor mounted on the underside of the PCB.		G +5V Signal
KY-034	**Automatic Flashing Color LED** LED will flash automatically when power is applied.		Signal +5V GND

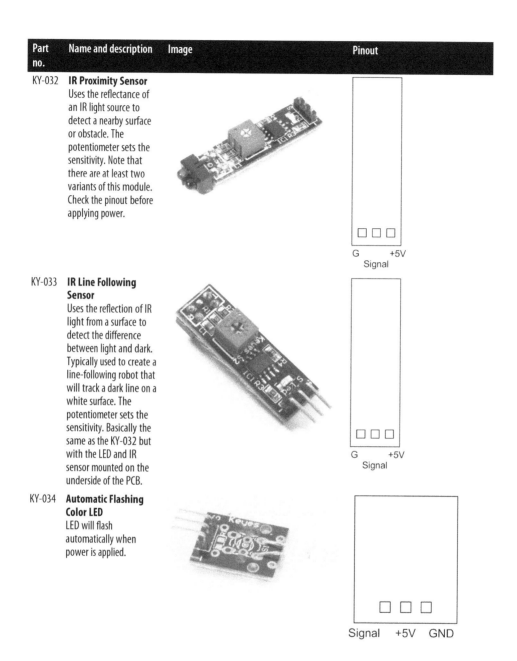

Part no.	Name and description	Image	Pinout
KY-035	**Hall Effect Magnetic Sensor** Similar to the KY-003 but with an SS49E linear Hall effect sensor.		GND +5V Signal
KY-036	**Conductive Contact Sensor** The output will change when something conductive (like a finger) touches the bare lead on the sensor.		GND +5V Signal
KY-037	**Sensitive Microphone Sensor** Microphone with a variable threshold. The larger microphone (as compared to the KY-038 module), gives this unit slightly more sensitivity.		AO GND +5V DO

Part no.	Name and description	Image	Pinout
KY-038	**Microphone Sensor** Microphone with a variable threshold. Due to the smaller microphone (as compared to the KY-037 module) this module is less sensitive.		AO GND +5V DO
KY-039	**LED Heartbeat Sensor** Basically the same thing as the finger clip popular in hospitals and doctor's offices for measuring a patient's pulse rate. A phototransistor is used to detect the slight variations in the light from an LED as blood flows through a finger.		Signal +5V GND
KY-040	**Rotary Encoder** A digital continuous rotary encoder. Refer to "Digital rotary encoders" on page 329 for more about these devices.		GND +5V SW DT CLK

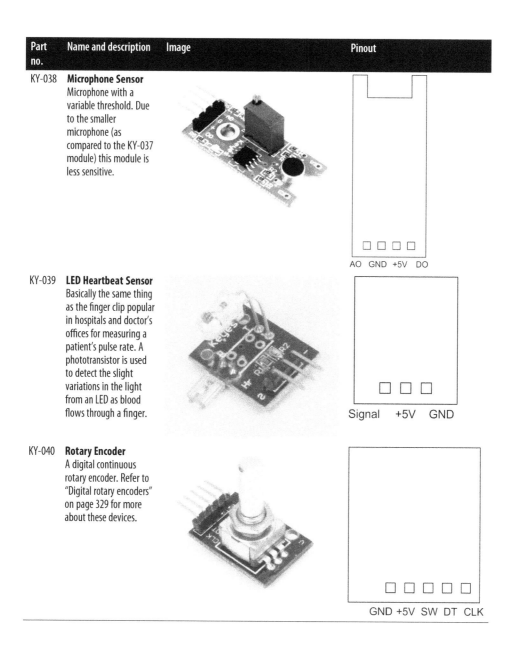

SainSmart modules

Table 9-8. SainSmart sensor and output modules

Part no.	Name and description	Image	Pinout
	Relay Module 5V relay with 10A contacts.		GND +5V Signal
20-019-100	**Ultrasonic Distance Sensor HC-SR04** Uses a pair of ultrasonic transducers to emit a signal and then receive the echo. The time between the trigger and the echo is proportional to the distance from the sensor to the reflective surface.		+5V Trig Echo G
	Active Buzzer Emits a tone when power is applied.		Signal +5V GND

Part no.	Name and description	Image	Pinout
	Passive Buzzer Responds to a square wave input, which allows this module to emit a programmable tone.		Signal +5V GND
20-011-983	**Hall Effect Sensor** Utilizes a linear Hall effect sensor. The small potentiometer sets the sensitivity threshold.		+5V G AO DO
	Infrared Receiver Responds to pulses from a remote control device to generate a digital signal.		GND +5V Signal
20-011-946	**Water Sensor** Detects the presence of water by responding to a change in the conductivity across the metallic fingers on the PCB. Can be used as a rain detector or as a splash detector.		+5V G Signal

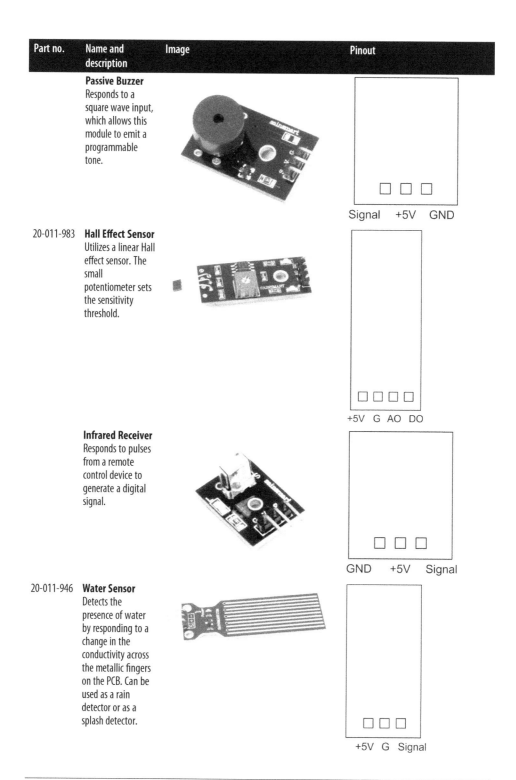

Part no.	Name and description	Image	Pinout
20-011-985	**Touch Sensor** Detects the touch of a finger.	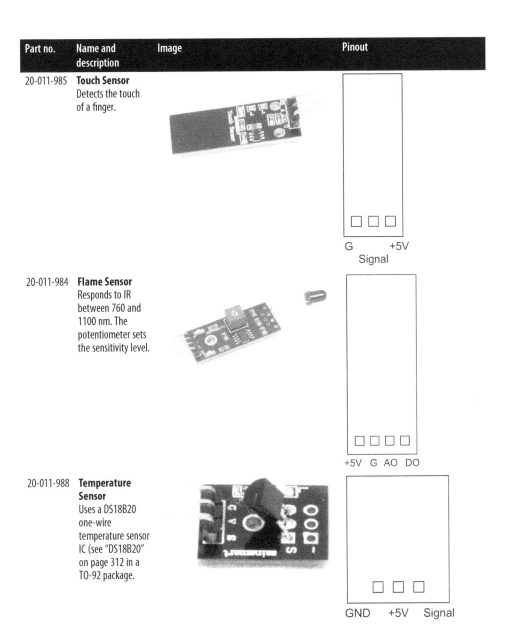	G +5V Signal
20-011-984	**Flame Sensor** Responds to IR between 760 and 1100 nm. The potentiometer sets the sensitivity level.		+5V G AO DO
20-011-988	**Temperature Sensor** Uses a DS18B20 one-wire temperature sensor IC (see "DS18B20" on page 312 in a TO-92 package.		GND +5V Signal

20-011-986	**Temperature & Relative Humidity Sensor** Uses a DHT11 temperature and humidity sensor. Refer to Figure 9-14 for more information on this part.		Signal +5V GND
20-011-982	**Vibration/Shock Sensor** Senses sudden movement (shock or vibration) with a sealed sensor, similar to the KEYES KY-002.		+5V G AO DO
20-011-987	**Tracking Sensor** Employs an IR LED and a photosensor to detect reflectance.		+5V G Signal

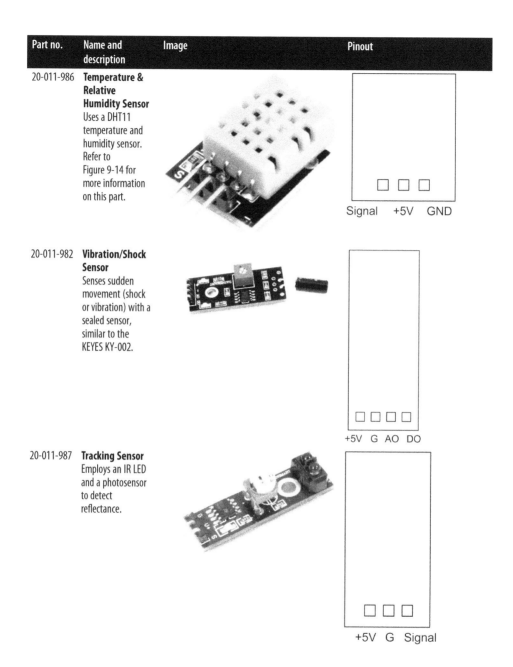

Part no.	Name and description	Image	Pinout
20-011-981	**Photosensitive Sensor** A basic LDR circuit (see "Photocells" on page 320) with variable sensitivity.		+5V G AO DO
20-011-944	**Joystick Module** A two-axis linear potentiometer joystick. Includes a large knob.		GND +5V X Y SW

TinkerKit modules

Table 9-9. TinkerKit I/O modules

Part no.	Name and description	Image
T000020	**Accelerometer Module** Based on the three-axis LIS344AL IC. Includes two signal amplifiers.	

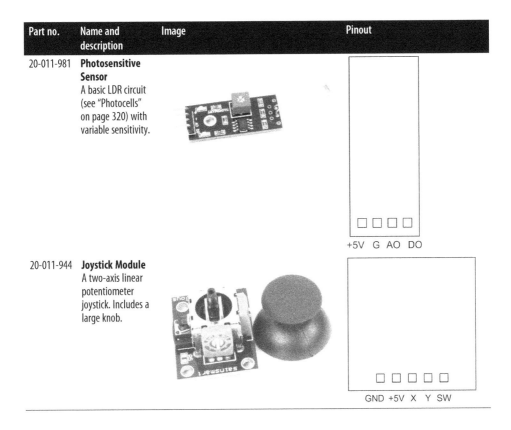

T000030 **Joystick Module**
Two potentiometers mounted on a two-axis gimbal.

T000070 **Hall Effect Sensor**
Outputs a voltage that is dependent on the strength of a local magnetic field.

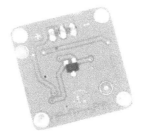

T000090 **LDR Module**
Light-dependent resistor (LDR) with an amplifier, outputs a voltage proportional to the light level.

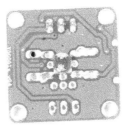

T000140 **Rotary Potentiometer Module**
4.7K ohm rotary analog potentiometer.

Part no.	Name and description	Image

T000150 **Linear Potentiometer Module**
4.7K ohm linear (slider) analog potentiometer.

T000180 **Pushbutton Module**
Simple pushbutton module, normally open, outputs +5V when pressed.

T000190 **Tilt Module**
Uses simple tilt sensor with internal metal ball.

T000200 **Thermistor Module**
Uses a thermistor and an amplifier to output a voltage proportional to temperature.

T000220   **Touch Sensor Module**

Utilizes a QT100A single touch controller to produce 5V when sensor is touched. The touch pad is part of the front of the PCB, which appears flat in the image.

T010010   **Relay Module**

Incorporates a 5V relay with 250V 10A contacts, a driver transistor, and a screw-type terminal block.

T010020   **Mosfet Module**

Switches up to 24V DC using an IRF520 MOSFET. Fast enough to be used with PWM for DC motor control.

T010110   **High Power LED Module**

Provides five ultra-bright LEDs. Requires significant current, so you might want to use with a T010020 and an external power supply.

Part no.	Name and description	Image

T010111   **5 mm Blue LED**

A simple module with a single 5 mm blue LED.

T010112   **5 mm Green LED Module**

A simple module with a single 5 mm green LED.

T010113   **5 mm Yellow LED Module**

A simple module with a single 5 mm yellow LED.

T010114   **5 mm Red LED Module**

A simple module with a single 5 mm red LED.

T010115 **10 mm Blue LED Module**
A simple module with a single large 10 mm blue LED.

T010116 **10 mm Green LED Module**
A simple module with a single large 10 mm green LED.

T010117 **10 mm Yellow LED Module**
A simple module with a single large 10 mm yellow LED.

T010118 **10 mm Red LED Module**
A simple module with a single large 10 mm red LED.

Grove Modules

SeedStudio has amassed a large number of modules under the Grove interconnect system. Many of these are functionally similar to modules available from KEYES, SainSmart, and TinkerKit. Others are unique to the Grove product line.

The Grove modules are categorized into six groups: environment monitoring, motion sensing, user interface, physical monitoring, logic gate functions, and power control. The modules all use a standardized connection scheme involving a 4-pin modular connector with power, ground, and signal lines.

As with the TinkerKit modules, the advantage of the Grove modules is that you don't have to try to verify (or figure out) what a module's pins actually do. A shield that follows the Grove conventions (such as the Grove shields shown in Chapter 8) will connect to a Grove sensor or actuator module using a prefabricated cable. Seeed Studio also sells premade cables for connecting modules to an interface or control shield.

One feature of the Grove modules that I find appealing is the inclusion of mounting tabs in the PCB of most of the modules. This gets the mounting hardware out of the way of the circuitry (as opposed to mounting holes in the middle of the PCB layout) and the modular connectors are easy to use. An example Grove module is shown in Figure 9-10.

Figure 9-10. An example Grove module

At this time of writing, the TinkerKit modules appear to be fading, while the Grove modules appear to be very much alive and well. That's not to say that TinkerKit won't make a comeback, or that Grove will be around forever; things can and do change in the Arduino world on short time scales. As I pointed out earlier, this chapter and Chapter 8 are intended to provide examples of what is available, not serve as definitive references. The market is too volatile to allow for that.

If you are looking for modules that follow a convention of one sort or another, then I would recommend investigating the Grove products. If you want to use a different expansion or interface shield, you can create your own interface cables using readily available shells, pins, and sockets, as described in "Building Custom Connectors" on page 352. You can see a complete listing of the modules currently available at the Seeed Studio wiki (*http://bit.ly/seeed-grove*).

Sensor and Module Descriptions

The remainder of this chapter describes some of the different types of Arduino-compatible components, modules, and sensors that are available. I say "some of" because, as with shields, new sensors and modules appear constantly, and some older types vanish if there isn't enough of a market to justify continued production. Some modules are rather specialized one-off types, such as gas sensors or current monitors. Some sensors may disappear as they are replaced with newer types. You can find all the devices described here, and more, with very little effort using Google, browsing through the listings on Amazon.com, or checking the websites listed in "Sources" on page 355.

Table 9-10 lists the controls, sensors, actuators, displays, and modules described in this section. They are organized by function, class (input or output), and type (device, component, or module). Remember that the components on any module are also available as single items, so you can create your own module or incorporate the parts into something else.

Table 9-10. Sensors and modules index

Function	Class	Type	Description
Audio	Output	Device	Buzzer (annunciator) components and modules
Audio	Sensor	Microphone	Audio pick-up (microphone) modules
Communication	I/O	Module	315/433 MHz RF modules
Communication	I/O	Module	APC220 wireless modules
Communication	I/O	Module	ESP8266 WiFi transceiver
Communication	I/O	Module	NRF24L01 module
Contact	Sensor	Switch	Contact switch modules
Contact	Output	Switch	Relays and relay modules
Control	Input	Device	Keypads
Control	Input	Module	Joystick modules
Control	Input	Device	Potentiometers
Display	Output	Module	7-segment modules
Display	Output	Module	ERM1601SBS-2 16 × 1 LCD display
Display	Output	Module	ST7066 (HD44780) 16 × 2 LCD display

Function	Class	Type	Description
Display	Output	Module	ERC240128SBS-1 240 × 128 LCD display
Display	Output	Module	ST7735R 128 × 160 TFT display
Light emit	Output	Display	7-segment LED display
Light emit	Output	Display	LED matrix modules
Light emit	Output	Laser	Laser LEDs
Light emit	Output	LED	Single-color LEDs
Light emit	Output	LED	Dicolor LEDs
Light emit	Output	LED	Tricolor (RGB) LEDs
Light sense	Sensor	Photocell	LDR modules
Light sense	Sensor	Diode	Photodiode modules
Light sense	Sensor	Transistor	Phototransistor modules
IR sense	Sensor	IR	PIR sensor modules
Magnetic	Sensor	Solid state	Hall effect sensor modules
Magnetic	Sensor	Solid state	Magnetometer modules
Moisture	Sensor	PCB	Soil moisture sensor modules
Motion	Output	Actuator	DC motor control
Motion	Output	Actuator	Servo control
Motion	Output	Actuator	Stepper motor control
Motion	Sensor	Solid state	Gyroscope modules
Motion	Sensor	Solid state	Accelerometer modules
Pressure	Sensor	Solid state	Barometric sensor modules
Range	Sensor	Module	Laser transmitter/receiver modules
Range	Sensor	Module	LED object sensor modules
Range	Sensor	Module	Ultrasonic range finder modules
Rotation	Sensor	Control	Digital rotary encoder modules
Signal	Output	Module	Waveform generator modules
Temperature	Sensor	Solid state	DS18B20 temperature sensor modules
Temperature	Sensor	Module	DHT11/DHT22 temperature and humidity sensor modules
Temperature	Sensor	Module	Thermistor temperature sensor modules
Tilt	Sensor	Switch	Single-axis tilt sensor modules
Tilt	Sensor	Switch	Dual-axis tilt sensor modules
Time	Support	Module	DS1302 RTC modules
Time	Support	Module	DS1307 RTC modules
Time	Support	Module	DS3231 RTC modules
Time	Support	Module	PCF8563 RTC modules
Voltage	Output	Module	DAC modules
Water	Sensor	PCB	Water conductivity sensor modules

Sensors

While it is possible to connect many sensors directly to an Arduino, modules like the ones listed in the previous section are definitely easier to deal with. But modules might not be a good choice if you want to create custom hardware for a specific application. In that case you will want to use just the sensor component and place it exactly where it needs to be.

A sensor is always an input device that acquires data from the physical environment and converts it into a form that a microcontroller can process. As the name implies, it senses something, where that something might be temperature, humidity, magnetic fields, visible light, heat, infrared, sound, or physical motion.

Temperature, Humidity, and Pressure Sensors

There are multiple choices available for environmental sensors that can be used with an Arduino. From simple continuity-based water detection sensors to sensitive temperature and humidity sensors, chances are there is a sensor that can measure how hot, how cold, how wet, or how dry the environment happens to be.

DS18B20

The DS18B20 is a so-called "one-wire" temperature sensor that returns a stream of binary data containing the current temperature at the sensor. Figure 9-11 shows a commonly available module with a DS18B20, some passive components, and an LED.

Figure 9-11. Typical DS18B20 module

Figure 9-12 shows the module mounted on a prototype shield along with a relay and a potentiometer. This, by the way, is a simple digital thermostat that was cobbled together to replace a dead bimetallic strip electromechanical thermostat in a small portable electric heater. The old-style mechanical thermostat never worked very well, but this simple digital replacement does an excellent job. The Arduino that it connects to also provides the ability to log temperature data and perform other functions. A cheap heater in the Internet of Things? Sure, why not?

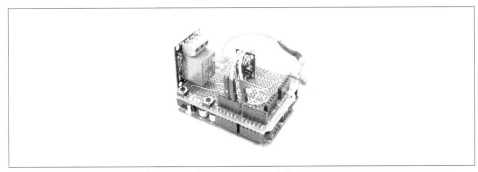

Figure 9-12. Example application of DS18B20 module

DHT11 and DHT22 sensors

The DHT11 and DHT22 temperature and humidity sensors come in three-terminal plastic packages, as shown in Figure 9-13. They can also be found already mounted on a small PCB, as shown in Figure 9-14.

Figure 9-13. DHT11 package

Figure 9-14. DHT11/DHT22 module

The DHT11 and DHT22 sensors differ in terms of resolution and serial data rate. The DHT11 is a basic device with a temperature sensing accuracy of +/- 2 degrees C, and a humidity resolution of +/– 5% relative humidity. It has a working range of 20 to 90% relative humidity, and 0 to 50 degrees C. The DHT22 features +/– 0.2 degrees C temperature resolution and +/– 1% RH sensing capability, and a faster serial data timing rate. The DHT22 has a wider sensing range than the DHT11: –40 to 80 degrees C and 0 to 100% RH. The pinout of the DHT22 is identical to the DHT11.

Both the DHT11 and DHT22 employ a nonstandard one-wire serial communications protocol. It works using a signal-response approach. The microcontroller pulls the single signal line low for a brief period of time, followed by allowing it to go high (via a pull-up resistor, either added externally or on the PCB module). The DHT11/DHT22 responds with 40 bits of serial data, organized as five 8-bit data values, including an 8-bit checksum.

Thermistors

A thermistor is a temperature-controlled resistor. They are available with either a negative or positive temperature coefficient. A negative temperature coefficient (NTC) device will exhibit a lower resistance as the temperature increases. A positive temperature coefficient (PTC) device has the opposite behavior. NTC thermistors are the most common types found in sensor applications, and PTC types are often used as current inrush limiters.

Figure 9-15 shows one way to connect a thermistor to an Arduino, but be aware that the response curve of the thermistor is not linear. Some circuits replace the fixed resistor with a constant current source.

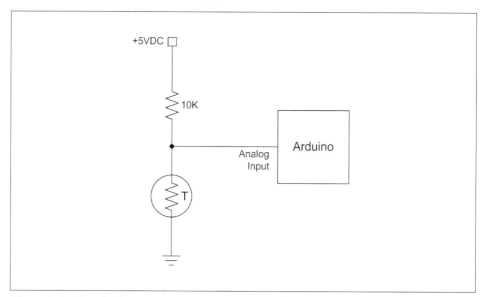

Figure 9-15. Simple thermistor connection to an Arduino

Although it is possible to connect a thermistor directly to an Arduino, an easier way is to use a module that includes the passive components necessary to create a voltage divider for the thermistor, such as the KY-028 shown in Table 9-7. You could also use the op amp circuit shown in Figure 9-26 to increase the sensitivity of the thermistor. Note that since most thermistors used for temperature sensing are NTC types, the voltage that appears on the analog input of the Arduino will drop as the temperature increases.

Water sensors

A water sensor is useful for many applications, from a flood sensor in a basement to a rain detector for an automated weather station. Table 9-8 shows one type of water detector available from SainSmart (the 20-011-946). This sensor incorporates an NPN transistor that will pull the output low when the thin traces on the PCB are connected by a water drop, or basically anything wet enough to cause the transistor to conduct.

Figure 9-16 shows the schematic for this sensor. As you can see, it's not complicated, and could be used with just about any conductive wires or probes. Connect this circuit to a pair of steel wires, mount the wires so they are about 1/4 inch or so (or about a centimeter) above the basement floor, and it could be used to trigger an alarm upstairs when water starts to flood the basement.

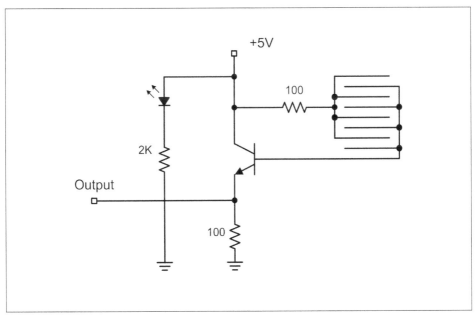

Figure 9-16. Water sensor schematic

Soil moisture sensors

In its simplest form a two-prong soil moisture sensor is really nothing more than a conductivity probe. The probe is configured to act as a component in a simple voltage divider, and the voltage that appears across it will be a function of the conductivity of the soil between the probes. You can purchase a kit consisting of a moisture probe, a small interface module, and a cable from suppliers like SainSmart. Figure 9-17 shows all three parts of such a kit.

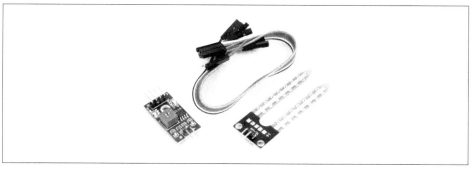

Figure 9-17. Soil moisture probe kit with interface module

The same effect could be achieved using almost any conductive material for the probes. For long-term use something like stainless steel or carbon rods might be a better choice where corrosion is a concern, and some type of amplifier or buffer is essential to get consistent readings without dumping a lot of current into the probe (which can cause some interesting side effects, and also help to corrode the probe's electrodes).

Another variation on the soil moisture sensor, a Grove 101020008 from Seeed Studio, is shown in Figure 9-18. This sensor includes an on-board NPN transistor to boost the voltage drop to a level an Arduino AVR ADC can work with. The copper layer on the prongs of this probe have been plated with a thin layer of gold to help resist corrosion.

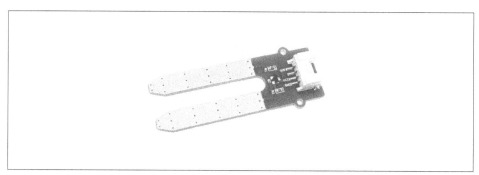

Figure 9-18. Self-contained soil moisture probe

The schematic for the self-contained moisture probe is shown in Figure 9-19. It is essentially the same as the circuit used with the water sensor, shown in Figure 9-16. Note that this design utilizes a four-terminal connector instead of the pin connections used on the probe in Figure 9-17. This is definitely more convenient, but pay attention to how the terminals in the connector are wired.

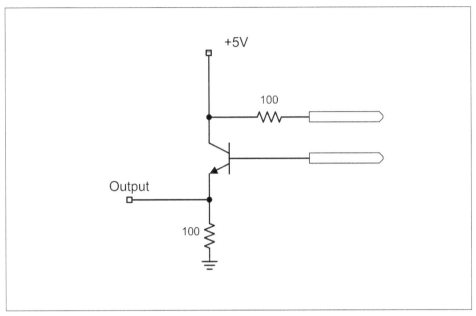

Figure 9-19. Self-contained soil moisture probe schematic

Barometric sensors

With a barometric pressure sensor, like the one shown in Figure 9-20, and the DHT11 or DHT22 modules described earlier ("DHT11 and DHT22 sensors" on page 313), an Arduino can be used to build a compact weather monitor. The sensor shown here is based on an MPL115A2 sensor with an I2C interface.

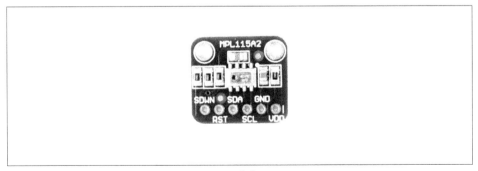

Figure 9-20. Barometric pressure sensor module

This particular module isn't accurate enough to be used as an altimeter, but it's fine for acquiring and logging weather data. Other modules based on sensors like the MPL3115A2 or BMP085 are accurate enough to be used as altimeters.

Tilt Sensors

A tilt sensor is typically nothing more than a small sealed capsule with a set of internal contacts and a small metal ball or bead of mercury inside. If the device is moved from its "neutral" orientation (perpendicular to the local gravitational pull), the ball or bead will move and close a circuit across the contacts. The operation is identical to closing a switch.

Bear in mind that a tilt sensor is not a proportional sensor. It is either tilted or not, so it is either open or closed electrically. There is nothing in between.

Single-axis tilt sensors

Figure 9-21 shows a module with a mercury bead tilt sensor, which happens to be a KY-017. This particular module will only sense tilt in one direction. You'll need to use two or more of them to sense a tilt along either end of an axis.

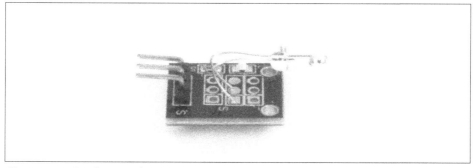

Figure 9-21. Single-axis tilt sensor

Dual-axis tilt sensors

The key to using a tilt sensor effectively is to determine the neutral position in a particular axis. Once this is known the sensor can be oriented to sense tilt along a particular horizontal axis. Figure 9-22 shows two tilt sensors on a single base to sense tilt in the x- and y-axes. Note that this arrangement will only sense tilt in one direction (either up or down, depending on how they are mounted).

If you want to sense up or down tilt in both axes, then four tilt sensors can be arranged at right angles to each other. This will detect tilt in both the +/– x and +/– y directions. Unlike a solid-state gyroscope (like those described in "Gyroscopes" on page 326), this type of circuit does not require a starting reference, and it will always work as long as there is gravity. The down side is that there is no in between: the tilt sensors are either on or off.

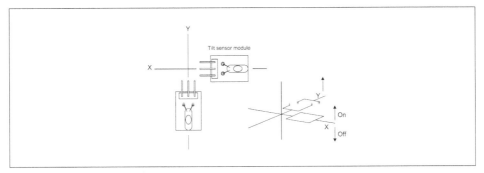

Figure 9-22. Dual-axis tilt sensor

Audio Sensors

A microphone can be used in a number of interesting ways. One way is to incorporate the microphone as part of a security system to detect loud noises such as breaking glass, someone kicking in a door, or the sound of a gun being discharged. If it is sensitive enough it can even be used to detect footsteps.

A contact microphone can be used to collect diagnostic data from an internal combustion engine or even an electric motor while it is running. It is possible to detect noisy bearings and loose components this way. When combined with an optical sensor an omnidirectional microphone can be used to build a lightning range detector (the thunder arrives about 4 seconds after the flash for every mile of distance from the observer to the lightning bolt).

There are small modules available with microphones and an IC, like the one shown in Figure 9-23. The miniature potentiometer is used to set the trip threshold of the circuit. This module uses a circuit similar to the one shown in Figure 9-8.

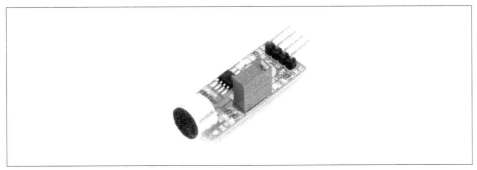

Figure 9-23. Audio pick-up module

It is also possible to connect a microphone directly to one of the analog inputs of an Arduino, although you wouldn't have any control over the sensitivity. A simple op-amp circuit like the one shown in Figure 9-26 can be used to improve the sensitivity.

Light Sensors

Sensors for detecting light come in a variety of styles and types. Some, like infrared (IR) sensors, can detect heat; some respond to the IR emitted by flames, others respond to visible light. A light sensor may employ a resistive element that changes its intrinsic resistance in response to the amount of light that falls on it. Others employ a semiconductor for increased sensitivity and fast response.

Photocells

A photocell, also known as a light-dependent resistor (LDR), is exactly what the name implies: a component in which the resistance changes as a function of the amount of light impinging upon it. Most look like the one shown in Figure 9-24. While these devices aren't all that fast by photodiode or phototransistor standards, they are still fast enough to carry audio in an amplitude-modulated beam of light. They are useful as ambient light level detectors, simple low-speed pulse-coded optical data links, beacon sensors for a robot so it can find a recharging station, and solar position trackers for a solar cell array.

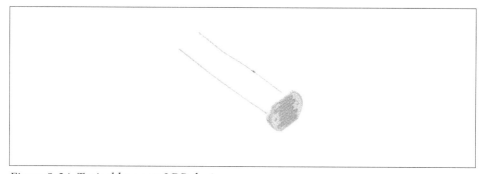

Figure 9-24. Typical low-cost LDR device

Photodiodes

Although most diodes have some degree of light sensitivity, a photodiode is built to enhance this effect. A photodiode is, as the name implies, a diode that has been manufactured such that it will go into conduction when light impinges on it, and a common type of photodiode is a PIN diode. The "I" stands for the layer of "intrinsic" silicon material between the P and N silicon parts of the diode, and that layer of intrinsic material makes a PIN diode a good light detector. Because they respond very quickly they are useful for optical data communications links and as position sensors for rotary mechanisms. PIN diodes are also found in high-frequency radio circuits, where they serve as switches. For more information about diodes and other solid-state devices, refer to the references listed in Appendix D.

Figure 9-25 shows how to connect a photodiode to an Arduino. Notice that the diode is reverse biased—in other words, it won't normally conduct current until it is exposed to light. While this circuit will work with a sufficiently bright light source, it may struggle with low illumination levels.

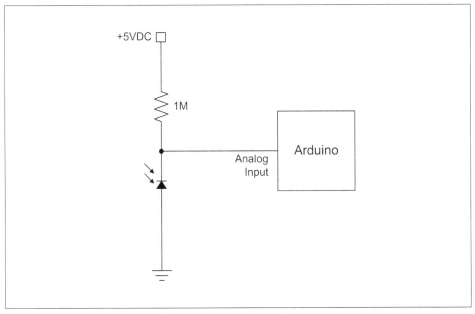

Figure 9-25. Simple photodiode connection to an Arduino

One way to improve the sensitivity is to use an op amp, as shown in Figure 9-26. In this case an LM358 op amp with a gain of up to around 10 is used to boost the small voltage change when the diode conducts to a level that the ADC in an Arduino can easily detect and convert. The trimmer potentiometer sets the gain of the circuit, so it can be adjusted to suit a particular application. This circuit can easily be assembled on a small solderless breadboard module, or the components can be mounted on a prototyping PCB (often called a "perf board") for a more permanent arrangement.

You can browse the selection of photodiodes (and op amps) available from electronics distributors such as Digikey (*http://www.digikey.com*), Mouser (*http://www.mouser.com*), and Newark/Element14 (*http://www.newark.com*). Many surplus electronics suppliers also have stocks of photodiodes on hand.

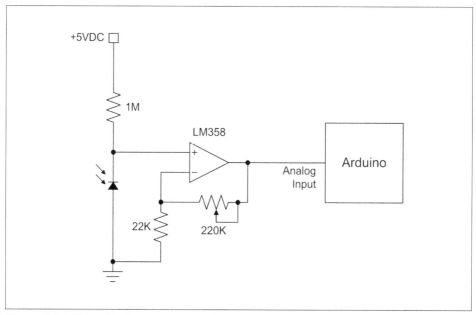

Figure 9-26. Photodiode connection with an op amp

Phototransistors

As the name implies, a phototransistor responds to light by varying the amount of current passing through the device, just as the base input lead would otherwise do. The circuit shown in Figure 9-8 will work with a phototransistor. Some modules, like the KY-039, simply bring out the transistor's leads to the connector pins (you can clip off the LED and use the module as a phototransistor sensor module, by the way).

You can expand on the basic circuit by adding an op-amp for some gain. Figure 9-27 shows how this can be done. Just about any garden-variety NPN phototransistor will work, but I happen to like the BFH310, mainly because I got a great deal on a large bag of the things (check with an electronics distributor such as DigiKey or Mouser for availability).

Common optical interrupters, like the one used in the KY-010 module, employ a phototransistor to sense the output of an LED. When something blocks the light by entering the gap in the component, the transistor will cease to conduct. Optical isolators (also called optocouplers or opto-isolators) also utilize an LED–phototransistor pair to couple a signal from one circuit to another without a direct electrical connection. You can build your own coupler with an LED, a phototransistor, and a section of black heat-shrink tubing to cover it all and keep out stray light.

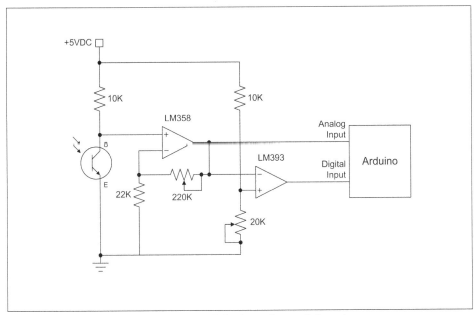

Figure 9-27. Phototransistor circuit with both analog and digital outputs

PIR sensors

A PIR (passive infrared) sensor measures the amount of infrared in its field of view. These are popular for security systems because they can usually detect very small changes in the ambient IR "glow" of a room. If the IR level deviates from the ambient baseline (such as when a warm human enters the room), the device will emit a signal. A PIR sensor can also be used to obtain a rough measurement of the temperature of whatever is in the field of view. Figure 9-28 shows a commonly available PIR module.

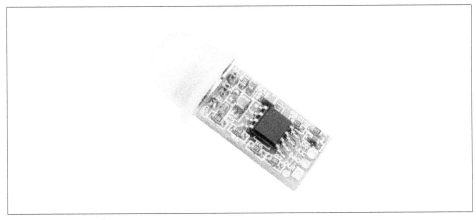

Figure 9-28. PIR detector

This module is available for around $2 from Banggood (*http://www.banggood.com*) and other distributors. It uses three connections: +5V, ground, and output. The output goes high when the sensor detects a change in the ambient IR. Combine this and an audio sensor, like the one shown in Figure 9-23, with the basic security system code described in Example 5-5, and you'll have a complete burglar alarm system.

Magnetic Sensors

One of the areas where semiconductor technology has made considerable strides is in the detection of static magnetic fields. Detecting an oscillating magnetic field, such as that produced by a coil, is relatively easy since all that is needed is another coil. Detecting a static magnetic field, such as that around a permanent magnet or the Earth's magnetic field, is a bit more challenging. Prior to the development of electronic devices magnetic field sensors often incorporated magnets, coils, mirrors, and other components. A standard camping compass is an example of old-style technology, and it works fine for a hiking trip or for detecting the magnetic field around a wire carrying a continuous direct current, but collecting data from it is rather tedious. These days you can build a magnetic field detector or an electronic compass with no moving parts that will interface directly with a microcontroller.

Hall effect sensors

A Hall effect sensor can detect the presence of a magnetic field. Some types, like the A3144 used in the KY-003, are designed to be on/off-type devices: there either is or is not a magnetic field present. Other types, like the SS49E, are linear types, with an analog output proportional to the sensed magnetic field. The KY-024 is an example of a module with a linear Hall effect sensor.

Both the A3144 and the SS49E look like small plastic body transistors, so I won't show them here. In the case of the A3144 and similar devices, you can connect it directly to an Arduino, which is essentially what the KY-003 module does. The KY-024 module uses the common comparator circuit shown in Figure 9-8.

Magnetometer sensors

Another form of magnetic sensor that can be useful with an Arduino is a compass, like the one shown in Figure 9-29. This unit from Adafruit uses an HMC5883L three-axis magnetometer. It features an I2C interface.

Vibration and Shock Sensors

Vibration and shock sensors are usually based on the detection of movement in a sensing mass of some sort. The sensing mass can be as simple as a mechanical arm with a small mass on the end and some contacts arranged so that the arm will close a circuit with one or the other as it deflects. Another variation might use an optical sensor that will change its output state when the arm breaks a beam of light. It might also be a spring-loaded sliding mass with contacts, optical sensors, or even a magnetic sensor. One type that is also used employs a piezo-electric sensor to detect movement in a sensing mass.

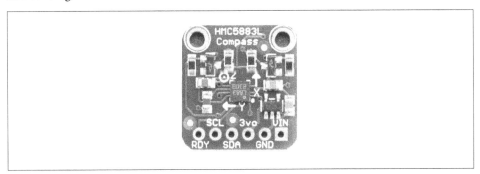

Figure 9-29. Magnetometer compass module

A very inexpensive type of vibration sensor uses a small conductive mass in a sealed enclosure. When the mass moves it will break and make contact with conductive sleeves at each end of the enclosure. The KY-020 is a typical module with a sensor of this type.

A shock sensor is similar to a vibration sensor, but in some cases shock sensors are designed to respond to specific input force levels in terms of some multiple of g (1g = the force of gravity at Earth's mean sea-level surface). The KY-031 is an example of a small low-cost shock sensor module.

You can build your own shock or impact sensor using a metal ball (a BB works well), a spring (perhaps from a ballpoint pen), a short section of plastic tubing, and some fine-gauge wire. Figure 9-30 shows how all the parts are assembled.

If you need more precision, industrial-grade shock sensors are available that are calibrated to specific force levels. These are used for things like automobile impact testing and testing the impact tolerance of shipping cases for delicate devices. They aren't cheap, however.

Motion Sensors

The ability to sense angular changes in position at varying rates is key to keeping things stable in three-dimensional space. Many quadcopters (or drones, as they are sometimes called) incorporate some form of multiaxis motion sensing to simulate the operation of a true mechanical gyroscope or inertial management unit (IMU). These types of devices are also popular with RC airplane and helicopter enthusiasts, and some adventurous souls have even incorporated them into model rockets to track and log the motion of the rocket during powered flight.

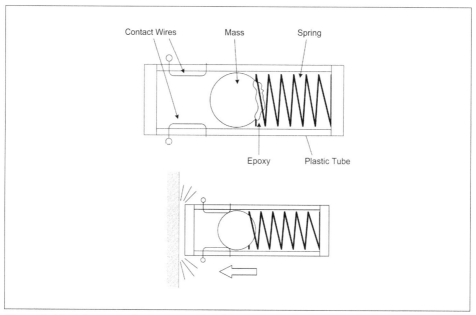

Figure 9-30. A homemade impact or shock sensor

Recent advances in technology and manufacturing techniques have driven down the prices of these devices to previously unimaginable levels. A solid-state rate gyroscope or accelerometer that once cost upwards of $50 can now be had for around $10. With the low cost, and the fact that the IC components are very small surface-mount packages with fine pitch leads, it makes more sense to buy a module rather than attempt to assemble something from scratch—unless, of course, you want to use it as part of something larger and you have the ability to deal with surface-mount parts.

Gyroscopes

The term "gyroscope," or "gyro," is something of a misnomer when applied to digital sensing devices. Unlike true mechanical gyroscopes, these devices are more like angular rate sensors intended to sense motion around an axis. They don't inherently reference an inertial starting position like a true gyroscope. An IMU can do that, but most electromechanical IMU devices are large, heavy things with three (or more) internal high-speed gyroscopes in precision ball-bearing gimbal mounts along with electric motors and position sensors. They also tend to be very, very expensive. However, with some clever programming and the use of multiaxis accelerometers (discussed next) it is possible to simulate an IMU. Figure 9-31 shows a three-axis rate gyro module, available from DealeXtreme, Banggood, and other suppliers.

Figure 9-31. Three-axis rate gyroscope module

Accelerometers

An accelerometer senses a change in velocity along a particular linear axis. When the accelerometer is moving at a constant velocity it will not detect any change, but when the velocity changes due to either acceleration or deceleration, the sensor will generate an output. A three-axis accelerometer arrangement will detect velocity changes in the x-, y-, and z-axes. Figure 9-32 shows an inexpensive MMA7361-based single-axis accelerometer module from DealeXtreme, but modules like this can also be obtained from Adafruit, SparkFun, and other suppliers.

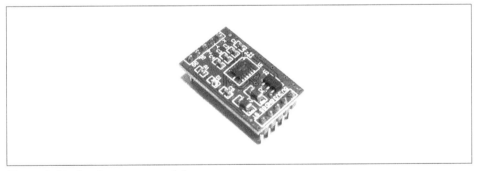

Figure 9-32. Accelerometer module

Contact and Position Sensors

Contact sensors are found in all manner of applications, from beverage bottling machines to computer-controlled tools in a machine shop. The heavy-duty pushbutton on an effects box like those used by musicians is a type of contact sensor. But regardless of how they are made, contact sensors all work on the same principle: either they are in physical contact with something, or they are not.

Position sensors, as the name implies, are used to sense the position of something. Unlike a contact sensor, a position sensor will usually have the ability to sense the degree of closeness or some amount of angular rotation. Some position sensors employ reflected light, others utilize sound, and still others incorporate a specially designed rotor and a light beam to measure angular motion. One other form of position sensor, called an absolute encoder, uses an internal glass disk with very fine marks to determine the precise degree of rotation of a shaft. Absolute encoders aren't covered here.

Contact switches

A contact switch can be as simple as a wire "whisker" made from a spring with a conductive post through the center, like those found on the little robotic bugs sold as children's toys. When the whisker is bent it causes the spring to make contact with the conductive post and closes the circuit. In other words, it's just a switch. Figure 9-33 shows a close-up view of this type of contact sensor.

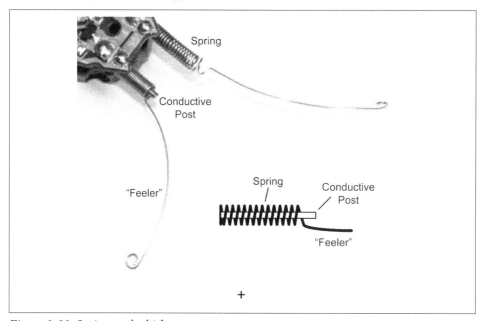

Figure 9-33. Spring and whisker contact sensor

At the other end of the spectrum are so-called snap-action switches like the one shown mounted on a module in Figure 9-34. This happens to be a Meeeno crash sensor module. These types of switches are often found in applications such as limit sensors for robotics or computer-numeric controlled (CNC) machine tools.

A pushbutton switch can also serve as a contact sensor. The old-style push switches used in automobiles to turn on the interior light when the door is opened make excellent contact sensors (although they might require a fair amount of force to activate them—more than the typical little robot can deliver). Even a copper strip and a screw can be used to sense physical contact, as shown in Figure 9-35. The main thing is that the switch closes (or opens, as the case may be) the circuit, which an Arduino can sense and then respond to.

Figure 9-34. Typical snap-action switch

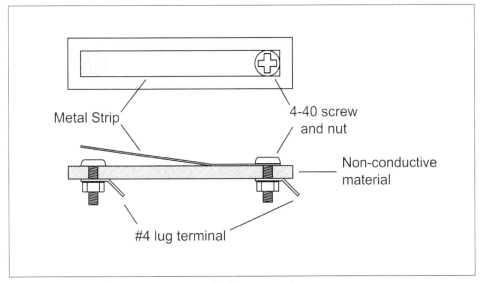

Figure 9-35. Simple contact sensor made from a metal strip

Digital rotary encoders

A digital rotary encoder, like the units shown in Figure 9-36, will generate pulses or emit a numeric value as a shaft is rotated. The KY-040 module also uses a rotary encoder. Some rotary encoders have detents on the shaft so that the operator will feel slight bumps as the shaft is turned. For applications that don't involve someone turning a knob, the shaft just moves freely and continuously. See Chapter 12 for an example of how to create software to read a rotary encoder like the KY-040.

Figure 9-36. Digital rotary encoders

Old-style computer mice used a coated metal ball instead of an LED, and two rotary encoders to detect the motion of the ball. These are becoming scarce, but they have some interesting parts inside. If you take one apart you can see the plastic wheels with evenly spaced slots. As the wheels move, the beam of light from an LED is interrupted. By sensing the timing of the pulses the small microcontroller in the mouse could generate numeric values indicating how far the mouse had moved in the x and y directions across a surface. Figure 9-37 shows the insides of a typical ball mouse.

Figure 9-37. Old-style mouse with a ball and rotary encoders

In this particular mouse the LED sender and phototransistor (or photodiode, perhaps) are two separate components. Other models of these old ball mice use optical interrupter components, like those used on the KY-010 module shown in Table 9-7. The interrupt wheels are driven by shafts that contact the sides of the metal ball. If you've ever used a ball mouse for any period of time you know that you must occasionally remove the ball and clean any accumulated stuff from the shafts (and the ball, as well). You can remove the interrupter wheels and the optical sensors and reuse them in something else—frankly, the interrupter wheels are probably the most useful parts in a ball mouse.

Laser transmitter/receivers

The module shown in Figure 9-38 is a short-range laser transmitter and receiver designed primarily for obstacle detection or any other task where reflectance can be used to sense an object. It could possibly be used as a data link, but the optics would need to be refined to achieve any significant range.

Figure 9-38. Short-range laser object detector

Range Sensors

The ability to both detect an object and determine its distance from the sensor is a key function in many robotics applications. A range sensor typically employs reflection, be it of light, sound, or radio waves in the case of radar. We'll cover light and sound here, as radar sensors tend to be rather pricey and work better over extended distances. For short-range sensing an optical or acoustic sensor works fine, and they're very inexpensive.

LED object sensors

An LED object sensor works by measuring the light from an LED (either optical or IR) reflected from a surface. These devices typically have an IR LED and a detector situated side-by-side, like the module shown in Figure 9-39. They don't measure the travel time of the light from the emitter back to the sensor, because at short ranges the speed of light makes that extremely difficult to do. If you want to measure the distance from the Earth to the Moon using one of the retroreflectors left by the Apollo missions, then a pulsed laser, and a big telescope, and a good high-precision timer will do the job. But for an Arduino, a reflective sensor will serve to keep a small robot from colliding with a wall or allow it to follow a line on the floor.

Ultrasonic range finders

If you're old enough, you might remember the old "instant" cameras that came with an ultrasonic range finder for automatic focusing and ejected a photograph that you could watch develop. The range finder typically consisted of a pair of piezoelectric sensors, one wired as an emitter and the other as a receiver. The emitter generated a short pulse of ultrasonic sound and the echo was detected by the receiver. With this kind of range finder, the time between the output of the pulse and the echo return is determined by the distance to whatever the sensors are pointing toward. This works because sound moves relatively slowly, so obtaining the time between the pulse and the return is not a particularly difficult thing to do with fast enough logic.

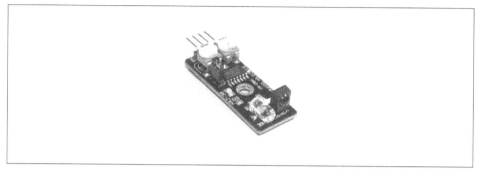

Figure 9-39. Reflective object sensor module

These days you can pick up an ultrasonic range finder for a few dollars that can be connected to an Arduino. An example of a readily available ultrasonic sensor is the 20-019-100 module shown in Table 9-8.

Communications

There are numerous modules for communications applications, ranging from plain RS-232 adapters to wireless communications and laser transmitter/receiver modules.

APC220 Wireless Modules

APC220 transceiver modules operate at 418 to 455 MHz. A complete digital link consists of two modules and an optional USB adapter. This allows for one module to be connected to a PC, and the other to be attached to an Arduino. The APC220 can transfer data at 19,200 bits per second with a range of up to 1,000 meters. The multifunction shield described in "Miscellaneous Shields" on page 268 comes with a connection point for an APC220 module, as does the 16 × 2 LCD shield from DFRobot (also covered in Chapter 8). Figure 9-40 shows a pair of APC220 modules.

315/433 MHz RF Modules

With a range of up to 500 feet (150 meters) these modules are a low-cost alternative to the APC220. The downside is that they are not transceivers, but come as a set consisting of a transmitter and a receiver as shown in Figure 9-41. They are available preset to either 315 MHz or 433 MHz. Note that you will need to add your own antenna.

Figure 9-40. APC220 RF transceiver modules (image source: DFRobot)

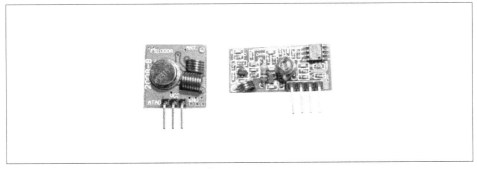

Figure 9-41. 433 MHz transmitter and receiver

ESP8266 Transceiver

This highly integrated WiFi module, shown in Figure 9-42, supports the 802.11 b/g/n protocols and uses a serial interface to communicate with an Arduino. The on-board 32-bit MCU has a TCP/IP network protocol stack in its firmware. It handles the low-level details of establishing and maintaining a digital link, so all the Arduino needs to do is specify an address to connect with, or wait for some other module to connect with it.

Figure 9-42. WiFi transceiver module

NRF24L01

The NRF24L01 module, shown in Figure 9-43, is a low-power transceiver operating at 2.4 GHz with about an 800 foot (250 meter) range. It uses an SPI interface to communicate with an Arduino. These modules can be purchased for around $3 from multiple sources.

Figure 9-43. NRF24L01 RF transceiver

RS-232 adapter

The AVR MCU devices used in Arduino boards have a built-in UART (or USART, if you want to follow Atmel's terminology), but it doesn't generate standard RS-232 signals. Rather than build a custom converter, an RS-232 adapter module provides the converter, IC, and a DB-9 connector, as shown in Figure 9-44. The RxD and TxD pins from the MCU connect directly to the module.

Figure 9-44. RS-232 adapter module

Output Devices and Components

An output from an Arduino can be an LED, a servo motor, a relay, or some other module, component, or device that responds to a signal or command from the AVR microcontroller on an Arduino. This section starts with light sources, followed by relays, motors, and servos. It also covers sound sources, like the KY-006 pulse-responsive speaker shown in Table 9-7. User input and output components are covered in later sections.

Light Sources

Light sources range from old-fashioned light bulbs in a wide range of sizes and types to LEDs, solid-state devices that act like diodes but emit a bright glow when current flows through them. We'll focus on LEDs in this section, mainly because they're inexpensive, they last a long time, and they don't always need a driver circuit. Even a tiny incandescent light bulb can draw a significant amount of current, and they tend to burn out and can get hot. That being said, there's no reason you can't use incandescent bulbs; just be prepared for additional circuit complexity (and cost) and the need to occasionally replace a dead bulb.

LEDs come in a wide range of styles, sizes, and colors. An AVR microcontroller can supply the 5 to 10 mA necessary to operate an LED, but it is generally not a good idea to directly connect a lot of LEDs or attempt to drive something like a high-output LED module. For that, a driver of some sort is the best way to go. The following images show some of the types of LEDs that are available, and they can also be found already mounted on module PCBs, like those shown in Table 9-7 and Table 9-9.

Single-color LEDs

Single-color LEDs range in size from miniscule surface-mount components like the D13 LED on an Arduino board to the huge devices used for lighting and illumination applications. Figure 9-45 shows a selection of some of the various types available.

Bicolor LEDs

A bicolor LED is basically just two LEDs mounted in a single package, which generally looks like a regular LED with an internal connection arrangement like that shown in Figure 9-46. The internal LEDs are connected backward with respect to one another. When current flows in one direction one of the LEDs will glow, and when the current is reversed the other LED will glow.

Another available type of bicolor LED employs three leads. One lead is common; the other two each connect to one of the LED chips inside the device's plastic package.

Tricolor (RGB) LEDs

A tricolor LED is comprised of three separate LED chips in a single molded plastic package. Figure 9-47 shows a surface-mount part. Large numbers of these can be mounted in an array on a single PCB, and multiple PCBs can be mounted side-by-side to create a large full-color LED display.

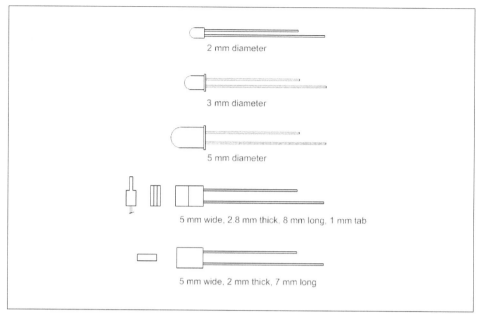

Figure 9-45. Common single-color LEDs

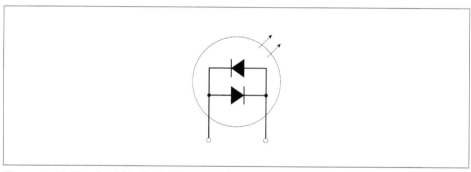

Figure 9-46. Two-lead bicolor LED internal connections

Figure 9-47. Surface-mount tri-color LED

A tricolor LED can produce an approximation of any visible color by varying the intensity of the output of each of the internal LEDs. Because the LED die (the individual LED chips) are physically separate, the colors are blended with a diffuser of some sort. From a distance it can look convincing. High-power, high-output RGB LEDs have been used to create huge color displays, like those seen on the sides of buildings or in large sports stadiums.

LED matrix

An LED matrix is useful for a variety of applications. An 8 × 8 matrix, like the one shown in Figure 9-48, can be used to display letters or numbers. If you arrange a lot of these modules side-by-side you can create a moving text display.

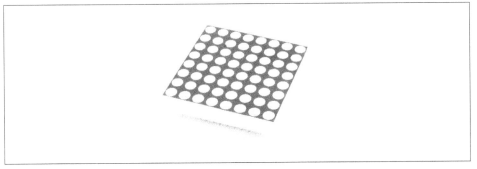

Figure 9-48. 8 × 8 LED matrix

Notice that the module is designed such that there is only a small amount of space at the side between the LED "pits" and the edge of the module. When these types of matrix modules are mounted side-by-side the gap between the last column or row on one module and the adjoining column or row on the next module is the same as the gap between the LEDs in the center of the module. This preserves the spacing when many modules are used to create large displays.

7-segment LED display

The venerable 7-segment display has been around for a long time. Other than single LEDs, this was the first viable application for LED technology, and by the late 1970s 7-segment and alphanumeric LED digit display modules were starting to show up in all types of applications. Figure 9-49 shows a typical four-digit display module.

As with the LED matrix module shown in Figure 9-48, the spacing gaps at the ends of this module are small so that multiple modules may be placed end-to-end. You could easily create a 12-digit floating-point display with three of these parts.

Figure 9-49. Typical 4-digit 7-segment display

7-segment LED modules

7-segment display modules are available with a built-in SPI or I2C interface. The module shown in Figure 9-50 is one such example. This particular module can be purchased from Tindie (see Table 9-11). There are also modules available with multiple displays, and they can be had in red, green, yellow, and blue.

Figure 9-50. 7-segment display module with SPI interface

Lasers

Some types of diode light sources are also lasers, like those found in laser pointers, while others are powerful enough to cut plastic or wood. A solid-state laser is essentially an LED with some internal tweaks to make it produce coherent light. Laser LEDs are available with output wavelengths ranging from infrared to blue. Without these devices things like laser levels for construction work, pointers for lecturers and instructors, surface profilers for 3D modeling, CD and DVD players and recorders, and some types of industrial cutting tools would not be possible. A typical small LED laser module is shown in Figure 9-51 (this is a KY-008). The laser LED can be purchased as a separate component from electronics distributors.

Figure 9-51. Typical low-power red LED laser

Relays, Motors, and Servos

Many small relays can draw more current to energize the internal coil than the AVR microcontroller on an Arduino PCB can safely deliver. Motors and servos are examples of very useful devices that often need a driver component of some sort to deliver the current they need to operate.

Relays

Relays come in a wide variety of sizes and shapes, ranging from small packages that look like a 14-pin DIP IC to huge things for controlling high-current loads in industrial equipment. For most Arduino applications a small relay is all that is needed. A small relay can operate a larger relay, which in turn can operate an even larger relay, and so on. The KY-019 is an example of a relay module that can be connected directly to an Arduino.

A relay driver can be as simple as a 2N2222 transistor, or it might be an IC designed specifically to deal with the current and reverse spikes encountered with relays. The circuit in Figure 9-5, shown earlier, uses an NPN transistor to control a small PCB-mounted relay.

Figure 9-52 shows a module with four relays. The PCB also includes the driver transistors, resistors, and diodes needed. All that is required is a source of 5V DC to drive the coils in the relays and standard logic signals to control them.

Servo control

The term *servo* typically refers to a small device something like a motor, although at one time it used to refer to large bulky actuators used for things like positioning guns on a naval vessel or performing analog calculations. Typical small servos, like the ones shown in Figure 9-53, are intended for use in radio-controlled cars and aircraft, as well as hobbyest robotics.

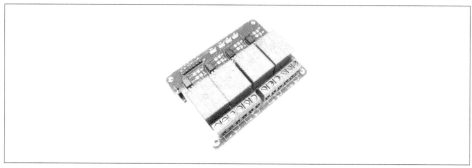

Figure 9-52. Relay module with four PCB-mounted relays

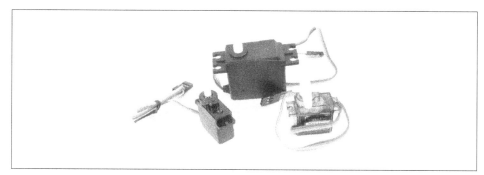

Figure 9-53. A selection of small servos

Although shields are available for use with servos (see Chapter 8), an AVR microcontroller can drive these devices directly with its PWM outputs. A servo like those in Figure 9-53 rotates a drive shaft through 180 degrees, with the amount of rotation determined by the pulse width and frequency of a control signal.

DC motor control

DC motors are commonly controlled with what is called an "H-bridge" circuit, and Figure 9-54 shows a simplified diagram. An H-bridge can be used with continuous DC or a PWM signal, and depending on how it is driven the motor can run in either forward or reverse.

I wouldn't recommended building a motor control circuit from scratch (unless you really want to, of course). A shield like the one shown in Figure 9-55 has everything needed to control a DC motor. It also has a heat sink to dissipate the heat generated with high current loads. This particular shield is from Seeed Studio (*http://bit.ly/seeed-motor-v2*). Note that the shield will control two DC motors or one stepper motor.

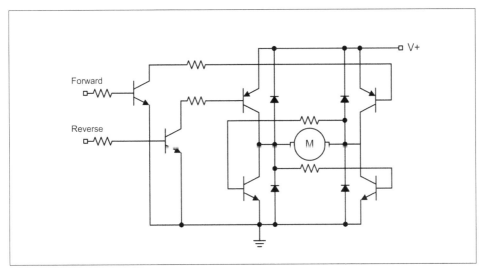

Figure 9-54. Simplified H-bridge

Figure 9-55. Motor control shield

Stepper motor control

Stepper motors are relatively simple to control, provided that the necessary electronics are in place to generate the pulses that will cause the motor shaft to rotate. One requirement is current, and while an IC like the ULN2003A provides everything necessary to drive a small stepper motor, it won't handle large motors with high current demands. The ULN2003A is basically just an array of eight Darlington transistors, so the Arduino has to take care of all of the timing for the motor pulses.

There are shields available (see Chapter 8) that contain circuitry for one to four stepper motors, along with connectors to make the job of wiring them up a bit easier. As with DC motors, using something that is already built is much easier than building it from scratch, and if you take your time into account, it's probably less expensive as well.

Analog Signal Outputs

Analog signals refer to the continuously variable cyclic phenomena we often refer to as sound, as well as those signals well above the range of human hearing, such as radio. There are multiple ways to use an Arduino to produce sine wave–type analog signals, all of which require some additional external components. An AVR microcontroller does not incorporate a digital-to-analog converter (DAC) in its design, so if you want anything other than square waves from a PWM or timer output, you'll need something to generate the signals.

Buzzers

Buzzers can be simple things that emit a fixed-pitch tone when active, or they can be a bit more sophisticated and generate a programmable pitch. The KY-006 and KY-012 are examples of these types of audio sources.

DAC modules

One way to give an Arduino digital-to-analog capability is with a DAC module, like the unit shown in Figure 9-56. This particular item is from Adafruit, and it is based on the MCP4725 IC, which is a single-channel 12-bit DAC with an I2C interface.

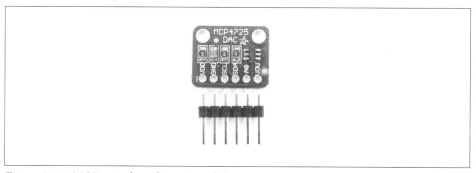

Figure 9-56. MCP4725-based DAC module

A DAC is useful when there is a need for a continuously variable signal, such as a control voltage for some other circuit or device. A DAC can also be used to generate a waveform such as a ramp or sine wave.

With the 100 Kb/s communications rate of standard mode I2C used with an AVR microcontroller it isn't possible to achieve the update rate possible with faster I2C interface types, but a DAC can still generate a respectable low-frequency sine wave. The downside is that the AVR will usually be doing nothing else but updating the DAC to produce a waveform.

Waveform generators

While it is possible to generate low-frequency waveforms directly with an Arduino and a DAC module, for quality waveforms beyond about 1 KHz an outboard circuit is necessary. One such device is the direct digital synthesis (DDS) module shown in Figure 9-57.

Figure 9-57. AD9850-based DDS module

The AD9850 IC can generate both square and sine waves from 1 Hz to 40 MHz. You can download the datasheet for the AD9850 from Analog Devices (*http://bit.ly/ ad9850-data*). The AD9850 uses its own unique interface, and can be controlled using either an 8-bit parallel or a serial interface. Arduino libraries for the AD9850 are readily available.

User Input

Sometimes its is necessary for a human to interact directly with an Arduino project, and this means pushbuttons, knobs, keypads, and joysticks. In addition to the modules described in this chapter, you can also purchase the bare components and mount them as you see fit.

Keypads

The term "keypad" usually refers to a small arrangement of switches with key caps, typically in a 3 × 3, 3 × 4, or 4 × 4 grid. It can also refer to a so-called membrane keypad, which is an array of thin membrane switches on a PCB. The keys are typically marked with letters and numbers, like the examples shown in Figure 9-58. Keypads aren't limited to small rectangular arrays, but can be found in a wide range of styles and layouts. In fact, a computer keyboard is just a large keypad.

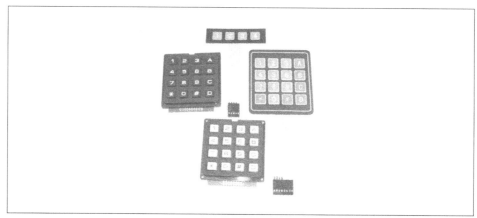

Figure 9-58. An assortment of keypads

Joysticks

Joysticks typically come as one of two types: continuous analog and discrete digital. A continuous analog joystick, like the one shown in Figure 9-59, uses two potentiometers, each connected to the x- and y-axes. The values read from the potentiometers indicate how far the joystick has moved and what position it is currently in.

Figure 9-59. Analog joystick module

A discrete digital joystick uses small switches or some other type of detector to sense when the joystick has moved to its maximum extent in either or both of the x or y directions. These types of joysticks were used with early low-cost consumer games and personal computers. They are all-or-nothing devices, but they are cheap to manufacture and they don't have the issues with dirt and wear that can affect an analog joystick. Many LCD shields incorporate a discrete joystick.

Potentiometers and Rotary Encoders

A potentiometer is a variable resistor, typically used for control input. A potentiometer, or pot, can be used in a light dimmer module, as a volume control, as a control input for a test instrument, or as an input for a wide variety of analog circuits. The TinkerKit T000140 and T000150 modules are examples of potentiometers.

A rotary encoder, like the one shown earlier in Figure 9-36, can also be put to use as a user input device. Instead of producing a variable voltage that must be converted into digital numbers so an AVR microcontroller can use it, the rotary encoder produces digital output that can be used directly.

User Output

The ability to display information to a user allows you to make something truly interactive. It might be simple status conveyed by LEDs, or complex messages or images on an LCD or TFT display screen. Whatever form they take, output devices give the user immediate feedback in response to command inputs.

There are a variety of displays available that can be used with an Arduino. The LCD display shields described in Chapter 8 utilize these same components, but in a more convenient form. There are, however, situations where a shield may not be appropriate, and in these cases an LCD display component that can be mounted in a specific fashion may be the better choice.

More examples of displays in the form of shields for Arduino boards are described in Chapter 8. Be sure to look there as well. Unless you absolutely must have a display with bare pins, a shield is a much easier way to go.

Text Displays

Some of the more common and inexpensive text-only displays have anywhere from 1 to 4 lines, with each line capable of displaying 8, 16, 20, 24, 32, or 40 characters. Of course, as the display density increases, so does the price. A 1-line 8-character display can be had for around $2, while a 4-line 40-character display goes for somewhere in the neighborhood of $18.

ERM1601SBS-2

The ERM1601SBS-2 LCD display is a 16 × 1 display with white characters on a blue background. A typical module is shown in Figure 9-60. These displays utilize an HD44780 or KS066 controller chip and LED backlighting, and this particular unit sells for around \$3. Similar products are available with black letters on a yellow-green background and black letters on a white background.

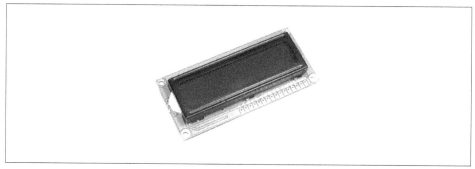

Figure 9-60. ERM1601SBS-2 display module

ST7066 (HD44780)

This is a 16 × 2 line LCD display with a simple parallel interface that uses either an HD44780 or an ST7066 controller. It is the same part as found on the LCD display shields from Adafruit, SparkFun, SainSmart, and other sources. Figure 9-61 shows an example of this type of display. These sell for about \$10.

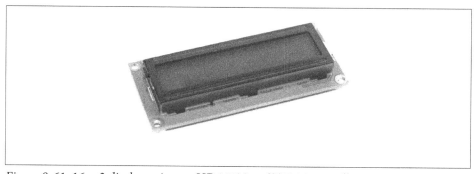

Figure 9-61. 16 × 2 display using an HD44780 or ST7066 controller

You may notice that the ERM1601SBS-2 shown in Figure 9-60 looks a lot like the display shown in Figure 9-61. This is because they both use the same ICs to drive the LCD. The only major difference is that one has a single-line display and the other has a two-line display.

Other types of both nongraphical and graphical LCD displays are available from a variety of sources. Occasionally you can find new surplus displays at very low prices, and if they use a standard controller chip they can usually be easily integrated into an Arduino project. The downside is that these surplus displays often have special-purpose symbols built in—so you might get a great deal but have a display with symbols for a microwave oven or a lawn sprinkler system that you don't need. (Or then again, maybe you do...)

Graphical Displays

Graphical displays are available using LCD, TFT, or OLED technologies in both monochrome and color formats. The TFT and OLED devices look much better than a simple dot-addressable LCD, but the prettiness comes at a price.

ERC240128SBS-1

The ERC240128SBS-1 display shown in Figure 9-62 is a 240 × 128 dot-addressable LCD display with an 8-bit parallel interface. It is available from BuyDisplay (*http://www.buydisplay.com*).

Figure 9-62. 240 × 128 color TFT display

ST7735R

An example of another type of display is the 128 × 160 TFT unit sold by Adafruit and shown in Figure 9-63. This display has a 1.8 inch (4.6 cm) diagonal size display and is capable of 18-bit color selection for 262,144 different shades. It uses an ST7735R controller with an SPI interface, and even includes a microSD card carrier on the back side of the PCB.

Support Functions

Most modules provide input or output functions. Not many have functions that could be classified as non-I/O support. Those that do generally fall into the categories of clocks and timers.

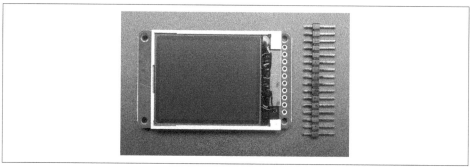

Figure 9-63. 1.8" color TFT display module

Clocks

There are multiple types of real-time clock (RTC) ICs available, including the DS1302, DS1307, DS3231, and PCF8563. They all do basically the same thing: keep track of the time and date. Some have on-board EEPROM and some do not. Some use a nonstandard interface, some use SPI, and others use I2C. Other than the obvious interface differences, the primary differences are largely a matter of accuracy as a function of stability over time, temperature sensitivity, power consumption, and cost.

Four common RTC modules encountered are those based on the DS1302, DS1307, DS3231, and PCF8563 ICs. The modules will usually have a holder for a "coin cell"–type battery, often a CR2032 or an LIR2032. Independent testing has shown that the DS3231 has the best overall long-term stability, but the other RTCs are perfectly usable. The modules with external crystals can suffer from temperature-induced drift, and all of them will deviate to some degree over extended periods of time.

DS1302 RTC module

The DS1302 RTC IC (Figure 9-64) uses a nonstandard serial interface. It's not SPI, but it is clocked. One line carries data, one is a clock signal, and another is the chip enable (CE) line. This is what Maxim (née Dallas Semiconductor) refers to as a three-wire interface. Figure 9-64 shows a typical DS1302 module. More information about the DS1302 is available from Maxim Integrated (*http://bit.ly/maxim-ds1302*).

DS1307 RTC module

The DS1307 RTC is an I2C device. It is not code- or pin-compatible with the DS1302, but the end result is the same. Refer to the datasheets from Maxim (*http://bit.ly/maxim-ds1307*) for details. Figure 9-65 shows a DS1307 RTC module from Tronix-labs (*http://tronixlabs.com*).

Figure 9-64. A DS1302 RTC module

Figure 9-65. A DS1307 RTC module

DS3231 RTC module

Like the DS1307, the DS3231 RTC uses an I2C interface, and it is code-compatible with the DS1307. The main difference is the accuracy. The DS3231 uses an internal crystal. This makes it less temperature sensitive. Figure 9-66 shows a typical DS3231 RTC module.

Figure 9-66. A DS3231 RTC module

RTC module using PCF8563

The PCF8563 is another RTC IC with an I2C interface. It is an NXP (*http://www.nxp.com*) part, and its internal registers are completely different from the Maxim DS1307 or DS3231. Figure 9-67 shows a typical module based on the PCF8563.

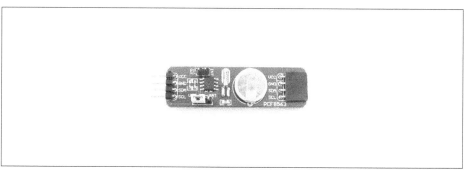

Figure 9-67. A PCF8563 RTC module

Timers

Although the AVR MCU contains a built-in watchdog timer, there are modules available that perform essentially the same function. One potential application for an external watchdog, or resettable countdown timer to be more precise, is when it makes sense for the timer reset signal to come from an external device instead of from the AVR itself. Consider a rotating mechanism fitted with a magnetic sensor that will emit a pulse for every revolution of the shaft. If the pulses are used to reset an outboard watchdog timer it can detect when the mechanism fails and stops turning. By either using an interrupt or just polling the status of the countdown timer the MCU can detect the fault condition and take appropriate action.

Many of the outboard timer modules seen in the wild use a 555 timer. A MOSFET is used to discharge the timing capacitor and reset the timer whenever a reset pulse is applied. Some other timer modules from Asian sellers use a black blob of epoxy to hide whatever is doing the actual timing on the PCB. I would avoid these, since there is no easy way of knowing what is inside without destroying the IC under the blob.

If you are interested in external countdown timer modules, I would suggest picking up a copy of Howard Berlin's book *The 555 Timer Applications Sourcebook* (see Appendix D). The original 1979 edition is out of print, but you can still find copies of it on Amazon, and there is newer, but slightly different, version available. The Australian company Freetronics (*http://www.freetronics.com.au*) sells an inexpensive watchdog timer module based on a 555 timer.

Connections

Over the past several years a trend has begun to emerge in module and shield inter-connect methods where modular connectors are replacing the connector pins and sockets once found on both shields and modules. Often referred to as *systems*, these involve a set of modules and some form of interface shield that all use the same connector types and pinouts for voltage, signal, and ground.

The TinkerKit modules (listed in Table 9-9) are just one example of a module connection system. Another is the Grove line of modules and associated base interface shields from Seeed Studio. Other shields feature three- and four-pin connectors for use with prefabricated cables with mating connectors at each end, such as the passive patch shield kit shown in Figure 8-21.

Working with Naked Jumper Wires

Let's say that you plan on permanently connecting a single module to an Arduino (perhaps it will be embedded somewhere and left to do its job for extended periods of time), and you don't want to take the time to build your own connector. That's fine; there is nothing wrong with using jumpers so long as you take some simple steps to make the connections more physically reliable.

The crimp socket (or pin) terminals used in jumper wires are the same terminals used in modular connectors. The difference is that a jumper wire has just one terminal, whereas a connector will have two or more. The more crimp terminals there are in a connector, the more robust it will be. This is due to the increased mechanical friction of multiple socket terminals all working together in the same connector housing.

A single jumper wire can wiggle and flex, and it doesn't have the mechanical grip afforded by a gang of terminals in a single connector housing. One way to achieve a more reliable connection is to apply a small blob of silicon rubber (also known by the brand name "RTV") to hold the jumper connectors in place on a module's pins. It might not be as elegant or robust as a modular connector, but unless the module is operating in a high-vibration environment like an RC vehicle or a machine tool in a factory, it will hold up just fine.

Just don't go crazy with the silicon, as you might want to remove it at some point and replace the module. A sharp razor knife will cut the soft silicon rubber and not damage the jumper (if you are careful, of course).

Module Connection Systems

In general, a shield with connectors—whether the open frame style used by TinkerKit or the closed shell types used by the Grove components and others—can also be used with either individual jumpers or crimped socket headers, just like the shields and modules with bare pins. Prefabricated connectors make it easier for someone to connect a module and not worry about how the pins are wired, so long as the module is designed to connect to a particular shield with the same types of connectors. This is the approach that the TinkerKit and Grove modules have taken. (More information about the Grove modules can be obtained from the Seeed Studio (*http://www.seeedstudio.com*) website.)

Other systems, like the TinyDuino modules from TinyCircuits, utilize small surface-mounted multipin connectors. Figure 9-68 shows a few examples of these types of modules. While technically not modules in the sense of the modules described in this chapter, this approach shows just one of the many ways to deal with the interconnect problem. TinyCircuits also produces sensor modules with the same type of connector, and they have extension cables available. These things are very small (look at the USB connector on a Nano and then compare that to Figure 9-68), and the first thing that popped into my mind when I saw them was "model rocket." More information is available at the TinyCircuits (*https://www.tiny-circuits.com*) website.

Figure 9-68. Example modules available from TinyCircuits

The connectors are the real problem with using modules and shields made for a particular interconnect system. There is no consistent standardization across all the different manufacturers, and no guarantee that a module from one vendor will just plug in to an interface shield from somewhere else. One way to address this is to create both the base interface shield and a selection of modules to go with it, which is exactly what the folks behind the TinkerKit products elected to do. When considering a particular module, pay attention to the connection method it uses. You should also be prepared to purchase additional cables, and perhaps some prototyping modules in order to interface with components not made for a paricular connector system. Or you could elect to build your own custom cables, which is covered in the next section.

Building Custom Connectors

Connectors are not only easier to use, they are also more robust and reliable than jumpers. But not having connectors isn't all that bad. An interface shield like the sensor interface shield from SainSmart shown in Figure 8-3 has the I/O pins arranged in neat rows with 0.1 inch (2.54 mm) spacing. A connector header, like those shown in Figure 9-69, with the same spacing and holes for crimp terminals will mate with the pins and make a solid connection.

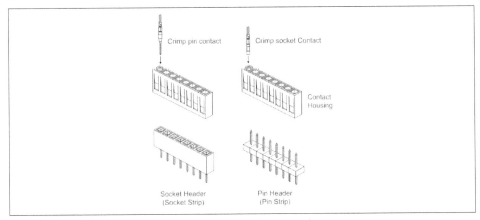

Figure 9-69. Pin and socket headers for crimped terminals

Connector shells and crimp terminals come in a variety of styles. The shells, or housings, are available with positions for one or more terminals. You may occasionally come across a jumper wire with single-position plastic housings at each end instead of the heat shrink insulation that is also used. I prefer the plastic shells, even if the jumpers do cost a bit more.

For those who just want to connect one or two modules and not worry about the jumpers coming loose and falling off, the socket header approach is an alternative worth considering. The downside is that you will need to invest in a crimping tool and a good pair of wire strippers. Figure 9-70 shows a crimping tool, and Figure 9-71 shows how it works.

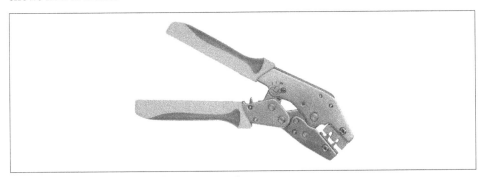

Figure 9-70. Typical terminal crimping tool

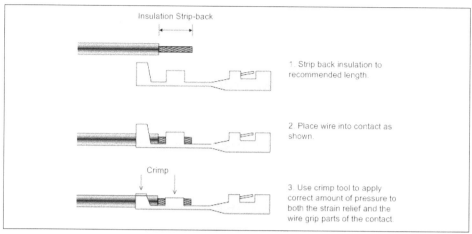

Figure 9-71. How a crimped terminal is attached to a wire

Figure 9-72 shows a three-position socket terminal header connected to a tempera-ture and humidity sensor module (a KEYES KY-015). This arrangement allows the sensor to be mounted where it is needed and connected to an interface shield. A small amount of clear silicon between the connecter housing and the module's PCB can be used to ensure that it won't easily come loose from the module.

Figure 9-72. Using a three-position header to connect a sensor module

Choosing a Connection Method

The trade-off boils down to either using modular connectors and prefabricated cables that are reliable and robust but require a matching set of components or making your own connections, either using jumper wires with crimped socket terminals or making your own socket header for a particular module. Which path you choose may depend on how much effort you want to expend to connect modules to an Arduino or a shield, and how "locked in" you want to be to a particular connection scheme.

You may have noticed that I didn't discuss one other available method for connecting modules: soldering. This is always an option, but unless a module is intended to become a permanent part of something it should be considered a last resort. Soldering a module directly to wires or onto a PCB makes a solid connection, but it's not easily undone and it can be ugly. It also means that you will have one less module to use in other projects.

That being said, you may have noticed in Figure 9-12 that the relay and temperature sensor modules are soldered directly onto a prototyping shield for the Arduino thermostat. The reasoning here is that the module will not be taken out of service any time soon, and it will operate in a somewhat nasty environment inside of an electric heater, with thermal fluctuations and fan vibrations. So, I elected to make it relatively permanent. While I may regret that at some point, for now it works just fine. You will need to make the soldering decision for yourself should the occasion arise.

Sources

The sources listed in Table 9-11 are just a sample of the companies selling Arduino-compatible components and modules. They are just the ones I happen to be aware of, and I've done business with most of them. There are many other sellers that I don't know about, but you may discover them if you look around on the Internet. And don't forget eBay.

Table 9-11. Parts sources

Distributor/vendor	URL	Distributor/vendor	URL
Adafruit	www.adafruit.com	Mouser Electronics	www.mouser.com
Amazon	www.amazon.com	RobotShop	www.robotshop.com
CuteDigi	store.cutedigi.com	SainSmart	www.sainsmart.com
DealeXtreme (DX)	www.dx.com	DFRobot	www.dfrobot.com
Seeed Studio	www.seeedstudio.com	Elecfreaks	www.elecfreaks.com
SparkFun	www.sparkfun.com	Elechouse	www.elechouse.com
Tindie	www.tindie.com	Freetronics	www.freetronics.com.au
Tinkersphere	tinkersphere.com	iMall	imall.itead.cc
Tronixlabs	tronixlabs.com	ITEAD Studio	store.iteadstudio.com
Trossen Robotics	www.trossenrobotics.com	KEYES	en.keyes-robot.com

Summary

This chapter has provided a survey of some of the various modules and components available for implementing input and output functions with an Arduino (or just about any modern microcontroller, for that matter). As was stated earlier, you can find many of the capabilities described here on a shield, but there may be times when a shield isn't appropriate. With an Arduino Nano and some sensors, input components, and some type of output, you can arrange things to suit your particular requirements.

With a web browser and access to the Internet you can easily locate hundreds of different I/O components that can be connected to an Arduino, or even just a bare AVR IC. Sources range from large electronics distributors like Digi-Key, Mouser, and Newark/Element14 to companies like Adafruit, SparkFun, SainSmart, and CuteDigi. There are also distributors that specialize in low-cost new and surplus components.

When it comes to purchasing sensors for an Arduino project, there are a few points to keep in mind:

1. Does the device use a simple interface (discrete digital, SPI, I2C, etc.)?
2. Does the device come with technical data (or is it readily available)?
3. Are Arduino software libraries available?

Depending on your programming skill level, item 3 may or may not be a dealbreaker for you. Personally, I put much more emphasis on items 1 and 2, mainly because I have better things to do than reverse engineer a complex interface for some really cool-looking thing I picked up for sale through eBay. For me, it makes more sense to buy the not-so-cool thing and get the technical information I need to get it up and running. The main thing is getting something that will do the job, and at a price that is within your budget.

Creating Custom Components

The more you work with Arduino devices in general, and the AVR microcontroller in particular, the more you may come to realize just how flexible and versatile they are. There seems to be a sensor or shield for almost any application you might imagine, and new shields appear on a regular basis. Even so, there are still a few applications for which there is no shield. There may be times when you spend hours online searching fruitlessly for a shield with specific capabilities, only to finally realize that it just doesn't exist. In that situation you basically have three choices: first, you could just give up and try to find another way to solve the problem; second, you might find someone to build it for you (for a fee, usually); or third, you could design and build your own PCB. This chapter describes two projects that illustrate the steps involved in creating a custom shield and an Arduino software–compatible device.

The first project is a shield, shown in Figure 10-1, that is intended for a specific range of applications. The GreenShield, as I'm calling it, is based on a conventional shield form factor. It utilizes surface-mount components in a layout that includes potentiometers, relays, and LEDs.

When coupled with an Arduino the GreenShield will be able to function as a standalone monitor and controller for gardening or agricultural applications. This shield can also serve as the foundation for an automated weather station, a storm warning monitor, or a thermostat (note that a programmable thermostat built using readymade modules and sensors is described in Chapter 12).

The second half of this chapter describes the Switchinator, an AVR ATmega328-based device that doesn't rely on the Arduino bootloader firmware, but can still be programmed with the Arduino IDE and an ICSP programming device. The Switchinator PCB is shown in Figure 10-2.

Figure 10-1. The GreenShield

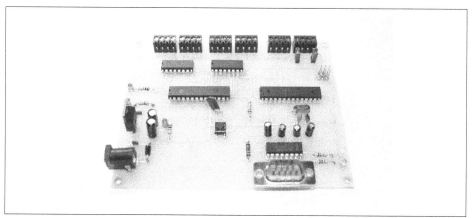

Figure 10-2. The Switchinator

The Switchinator can remotely control up to 14 DC devices such as relays or high-current LEDs, drive up to 3 unipolar stepper motors, or control AC loads using external solid-state relays. It uses an RS-232 interface and does not require a connection to a host PC via USB.

The Switchinator incorporates all of the essential components of an Arduino into a board of our own design. We won't need to worry about the socket header dimensions and layout considerations required for a shield PCB; the only constraints on overall size and shape will be those that we impose on the design.

With some patience and a plan you can easily create a custom PCB that doesn't look anything like an Arduino, but has the same ease of programming. Best of all, it will do exactly what you design it to do, and in a physical form that exactly meets your requirements.

Creating a PCB is not all that difficult, but it is a task that requires some knowledge of PCB design and electronics. There are low-cost/no-cost software tools available to handle PCB layout chores, and getting a circuit board fabricated is actually rather easy. The projects in this chapter will also require some soldering skill, particularly in the case of surface-mounted components. If you're already experienced in these areas, then you're almost there. If not, then learning how to create a schematic, work with PCBs, use a soldering iron, and select the right parts can be an enjoyable and rewarding experience.

For the shield PCB we will use the Eagle schematic capture and PCB layout tool, and for the Arduino-compatible PCB we'll use the Fritzing tool. Both of these are very common and capable tools. Best of all, they are free (well, Fritzing is free, and the entry-level version of Eagle, with limitations, is available at no cost).

 When creating custom shields or Arduino-compatible boards, you can do yourself a huge favor by keeping a notebook. Even if it's nothing more than a collection of printed or photocopied pages in a three-ring binder, you will be grateful if you need to look up some tidbit of information in the future for a similar project. Why not just save it all on disk in your PC? Because it can evaporate if the disk drive fails without a backup, or even get lost in the crowd if there are lots of things already stored on the drive (this happens to me more than I'd like to admit). Also, putting it into a notebook allows you to go back later and annotate things as you gain experience with testing, fabricating, and deploying your device. Red pens aren't just for high school English teachers.

This chapter also provides a list of resources for software, parts, and PCB fabrication. The only assumption I've made is that you may already have some electronics experience, or at least be willing to put in a little extra effort to learn some of the basics. Be sure to take a look at the brief overview of tools in Appendix A, and definitely check out the reading suggestions found in Appendix D. Finally, don't forget to peruse articles from websites like Hackaday (*http://hackaday.com*), Makezine (*http://makezine.com*), Adafruit (*http://www.adafruit.com*), SparkFun (*http://www.sparkfun.com*), and Instructables (*http://www.instructables.com*). You can also find tutorial videos on YouTube. Many others have been down these paths before, and many of them have been kind enough to document their adventures for the benefit of others.

 Remember that since the main emphasis of this book is on the Ardunio hardware and related modules, sensors, and components, the software shown is intended only to highlight key points, not present complete ready-to-run examples. The full software listings for the examples and projects can be found on GitHub (*https://www.github.com/ardnut*).

Getting Started

As with any endeavor worth devoting any significant amount of time to, planning is essential. This applies to electronics projects just as it applies to the development and implementation of complex software, building a house, designing a Formula 1 race car, or mounting an expedition to the Arctic. As the old saying goes, "Failure to plan is planning to fail."

Every project that is beyond trivial can be broken into a series of steps. In general, there are seven basic steps to creating an electronic device:

1. Define
2. Plan
3. Design
4. Prototype
5. Test
6. Fabricate
7. Acceptance test

Some projects may have fewer steps and some more, depending on what is being built. Here's some more detail about each step:

Define

In formal engineering terms this might be referred to as the requirements definition phase, or, more correctly, the functional requirements definition phase. A brief description of what the end result of the project will do and how it will be used is sufficient for simple things. A more detailed description will probably be needed for something complex, like a shield for use with an Arduino on board a CubeSat. But regardless of the complexity, putting it down in writing can help chart the course for the steps to come, and it can also reveal omissions or errors that may otherwise go unnoticed until it's too late to easily make substantive changes. Lastly, it's a good idea to write the project definition such that it can be used to test a prototype or a finished device. If the functional requirements don't clearly state what the device is supposed to do in such a way that someone could use this description to test the device, then it really isn't a good set of requirements and it doesn't define the desired device or system very well.

Writing functional requirements may sound like the equivalent of watching paint dry, but it's actually an essential aspect of engineering. If you don't know where you are going, how can you tell when you get there? A good requirement, of any type, should have four basic characteristics: (1) it should be *consistent* with itself and the overall design, (2) it should be *coherent* so that it makes sense, (3) it should be *concise* and not overly wordy, and (4) it must be *verifiable*. A functional requirement that states that "The device must be able to heat 100 milliliters of water in a 250 ml beaker to 100 degrees C in 5 minutes" is testable, but a statement like "The device must be able to make hot water" is not (How hot? How long? How much water?).

Plan

Sometimes planning and definition can happen in the same step, if the device is something relatively simple and well understood. But in any case, the planning step involves identifying the information necessary for the design step and getting as much essential information assembled in one place as possible—things like component datasheets, parts sources, identification of necessary design and software tools, and so on. The idea is to go into the design step with everything needed to make good decisions about what is available, how long it may take to get it, and how it will be used.

The planning step is also where you make some educated guesses as to how long it will take to complete each of the upcoming steps: designing, prototyping, testing, fabrication, and acceptance testing. I say "educated guesses" here because you should have some idea of what will be involved after collecting as much information as possible, but these are guesses because no one has a crystal ball that can let them see the future. Unexpected things can happen, and sometimes it takes longer to complete some aspect of the project than could be realistically anticipated. This is just how things go in the real world. It's also why people who do project management for a living multiply their time estimates by a factor of two or three. It's better to overestimate and get it done early than to underestimate the amount of time necessary and deliver it late.

A planning tool is handy both for establishing a realistic schedule and as a way to gauge progress. My favorite tool for quick and simple scheduling is the timeline chart, also known as a Gantt chart in fancier form. These can be complicated affairs created using project management software, or simple charts like the one shown in Figure 10-3.

For many small projects the fancy charts are not necessary, and the added complexity is just extra work. The important things are: (1) all the necessary tasks are accounted for in the planning, (2) the plan is realistic from both time and resource perspectives, and (3) the plan has a definite objective and ending. One other thing to notice about the simple timeline chart is that some tasks start before a preceding task is complete, and the chart does not show the dependencies that would indicate a critical path. I have found that for small projects involving one or just a few people, this overlapped scheduling more realistically reflects how things really happen.

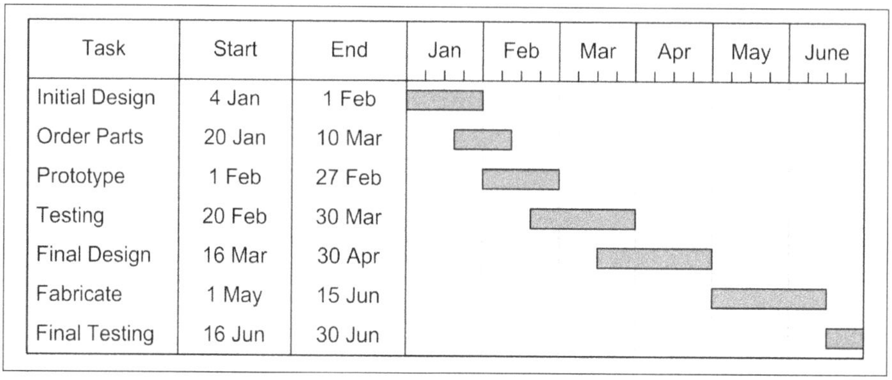

Task	Start	End	Jan	Feb	Mar	Apr	May	June
Initial Design	4 Jan	1 Feb	▨					
Order Parts	20 Jan	10 Mar	▨					
Prototype	1 Feb	27 Feb		▨				
Testing	20 Feb	30 Mar		▨				
Final Design	16 Mar	30 Apr			▨			
Fabricate	1 May	15 Jun					▨	
Final Testing	16 Jun	30 Jun						▨

Figure 10-3. Example timeline chart

Design

For a hardware project, the design step is where the circuit diagrams start to emerge and the design of the physical form takes shape. With the project definition and the planning information in hand, what needs to be done should be clear. While defining the circuitry is what most people think of when considering design, other significant activities in the design step include selecting components for form and fit, evaluating electrical ratings of components and environmental considerations (humidity, vibration, and temperature), and perhaps even potential RFI (radio frequency interference) issues.

When designing a new device there are always choices to be made. Sometimes the reason for choosing one method over another comes down to cost and availability of parts or materials. The intended functionality is another consideration, such as in the case where an input control may need to perform more than one function. Other times the choice might be based on aesthetics, particularly if there is no cost benefit of going one way over another. Lastly, in some cases it's just a matter of using something that is known and familiar rather than working with something unknown. This isn't always the best reason for making a choice, but it does happen quite often.

Regardless of the type of project (hardware, software, structure, or whatever), the design step is typically iterative. It's not realistic to expect that the design will just fall into place on the first attempt, unless perhaps it's something profoundly trivial (and even then, it's not a sure thing). Actually, design and prototyping (discussed next) work together to identify potential problems, devise feasible solutions, and refine the design. This is common in engineering, and while sometimes aggravating, iteration is an essential part of design refinement.

Changes and More Changes

During the development of the signal generator in Chapter 11, it went through several major revisions. The original plan called for the ability to have remote host control of all of the generator's functions via a serial RS-232 interface, and a parallel digital pattern output function. It turns out that there just aren't enough I/O pins on the Arduino to make all that happen without resorting to the use of I/O expansion and RS-232 shields. Other options, such as the use of a rotary encoder, could have been used to solve some problems, but each alternative brought challenges of its own into play. The instrument would still need pushbutton controls for some functions, and the LCD was becoming very cramped in terms of available display space.

So, rather than incorporate more complexity into both the hardware and software, I opted to keep it simple. The resulting design does not use interrupts in the software (no need for them), the LCD isn't overly crowded with cryptic data values, and only two main PCB components are needed: the Arduino board and the DDS module. The resulting device provides quite a bit of functionality as a signal generator, even if it doesn't have an arbitrary pattern output or remote host control.

Prototype

For simple things, such as a basic I/O shield with no active circuitry, building a prototype may not really be necessary (or even feasible). In other cases, a prototype can be used to verify the design and ensure that it performs according to the definition created at the start of the project. For example, it might not be a good idea to jump right into laying out a PCB for a device that combines an AVR processor, an LCD display, a Bluetooth transceiver, a multiaxis accelerometer, and an electronic compass, all on the same PCB. It might all work the first time, but if there are unforeseen subtle issues, they may not be apparent until after the PCB has already been made and paid for. Building and testing a prototype first can save a lot of aggravation and money later.

Problems identified with a prototype feed back into the design to help improve it. In an extreme case the prototype might even demonstrate that the initial design is just wrong, and you might need to start over. While this is annoying, it's not a disaster (it's actually more common than you might think). Building a hundred circuit boards only to find out that there is a fundamental design flaw—now that's a disaster. Prototyping, testing, and design revision can prevent that from occurring.

Test

The essence of testing is simply, "Does it correctly do what it's supposed to do, and does it do it as safely and reliably as intended?" The project definition created at the outset is the yardstick used to determine if the device has the desired functionality and exhibits the required safety and reliability. This is basic functional testing. It can get a lot more complicated, but unless you plan to send your design into space (or to the bottom of the ocean), or it will be controlling something that could cost a lot of money if something goes wrong (or damage something else, such as human beings), then basic functional testing should be sufficient for the prototype.

A word about the differences between "correct," "safe," and "reliable": just because something behaves in a correct way doesn't means it's safe (safe can also mean "operates without introducing unacceptable risk"), and something that behaves in a safe way isn't necessarily correct or even reliable. If a device won't turn on, it could be considered to be safe and reliable (it reliably will not do anything), but it definitely would not be correct. Lastly, to say that something is correct and reliable does not automatically mean that it's safe. A power tool such as a handheld circular saw may correctly and reliably cut lumber, but it will also correctly and reliably cut off a hand just as easily. An electrical device with an internal short circuit will reliably emit smoke (and perhaps even some flames) when power is applied, but the overall operation is neither safe nor correct.

Fabricate

After the prototype has been tested and has demonstrated correct behavior in accordance with the functional requirements, the design can move on to the fabrication stage. This step may involve fabricating a PCB and then loading it with parts. It might also refer to the integration of prefabricated modules and associated cables and wires into an enclosure or a larger system. In terms of possible delays, fabrication can be problematic. A supplier might be out of stock of a critical component, or if the assembly has been contracted, the assembly house might be having some problems or be overbooked. Custom-made parts might be late for any number of reasons.

Fortunately, if you are only making one or a few of something and you are doing all of the fabrication yourself, then you can avoid many of these potential problems. It still doesn't hurt to give yourself plenty of time, however. Making a run to the hardware store for a box of 3/8 inch 6-32 machine screws takes time, and if they're out of stock then it is going to take that much longer to try another store. Of course, if the planning and design steps were done with an eye toward what would be needed for fabrication, then you should have all the components, PCBs, nuts, bolts, screws, washers, connectors, wire, brackets, and glue you will need.

Acceptance test

This last step is also known as final testing, since it is the last thing to occur before the device is deemed ready to use or deploy. Some basic functional testing was already done with the prototype, but now it is time to test the final product. This is definitely not duplicated effort. It's all too easy to mount the wrong part on a PCB, or even install a part backward. Also, circuits on a PCB will sometimes behave differently than those built on a solderless prototyping block, so thoroughly testing the assembled device is always a good idea.

The testing is typically conducted in two steps. The first step verifies that the device or system behaves in the same manner as the prototype. The idea is to apply the same tests used with the prototype to verify that nothing has changed. In software engineering this is referred to as regression testing. The next step of testing involves additional tests to verify that the thing you've built works correctly in its final configuration with actual I/O. This is why this step is referred to as acceptance testing, and the main point is to answer the question, "Is the device or subsystem acceptable for its intended application?"

Custom Shields

There are basically three primary form factors to consider when designing an Arduino shield. These are the baseline, extended, and Mega pin layouts. The original or baseline (a.k.a. R2) layout, found on the Duemilanove, Uno R2, and other older boards, can be considered to be the standard for shield layouts, but that doesn't mean a shield can't be designed to utilize the extended (a.k.a. R3) layout found on later-model Uno and Leonardo boards. Shields can also been designed to utilize all of the pins on a Mega-style Arduino, and there's no reason why a shield has to have the same outline shape as an Arduino. Some shields, such as those with relays or large heat sinks, have a physical form factor suited to the components on the shield, rather than the Arduinos to which they are connected.

A novel approach that ignores size constraints is to create a large PCB with a set of pins arranged so that an Arduino can be connected in an inverted position. This might sound odd, but take a look at Figure 10-4. This is a board from a Roland SRM-20 desktop CNC milling machine. You can read more about it at Nadya Peek's *infosyncratic.nl* blog (*http://bit.ly/open-hardware-footbath*).

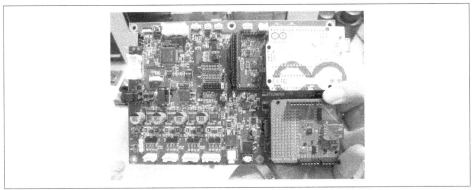

Figure 10-4. An inverted Arduino on a large PCB (image courtesy of Nadya Peek)

If you are designing a shield to sell commercially, then you may want to use the base-line layout as your template, since it will also work just fine with an Uno or Leonardo board, as well as a Mega-type PCB. Chapter 4 describes each of these board types and provides dimensions and pinout information.

The small form-factor Arduino boards such as the Nano, Mini, and Micro are unique in that they have all of the I/O pins on the bottom of the PCB, rather like a large IC. Adding a shield to one of these boards entails using an adapter, like the one shown in Figure 10-5, to bring out the signals to pin sockets for connecting a shield PCB.

Figure 10-5. Arduino Nano interface adapter PCB

You might notice in Figure 10-5 that pin socket headers have been added to the PCB. This was done largely as an experiment to see what would be involved in physically interfacing a Nano with a regular shield. With some creativity one could conceivably plug a shield into the board, except for the fact that the Nano sits too high on its socket headers. One solution would be to use extensions for the shield pins, which are just extended pin socket headers with some of the pin length removed. A more drastic solution would be to desolder the existing headers for the Nano and replace them with shorter types. I recommend the pin extension option. This chapter doesn't cover the steps involved in creating this type of adapter.

Using something like a Nano, Mini, or Micro Arduino with a shield actually doesn't make a whole lot of sense most of the time (although in Chapter 12 there is an example where this was done for a particular application). It *does* make sense to treat these small PCBs as if they were large ICs, and use them as components on a large PCB.

Physical Considerations

Be aware that on the Duemilanove PCB, and probably other boards as well, there are two surface-mounted capacitors that can collide with the extra pins found on the power and analog connector row of some shields designed for the R3 extended baseline layout. Some variants of the Uno also have components that can collide with shield pins.

Another consideration is that the type B USB jack found on the Duemilanove and Uno boards can potentially short out against a shield PCB. The DC power jack on these boards can also interfere with a shield PCB. Although it is plastic and won't short anything, it can prevent the shield from seating completely. Figure 10-6 shows this situation with a Duemilanove and an Ethernet shield.

For these reasons, it's a good idea to either size the length of the shield so that it won't interfere with the underlying Arduino, or design the component placement on the shield PCB to leave collision areas blank. See the next section, Chapter 4, and Chapter 8 for more on PCB dimensions and shield stacking.

When designing a shield you may also want to consider what will no longer be accessible on the Arduino under the shield and what can, or cannot, be mounted on it. This includes the reset button, the surface-mounted LEDs, and the ICSP pin group. Some shields deal with this by simply replicating the pinout of the Arduino board. Others, like most LCD shields, may not replicate the Arduino's pins to the top side of the shield PCB, but sometimes provide a reset button. With an LCD shield this makes sense, of course, since it would be the top shield in a stack in any case.

Figure 10-6. Duemilanove with shield

Stacking Shields

One of the nice things about the Arduino form factor is the ability to stack shields. You could create a stack containing a base Arduino board (an Uno, for example) with an SD memory card shield on top of it, followed by an input/output shield of some sort, and then an LCD or TFT display shield on top of all of that. Presto, you now have a basic data logging device.

What can be mounted above a shield should always be a consideration, be it another shield or perhaps some type of sensor module. When selecting components for a shield it is wise to consider the height of the various parts. If the parts are too high and another shield cannot be physically attached without interfering with something, then that shield will always need to be on the top of the stack.

There are basically two ways to allow for shield stacking: staggered socket and pin headers, and extended pin socket headers. Staggered, in this context, means that the upper connectors (socket headers) are offset from the lower pins (pin headers) by some amount, with both the upper digital I/O and power/analog headers shifted to the same side by the same amount.

In Figure 10-7 you can see a modular I/O shield on a Duemilanove. There are a few things to notice here. First, the modular I/O shield uses the staggered connector approach, so it is not vertically edge-aligned with the underlying Arduino board. This may need to be taken into account if the assembly will be mounted in some type of enclosure that might be subjected to shock or vibration—it may not be possible to secure the shield with nuts and bolts. Secondly, the shield does not bring out the ICSP pins, so that functionality is effectively lost (which may, or may not, be a big deal). Lastly, the pins of the connectors along the edge of the shield above the Arduino's power and USB connector can (and do) collide, so some type of insulating shim is necessary.

Figure 10-7. Example of a staggered shield

Extended pin connector sockets are a common variation on the 0.1 center connectors that have a long pin that goes through the PCB. The pin is long enough to make a solid connection with an underlying board, and the connector sockets provide for another shield board to mount on top and be in alignment with the shield below it. These are found on many shields, and you should look for them when selecting a shield.

A big consideration in shield design is Arduino pin usage. This is in addition to stacking concerns. The pins used by a shield determine what else can be used with that shield in a stack. Avoiding exclusive use of the SPI and I2C pins means other shields can be used in a stack, including SD and microSD flash memory, I/O expander, Bluetooth, ZigBee, Ethernet, and GSM shields. In other words, don't repurpose the SPI or I2C pins unless the shield design absolutely needs them.

Sometimes you may encounter shields that someone obviously thought were a good idea, but which don't always work out so well in practice. I/O shields often have multiple blocks of connector pins that are rendered inaccessible if another shield is placed on them. Other shields have solder pads that interfere with the ICSP pins on the Arduino PCB. Problems like these are not uncommon. Unfortunately it's not always possible to know in advance if there are problems with a shield, and sometimes the only way to know if a particular shield might have physical mounting issues is to purchase one and try it.

Electrical Considerations

If your custom shield has nothing but passive components (i.e., connectors, switches, resistors), power supply requirements probably won't be an issue for the board. However, if it has LEDs or active circuitry, it is a good idea to consider how much power it will need, and where it will get it. Even something as simple as an LED draws some power, and enough of them can overload an AVR processor and do some damage.

As a general rule, if a shield has one or more relays or connectors to attach things that could draw more than a few milliamps each, then some type of driver circuit or IC should be considered. It is possible to operate a shield from a separate power supply, but passing signals between the Arduino and the shield can sometimes be tricky.

If a shield has its own DC power source, then ground might also be something to consider. There are three ground sockets on a baseline Arduino: two on the side with the analog inputs and one on the digital I/O side. If you use only one of the Arduino's ground sockets for the signal ground reference for a shield, you can avoid potential problems with induced noise and ground loops. Although these situations are very rare, they are still a possibility, particularly for shields that may incorporate high-gain operational amplifiers or high-frequency circuits. Applying good design practices can help avoid strange and hard-to-diagnose problems in the future.

The GreenShield Custom Shield

In this section we will create a custom shield to illustrate the steps involved in the design and fabrication of an Arduino shield. The shield will be a humidity, temperature, ambient light, and soil moisture monitor. It is intended mainly for use in a greenhouse, although with the correct enclosure, some solar cells, and a wireless transceiver of some sort it could be put into a field to monitor the turnips (or whatever). As water gets scarcer in some parts of the world (including the Western United States), keeping an eye on the soil moisture content as well as temperature and humidity can help to minimize watering times and volumes while still keeping the crops healthy. The farmer can sit in the living room and get a quick readout of how things are doing in the field from a smartphone, tablet, or desktop PC.

I'm calling this the GreenShield, for obvious reasons, and the definition and planning steps are combined into one step. The GreenShield is physically very simple, and the main hardware design challenge will be fitting some large components (two relays and a DHT22) onto a small shield PCB.

Objectives

The goal of this project is to create a shield that can be used as an autonomous remote monitor to sense temperature, humidity, soil moisture content, and ambient light level. Based on predefined sensor input limits, it will control two relays.

The relays can be used to control a water valve, and perhaps a fan or maybe some lights to compensate for cloudy days. It also has six LEDs: two for high and low humidity points, two for high and low soil moisture levels, and an LED for each relay to indicate activity.

The software will support a command-response protocol and maintain an internal table of automatic relay functions mapped to sensor inputs and limits. Alternatively, a host control computer can obtain the sensor readings on demand and control the relays directly.

Definition and Planning

The shield will incorporate a DHT22 combination temperature and humidity sensor, a light-dependent resistor (LDR) sensor (see Chapter 9) to detect the ambient light level, and a conductivity-based soil moisture probe. Two relays will provide automatic or commanded control of external devices or circuits.

Sensor inputs:

- Temperature
- Relative humidity
- Soil moisture (relative)
- Ambient light level

Control and status outputs:

- Two control relays, software function definable, 10A control capability

- Four LEDs to indicate soil moisture and humidity limits
- Two LEDs to indicate relay status

Electrical interface:

- Two-position terminal block for moisture probe input
- Two-position terminal block for LDR connection
- One three-position terminal block per relay (NC, C, and NO)
- +5V DC supplied by attached Arduino board

Control interface:

- Command-response protocol, control host driven
- Sensor readings available on demand
- Relay override by control host

All of the components will be placed on a standard baseline-type shield, with dimensions as described in Chapter 4. Most of the components will be surface-mount types, with the exception of the terminal blocks, the relays, and the DHT22 temperature and humidity sensor.

Design

The GreenShield is intended to be used without a display or user controls. In other words, it and an Arduino will operate as an autonomous remote sensor and controller. It can be connected to another computer system (the master host system) to receive operating parameters, return sensor data, and override relay operation.

Autonomous in this case means the GreenShield will be able to operate the relays automatically when specific conditions are met, such as humidity, soil moisture, or light levels. The software will accept commands from a master computer to set the various threshold levels and override the operation of the relays. It will generate a response on command containing the current temperature, humidity, light level, and relay states. All interactions between the control host and a GreenShield Arduino will be command-response transactions.

The GreenShield software will be developed entirely with the Arduino IDE. The host computer used to compile and upload the finished code will also serve as the test terminal interface when the code is running on the Arduino. Ideally one would want to create a custom interface program using something like Python, or, in a Windows environment, a terminal emulator like TeraTerm (*https://ttssh2.osdn.jp/index.html.en*). It includes an excellent scripting facility, and I highly recommend it.

Eagle Schematic and PCB Tool

For this project I'm using the Eagle schematic and PCB tool. If you don't already have it, you can download it from *http://www.cadsoftusa.com/download-eagle*. Most major Linux distributions have an older version available in their package repositories. The limitations of the free version of Eagle are given on the CadSoft website as:

- The usable board area is limited to 100 × 80 mm (4 × 3.2 inches).
- Only two signal layers can be used (Top and Bottom).
- The schematic editor can only create two sheets.

The baseline Arduino dimensions are approximately 69 mm × 53 mm, so there is no problem with using Eagle for a baseline- or extended-type shield. It cannot be used to create a shield for an Arduino Mega-type PCB, due to the size limitation. The two layer limitation is usually not a problem for most shield designs, but in some cases shields that process video or RF signals may need additional layers for ground and power planes.

The folks at SparkFun have some clear and concise tutorials online to help you get the Eagle software installed and running. You can find those at *http://bit.ly/sparkfun-eagle* and *http://bit.ly/sparkfun-using-eagle*.

You can also get Eagle PCB library components from SparkFun's GitHub repository (*https://github.com/sparkfun/SparkFun-Eagle-Libraries*). I found that the SparkFun libraries didn't work with the version of the Eagle package used by Kubuntu 12.04 (the latest version of Eagle available is 5.12), but the newest version from CadSoft (version 7.4.0) will install and run just fine. I did it the lazy way and installed 7.4.0 with 5.12 already installed, copied the old Eagle executable in */usr/bin* to *eagle.old*, and created a symbolic link in */usr/bin* to point at the newer version in */opt/ eagle-7.4.0/bin*.

Functionality

The first consideration is the physical design of the shield PCB. The block diagram shown in Figure 10-8 gives an overview of what types of functions will be on the shield.

Note that in Figure 10-8 none of the hardware functions interact directly. The sensors, LEDs, and relays are just extensions of the Arduino's basic I/O capabilities.

The humidity/temperature sensor is mounted on the PCB, while the photocell (a light-dependent resistor) and the soil moisture probe can be located off-board if desired. Miniature terminal blocks are used for sensor connections, so no soldering or connector crimping is required.

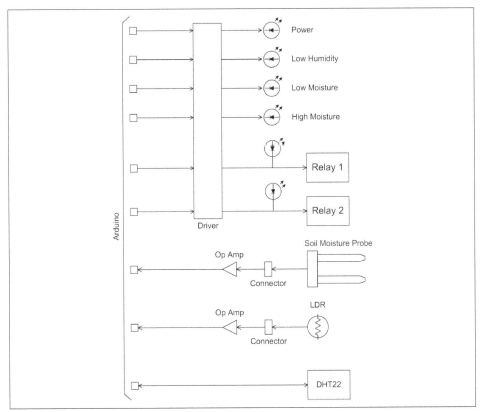

Figure 10-8. GreenShield block diagram

The GreenShield is intended to be the last (top) shield on a stack. This is due to the relays and the temperature/humidity sensor, all of which are tall enough to prohibit stacking another shield.

Hardware

Circuit-wise the GreenShield isn't very complicated, as can be seen in the schematic shown in Figure 10-9. A ULN2003A is used to drive status LEDs and two relays. A dual op amp is used to buffer the voltage level from a soil moisture sensor and an LDR for input to the AVR ADC.

The sensors connected to the op amp inputs are effectively variable resistors, and with the two trimmer potentiometers they form a voltage divider. The trimmers can be adjusted to achieve an optimal response from the op amp without driving it too far one way or the other voltage-wise.

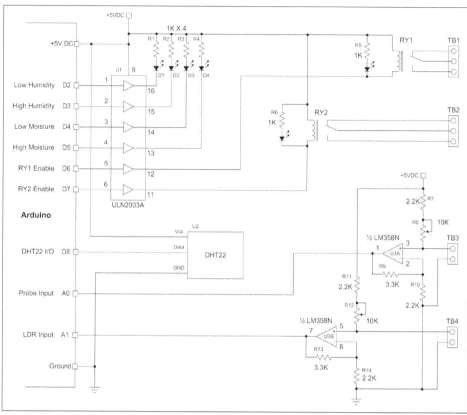

Figure 10-9. GreenShield schematic

The GreenShield has been designed to allow for extending the Arduino shield stack by avoiding critical digital and analog I/O pins. Table 10-1 lists the Arduino pins and assignments used for the GreenShield.

Table 10-1. GreenShield Arduino pin usage

Pin	Function	Pin	Function
D2	ULN2003A channel 1	D7	ULN2003A channel 6
D3	ULN2003A channel 2	D8	DHT22 data input
D4	ULN2003A channel 3	A0	Soil moisture sensor input
D5	ULN2003A channel 4	A1	LDR sensor input

Notice that the SPI pins—D10, D11, D12, and D13—are not used, so they are available for SPI shields. D0 and D1 are also available if you want to connect an RS-232 interface and forgo the USB. A4 and A5 are available for I2C applications.

Now that we have a schematic we can assemble a complete parts list, which is given in Table 10-2.

Table 10-2. GreenShield parts list

Quantity	Type	Description	Quantity	Type	Description
2	SRD-05VDC-SL-C	Songle 5A relay	4	2.2K ohm, 1/8W	Resistor
1	DHT22	Humidity/temperature sensor	2	3.3K ohm, 1/8W	Resistor
1	Generic	LDR sensor	2	10K trim	PCB mount potentiometer
1	SainSmart	Soil moisture probe	2	0.1" (2.54 mm)	3-position terminal block
6	3 mm	LED	2	0.1" (2.54 mm)	2-position terminal block
1	LM358N	Op amp	2	0.1" (2.54 mm)	8-position socket header
1	ULN2003A	Driver IC	2	0.1" (2.54 mm)	6-position socket header
6	1K ohm, 1/8W	Resistor	1	Custom	Shield PCB

Software

The GreenShield software is based on three primary functions: sensor input, command parsing and output generation, and relay function mapping. The first function is responsible for obtaining data from each of the four sensor inputs (temperature, humidity, soil moisture, and ambient light level) and storing the values for use by other parts of the software. The command parsing functions interpret the incoming command strings from a host PC and generate responses using the command-response protocol described here. The output functions control the relays based on the sensor inputs and preset limits defined by the various commands.

The command-response protocol used for transactions between the host computer and the GreenShield Arduino is shown in Table 10-3. Note that the GreenShield only responds to the host; it will never initiate a transaction on its own.

You can use the built-in serial terminal tool in the Arduino IDE, or you can exit from the IDE and connect directly to the USB port that the Arduino with the GreenShield happens to be using. This approach can be used to create a user interface application for setting up the GreenShield and monitoring its operation. In practice the idea is to configure the GreenShield software on an Arduino and then let it run unattended.

Status query commands

The GreenShield software provides four query commands. These are listed in Table 10-4. These commands allow a control host to get the current on/off state of either of the two relays, the last value read from either the LDR or the moisture sensor analog input, and the latest temperature and relative humidity reading from the DHT22 sensor.

Table 10-3. GreenShield command-response protocol (all commands)

Command	Response	Description
AN:*n*:?	AN:*n*:*val*	Get analog input *n* in raw DN
GT:HMX	GT:HMX:*val*	Get humidity max value
GT:HMN	GT:HMN:*val*	Get humidity min value
GT:LMX	GT:LMX:*val*	Get light max value
GT:LMN	GT:LMN:*val*	Get light min value
GT:MMX	GT:MMX:*val*	Get moisture max value
GT:MMN	GT:MMN:*val*	Get moisture min value
GT:TMX	GT:TMX:*val*	Get temp max value
GT:TMN	GT:TMN:*val*	Get temp min value
HM:?	HM:*val*	Return current humidity
RY:*n*:?	RY:*n*:*n*	Return status of relay *n*
RY:*n*:1	OK	Set relay *n* ON
RY:*n*:0	OK	Set relay *n* OFF
RY:A:1	OK	Set all relays ON
RY:A:0	OK	Set all relays OFF
RY:*n*:HMX	OK	Set relay *n* to ON if humidity >= max
RY:*n*:HMN	OK	Set relay *n* to ON if humidity <= min
RY:*n*:LMX	OK	Set relay *n* to ON if light level >= max
RY:*n*:LMN	OK	Set relay *n* to ON if light level <= min
RY:*n*:MMX	OK	Set relay *n* to ON if moisture >= max
RY:*n*:MMN	OK	Set relay *n* to ON if moisture <= min
RY:*n*:TMX	OK	Set relay *n* to ON if temp >= max
RY:*n*:TMN	OK	Set relay *n* to ON if temp <= min
ST:HMX:*val*	OK	Set humidity max value
ST:HMN:*val*	OK	Set humidity min value
ST:LMX:*val*	OK	Set light max value
ST:LMN:*val*	OK	Set light min value
ST:MMX:*val*	OK	Set moisture max value
ST:MMN:*val*	OK	Set moisture min value
ST:TMX:*val*	OK	Set temp max value
ST:TMN:*val*	OK	Set temp min value
TM:?	TM:*val*	Return current temperature

Table 10-4. GreenShield query commands

Command	Response	Description
RY:*n*:?	RS:*n*:*n*	Return status of relay *n*
AN:*n*:?	AN:*n*:*val*	Get analog input *n* in raw DN
TM:?	TM:*val*	Return latest DHT22 temperature
HM:?	HM:*val*	Return latest DHT22 humidity

Relay override commands

The relays on the GreenShield may be controlled via software commands. Four relay control commands allow an individual relay to be set to either on or off, or both relays may be set on or off at one time. Table 10-5 lists the relay override commands.

Note that when a relay is set using an override command any previous setpoint mapping is deleted. To use the relay again with a setpoint, one of the setpoint commands must be sent to the GreenShield.

Table 10-5. GreenShield relay commands

Command	Response	Description
RY:*n*:1	OK	Set relay *n* ON
RY:*n*:0	OK	Set relay *n* OFF
RY:A:1	OK	Set all relays ON
RY:A:0	OK	Set all relays OFF

Relay action mapping commands

The activation of either of the two relays may be mapped to a specific minimum or maximum setpoint condition for humidity, light level, soil moisture content, or ambient temperature. Table 10-6 lists the relay setpoint commands. When mapping an action to a relay, the most recent mapping command will override any previous command.

Setpoint commands

The minimum and maximum setpoints are defined using the ST commands, listed in Table 10-7. The setpoint value may be returned to the host control PC using the GT commands. The setpoint values may be modified at any time.

The relay function mapping associates a relay with a sensor input and a set of state change conditions in the form of upper and lower limits. A relay may be enabled if a sensor value is above or below a limit set by the host control system. Relay association is not exclusive, meaning that both relays could be assigned to the same sensor input and limit conditions. This might not make sense to do, but it can still be done.

Table 10-6. GreenShield relay setpoint commands

Command	Response	Description
RY:*n*:HMX	OK	Set relay *n* to ON if humidity >= max
RY:*n*:HMN	OK	Set relay *n* to ON if humidity <= min
RY:*n*:LMX	OK	Set relay *n* to ON if light level >= max
RY:*n*:LMN	OK	Set relay *n* to ON if light level <= min
RY:*n*:MMX	OK	Set relay *n* to ON if moisture >= max
RY:*n*:MMN	OK	Set relay *n* to ON if moisture =< min
RY:*n*:TMX	OK	Set relay *n* to ON if temp >= max
RY:*n*:TMN	OK	Set relay *n* to ON if temp <= min

Table 10-7. GreenShield min/max setting commands

Command	Response	Description
ST:HMX:*val*	OK	Set humidity max value
ST:HMN:*val*	OK	Set humidity min value
ST:LMX:*val*	OK	Set light max value
ST:LMN:*val*	OK	Set light min value
ST:MMX:*val*	OK	Set moisture max value
ST:MMN:*val*	OK	Set moisture min value
ST:TMX:*val*	OK	Set temp max value
ST:TMN:*val*	OK	Set temp min value
GT:HMX	GT:HMX:*val*	Get humidity max value
GT:HMN	GT:HMN:*val*	Get humidity min value
GT:LMX	GT:LMX:*val*	Get light max value
GT:LMN	GT:LMN:*val*	Get light min value
GT:MMX	GT:MMX:*val*	Get moisture max value
GT:MMN	GT:MMN:*val*	Get moisture min value
GT:TMX	GT:TMX:*val*	Get temp max value
GT:TMN	GT:TMN:*val*	Get temp min value

Although the GreenShield is currently configured for two relays, there is no hard limit on the number of relays that could be used. As can be seen from Table 10-1 the I2C pins (A4 and A5) are available, so an I2C digital I/O expander shield can be used to connect additional devices to the Arduino.

Prototype

To create the prototype for this project I'm using something called a Duinokit, which is shown in Figure 10-10. This clever thing has an array of sensors, LEDs, switches, and other accessories along with an Arduino Nano, all mounted on a large PCB with lots of socket headers. It also has a position for attaching a conventional shield (or stack of shields). It's like a modern take on the old all-in-one electronics project kits that were once popular.

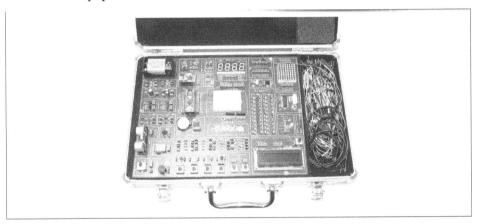

Figure 10-10. The Duinokit

An equally valid approach would be to assemble all the necessary components from a sensor kit and just about any Arduino, but the Duinokit keeps things neat and tidy, and it provides a nice development platform to use to create the software while waiting for the PCB and some of the other parts to show up. The Duinokit is available from *http://duinokit.com* and through Amazon.com.

The Duinokit has one DHT11 temperature/humidity sensor, which is a slower, lower-resolution version of the DHT22 that will be used with the GreenShield. In terms of software, the DHT11 and DHT22 are similar but not identical. The DHT22 uses a different data word (bit string) than the DHT11 to accommodate the improved accuracy of the DHT22 sensor.

The LDR and soil moisture sensor use an LM358 dual op amp, with one-half assigned to each input. I placed the LM358 on the solderless breadboard provided on the Duinokit's single large PCB. This also provided a place to mount the resistors, and the two 10K potentiometers on the Duinokit served as the input offset trim controls.

Prototype software

The software will be developed in prototype and final forms. The first step is to create software to run on the Duinokit prototype that will read sensor inputs, verify that the

LM358 op amp circuits are behaving as expected, and support some testing to determine initial input range limits. The next step is the development of the final software that will support the command-response protocol defined in "Software" on page 375 and the relay function mapping. The actual shield hardware will be used to develop the final software.

The prototype software is intended for reading data from the analog inputs and the on-board DHT11 sensor. This is what will be used for prototype testing when the initial input ranges are established. The output appears in the serial monitor window provided by the Arduino IDE. The prototype test software shown in Example 10-1 is contained in a single sketch file called *gs_proto.ino*.

 The functions shown in Example 10-1 for the various temperature conversions and dewpoint values aren't really necessary for the basic Greenshield. I've included them as examples if you want to use them.

Example 10-1. GreenShield sensor prototype software

```
// GreenShield prototype software
// 2015 J. M. Hughes
//
// Uses DHT11 library from George Hadjikyriacou, SimKard, and Rob Tillaart
//
// Repeatedly reads and outputs temperature, humidity, soil moisture,
// and light level. No relay setpoint functionality.

#include <dht11.h>

// Definitions

#define LHLED       2       // D2   Low humidity LED
#define HHLED       3       // D3   High humidity LED
#define LMLED       4       // D4   Low moisture
#define HMLED       5       // D5   High moisture
#define RY1OUT      6       // D6   RY1 enabled
#define RY2OUT      7       // D7   RY2 enable
#define DHT11PIN    8       // D8   DHT11 data
#define SMSINPUT    A0      // A0   SMS input
#define LDRINPUT    A1      // A1   LDR input

// Global Vars

int curr_temp   = 0;        // current (latest) temperature
int curr_hum    = 0;        // current humidity
int curr_ambl   = 0;        // current ambient light level
int curr_sms    = 0;        // current soil moisture
```

```
int cntr = 0;


// dht11 object is global
dht11 DHT11;


// Read data from DHT11 and store in curr_hum and curr_temp global
// variables for later use
void readDHT()
{
    if (!DHT11.read(DHT11PIN)) {
        curr_hum = DHT11.humidity;
        curr_temp = DHT11.temperature;
    }
}


// Read data from analog inputs via LM358 op amp circuits and store
// in curr_ambl and curr_sms for later use
void readAnalog()
{
    curr_ambl = analogRead(LDRINPUT);
    curr_sms  = analogRead(SMSINPUT);
}


// Celsius to Fahrenheit conversion
// From example code found at http://playground.arduino.cc/main/DHT11Lib
double Fahrenheit(double celsius)
{
    return 1.8 * celsius + 32;
}


// Celsius to Kelvin conversion
// From example code found at http://playground.arduino.cc/main/DHT11Lib
double Kelvin(double celsius)
{
    return celsius + 273.15;
}


// dewPoint() function NOAA
// reference: http://wahiduddin.net/calc/density_algorithms.htm
// From example code found at http://playground.arduino.cc/main/DHT11Lib
double dewPoint(double celsius, double humidity)
{
    double A0= 373.15/(273.15 + celsius);
    double SUM = -7.90298 * (A0-1);
    SUM += 5.02808 * log10(A0);
    SUM += -1.3816e-7 * (pow(10, (11.344*(1-1/A0)))-1) ;
```

```
    SUM += 8.1328e-3 * (pow(10,(-3.49149*(A0-1)))-1) ;
    SUM += log10(1013.246);
    double VP = pow(10, SUM-3) * humidity;
    double T = log(VP/0.61078);    // temp var
    return (241.88 * T) / (17.558-T);
}

// delta max = 0.6544 wrt dewPoint()
// 5x faster than dewPoint()
// reference: http://en.wikipedia.org/wiki/Dew_point
// From example code found at http://playground.arduino.cc/main/DHT11Lib
double dewPointFast(double celsius, double humidity)
{
    double a = 17.271;
    double b = 237.7;
    double temp = (a * celsius) / (b + celsius) + log(humidity/100);
    double Td = (b * temp) / (a - temp);
    return Td;
}


void setup()
{
    // Init the serial I/O
    Serial.begin(9600);

    // Set up the AVR's pins
    pinMode(LHLED,OUTPUT);
    pinMode(HHLED,OUTPUT);
    pinMode(LMLED,OUTPUT);
    pinMode(HMLED,OUTPUT);
    pinMode(RY1OUT,OUTPUT);
    pinMode(RY2OUT,OUTPUT);

    // Initial current data variables
    curr_temp = 0;
    curr_hum  = 0;
    curr_ambl = 0;
    curr_sms  = 0;
}


void loop()
{
    // Get DHT11 readings
    readDHT();

    // Get LDR and SMS readings
    readAnalog();

    // Print data to output
    Serial.println("\n");
```

```
    Serial.print("Raw LDR          : ");
    Serial.println(curr_ambl);
    Serial.print("Raw SMS          : ");
    Serial.println(curr_sms);
    Serial.print("Humidity (%)     : ");
    Serial.println((float)DHT11.humidity, 2);
    Serial.print("Temperature (oC) : ");
    Serial.println((float)DHT11.temperature, 2);
    Serial.print("Temperature (oF) : ");
    Serial.println(Fahrenheit(DHT11.temperature), 2);
    Serial.print("Temperature (K)  : ");
    Serial.println(Kelvin(DHT11.temperature), 2);
    Serial.print("Dew Point (oC)   : ");
    Serial.println(dewPoint(DHT11.temperature, DHT11.humidity));
    Serial.print("Dew PointFast (oC): ");
    Serial.println(dewPointFast(DHT11.temperature, DHT11.humidity));

    // Scroll up to align display
    for (int i = 0; i < 12; i++)
        Serial.println();

    delay(1000);
}
```

The setup() function simply initializes and opens the serial I/O, sets some pin modes, and clears the global variables for current data readings. Each time the readDHT() and readAnalog() functions are called they will obtain the latest values from the DHT11 and the analog inputs and place them into these variables.

The main loop reads the analog inputs and the DHT11, formats the data, and writes the current values to the USB serial monitor. It does not communicate with a control host computer and it doesn't do any relay setpoint mapping. Its purpose is to continuously obtain and display sensor data.

The prototype uses an open source library for the DHT11. The final version will use a custom library for the DHT22, but it's not needed for the prototype. The DHT11 library by George Hadjikyriacou, SimKard, and Rob Tillaart is available from the Arduino Playground (*http://bit.ly/apg-dht11*). The Fahrenheit(), Kelvin(), dew Point(), and dewPointFast() functions are from the same source.

Prototype testing

Using the Duinokit we can test the various sensor input functions of the GreenShield and fine-tune the operation. If there's a problem with the circuitry (which will be easy to resolve, given that it's so simple), this is where you would want to find it and fix it. Trying to fix a problem on a PCB after it has been fabricated and loaded with parts can be really frustrating, and there is always the risk of something being damaged in the process.

For the GreenShield we want to verify that the sensors work correctly, the software is able to derive sensible values for things like the soil moisture sensor and the LDR, and the humidity/temperature sensor is operating as expected. To do this we'll use some dry sand, water, a reliable digital thermometer, a refrigerator, an oven, and a sunny day.

The first thing to check is the humidity/temperature sensor. Using a external thermometer (in this case I used a digital thermometer with the sensor on a long lead), I started with an ambient reading. The next step was to put the Duinokit and the thermometer into a refrigerator. A small netbook PC provided power and displayed the temperature, and the USB cable was thin enough to allow the door of the refrigerator to close completely. Last, the Duinokit and the thermometer were placed in a warm oven that was at about 140° F (60° C). With three data points we can generate a rough calibration curve to compensate for variances in the temperature sensor.

Testing the humidity response is a bit trickier, but getting readings near the ends of the usable range isn't too hard to do. A short stay in the freezer section of the refrigerator will expose the sensor to a very low-humidity environment. Freezers are dry because moisture in the air condenses on the coils inside the freezer compartment. This is what causes "freezer burn," by the way, when food isn't properly sealed before being frozen. It's also the principle behind freeze-drying, although that is typically done at much colder temperatures (around –112° F, or –80° C), and in a partial vacuum. In the case of a kitchen freezer we would expect to see something like 5% humidity, or perhaps a bit lower.

Another method is the so-called "salt test." This technique uses water-saturated salt to establish a constant relative humidity in a sealed environment. You can read one way to perform a salt-base calibration at the Ambient Weather wiki (*http://bit.ly/wiki-aw*). If you elect to do this, be careful not to get any of the salt or water on the circuit components. This might not be very practical with a large item like the Duinokit, but it can be used with the finished Greenshield.

Once we have a low-humidity reading, the next step is to boil some water on the stove and use a small fan to blow the steam over the sensor. The resulting flow of air won't be fully saturated, but it will be in the 80 to 90% humidity range. These tests verify that the sensor is working, but we can't really use the data for anything beyond that because we have no reference to compare it to. If you happen to have an accurate humidity sensor available, then by all means use it and create a calibration curve like the one that was created for the temperature.

Testing the soil moisture sensor involves some clean, dry sand, a scale, and some water. First, get a large glass jar or ceramic bowl. Either will work; choose one that can hold a quart or so (or about 1 liter). Don't use a metal bowl for this test, because the moisture probe uses current flow and a metal bowl could create a false reading. First, weigh the container and record the value. We'll need this later on. Next, measure out about 1/2 pound (or about 225g) of sand into the container. Put it back on the scale and weigh it again. The actual weight of the sand is whatever the scale shows minus the weight of the container. You can leave the container on the scale for the rest of the test procedures if you want to.

Now insert the soil moisture probe into the dry sand and note the reading shown on the Arduino IDE's serial monitor output. Remove the sensor and add water until the weight is about one-quarter more than the original weight of the sand plus the weight of the container. Let it sit for a bit to allow the water to work through the sand. The sand should feel damp to the touch, but it shouldn't be wet or muddy.

Reinsert the sensor and observe the output. The sand is now about 50% saturated, and from these two readings, dry and damp, we can interpolate a point in between, which we will call the 25% point.

Lastly, there is the LDR photocell. The response of the photocell really isn't all that critical, but it is a good idea to establish a low-light trip point. On a cloudy day this is when the GreenShield can be used to turn on some auxiliary lighting, or it can simply be used to determine the difference between day and night. All that is needed to test the photocell is an interior room in your house (perhaps with the curtains partly drawn, and no lights on) and a nice sunny day outside. The direct sunlight outside is as much light as the photocell is ever likely to be exposed to, and an interior room in your house is roughly equivalent to the light level of a dim, cloudy day outside.

We need to record the data for the temperature/humidity sensor, the LDR, and the soil moisture probe. These will be our initial values when we set up the GreenShield for the first time, and since we now know what to expect we won't have to guess at appropriate minimum and maximum setpoint values to start off with.

Final Software

The prototype software only handles the sensor inputs. The final version will also handle the host control interface and relay setpoint function mapping. This involves input command parsing, along with data storage and lookup.

There are no output displays beyond the four humidity and temperature range status LEDs, and no manual control inputs. A simple USB serial interface is used for command-response transactions between the GreenShield Arduino and a host control computer. The bulk of the software involves interpreting the commands from the control host PC, and then applying the setpoint mapping to the relays.

Source code organization

The final version of the GreenShield source code is contained in multiple source files, or modules. When the Arduino IDE opens the main file, *GreenShield.ino*, it will also open the other associated files in the same directory. The secondary files are placed in "tabs" in the IDE as shown in Figure 10-11.

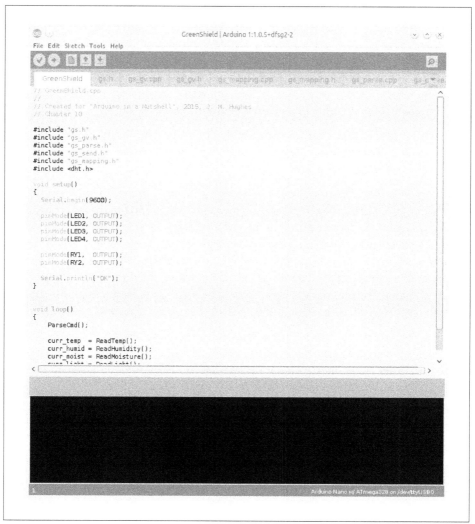

Figure 10-11. The Arduino IDE with the GreenShield files loaded

Table 10-8 lists the files in the GreenShield set. Two files are shared by all the source modules. These are *gs.h* and *gs_gv.h*. The global variables defined in *gs_gv.cpp* that would otherwise be found at the start of a conventional sketch are compiled separately and shared as necessary among the other modules.

Table 10-8. GreenShield source code modules

Module	Function
GreenShield.ino	Primary module containing setup( ) and loop( )
gs_gv.cpp	Global variables
gs_gv.h	Include file
gs.h	Constant definitions (#define statements)
gs_mapping.cpp	Function mapping
gs_mapping.h	Include file
gs_parse.cpp	Command parsing
gs_parse.h	Include file
gs_send.cpp	Data send (to host) functions
gs_send.h	Include file

Organizing a project in this manner makes it easier to deal with just one section at a time without wading through line after line of source code. Once the main section is done, then changes can be made to other modules without interfering with the finished code. This approach also helps in thinking about your software from a modular perspective, and this in turn makes it easier to understand and easier to maintain.

Software description

Figure 10-12 shows the flowchart for the main loop of the software. The loop( ) function begins with the "Start" block and it will continue until the Arduino is powered off. Note that there are three primary functional sections: command input and response processing, data acquisition, and minimum/maximum setpoint testing. Also note that the block labeled "Setpoint Test" is one instance of four blocks, one for each pair of min/max setpoints. In order to keep the size of the diagram reasonable, only one test section is shown.

The *GreenShield.ino* source file, shown in Example 10-2, contains the setup( ) and loop( ) functions. The complete GreenShield software can be found on GitHub (*https://github.com/ardnut*).

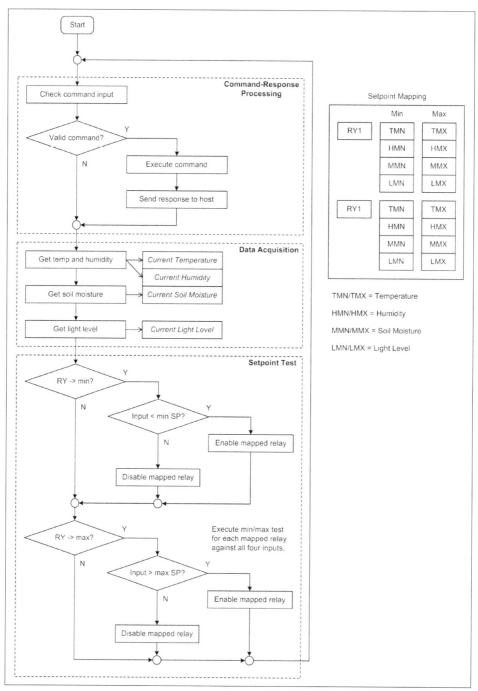

Figure 10-12. GreenShield flowchart

Example 10-2. GreenShield main source file

```
// GreenShield.ino
//
// Created for "Arduino: A Technical Reference," 2016, J. M. Hughes
// Chapter 10

#include "gs.h"
#include "gs_gv.h"
#include "gs_parse.h"
#include "gs_send.h"
#include "gs_mapping.h"
#include <dht.h>

void setup()
{
  Serial.begin(9600);

  pinMode(LED1, OUTPUT);
  pinMode(LED2, OUTPUT);
  pinMode(LED3, OUTPUT);
  pinMode(LED4, OUTPUT);

  pinMode(RY1,  OUTPUT);
  pinMode(RY2,  OUTPUT);

  Serial.println("OK");
}


void loop()
{
    ParseCmd();

    curr_temp  = ReadTemp();
    curr_humid = ReadHumidity();
    curr_moist = ReadMoisture();
    curr_light = ReadLight();

    ScanMap();
}
```

The Arduino IDE relies on the `#include` statements to determine which modules belong in the code set. Even if a source file isn't directly used by the top-level module, it must still be included.

The global definitions file *gs.h*, shown in Example 10-3, defines a set of constants used by the GreenShield source modules. The `#define` statements result in a smaller compiled object, as demonstrated in "Constants" on page 110.

Example 10-3. GreenShield global definitions

```
// gs.h
//
// Created for "Arduino: A Technical Reference," 2016, J. M. Hughes
// Chapter 10

#ifndef GSDEFS_H
#define GSDEFS_H

#define MAXINSZ      12      // Input buffer size

#define NOERR        0       // Error codes
#define TIMEOUT      1
#define BADCHAR      2
#define BADVAL       3

#define LED1         2       // LED pin definitions
#define LED2         3
#define LED3         4
#define LED4         5
#define RY1          6       // Relay pin definitions
#define RY2          7

#define DHT22        8       // DHT22 I/O pin
#define MPROBE       A0      // Moisture probe input
#define LDRIN        A1      // LDR input

#define MAXRY        2       // Maximum num of relays

#define MAP_NONE     0       // Mapping vectors
#define MAP_TEMPMIN  1
#define MAP_TEMPMAX  2
#define MAP_HUMIDMIN 3
#define MAP_HUMIDMAX 4
#define MAP_MOISTMIN 5
#define MAP_MOISTMAX 6
#define MAP_LIGHTMIN 7
#define MAP_LIGHTMAX 8

#endif
```

In Example 10-3 the definition of MAXRY is 2. This can be a larger value if the hardware to support additional relays is present, and the outputs don't have to be relays. The include file *gs_mapping.h*, shown in Example 10-4, declares the functions for reading the DHT22 and the analog inputs, setting the on/off state of each relay (or all relays), controlling the status LEDs, and performing a scan through the response conditions that will control the relays in the function ScanMap().

Example 10-4. GreenShield mapping functions

```
// gs_mapping.h
//
// Created for "Arduino: A Technical Reference," 2016, J. M. Hughes
// Chapter 10

#ifndef GSMAP_H
#define GSMAP_H

void ReadDHT22();
int  ReadTemp();
int  ReadHumidity();
int  ReadMoisture();
int  ReadLight();

int  RyGet(int ry);
void RySet(int ry, int state);
void RyAll(int state);

void LEDControl(int LEDidx);
void ScanMap();

#endif
```

The ScanMap() function in *gs_mapping.cpp*, shown in Example 10-5, is executed on each cycle of the loop() function in the *GreenShield.ino* main source file. It evaluates the analog inputs against a set of configurable limits, and either enables or disables the relays based on those conditions.

Example 10-5. The GreenShield function map scanner

```
// NOTE: There are no checks in this code to prevent multiple relays being
// mapped to the same operational mode.

// Each RY is mapped to one of 8 possible operational modes. Determine the
// mapping for a specific relay and see if the enable condition has been
// met. This is extensible to any reasonable number of relays.
void ScanMap()
{
    for (int i = 0; i < MAXRY; i++) {
        if (rymap[i] != MAP_NONE) {
            switch (rymap[i]) {
                case MAP_TEMPMIN:
                    if (curr_temp < mintemp)
                        { RySet(i, 1); } else { RySet(i, 0); }
                    break;
                case MAP_TEMPMAX:
                    if (curr_temp > maxtemp)
                        { RySet(i, 1); } else { RySet(i, 0); }
                    break;
```

```
            case MAP_HUMIDMIN:
                if (curr_humid < minhum)
                    { RySet(i, 1); } else { RySet(i, 0); }
                break;
            case MAP_HUMIDMAX:
                if (curr_humid > maxhum)
                    { RySet(i, 1); } else { RySet(i, 0); }
                break;
            case MAP_MOISTMIN:
                if (curr_moist < minmoist)
                    { RySet(i, 1); } else { RySet(i, 0); }
                break;
            case MAP_MOISTMAX:
                if (curr_moist > maxmoist)
                    { RySet(i, 1); } else { RySet(i, 0); }
                break;
            case MAP_LIGHTMIN:
                if (curr_light < minlite)
                    { RySet(i, 1); } else { RySet(i, 0); }
                break;
            case MAP_LIGHTMAX:
                if (curr_light > maxlite)
                    { RySet(i, 1); } else { RySet(i, 0); }
                break;
            default:
                // Do nothing
                break;
        }
    }
}
```

The source module *gs_parse.cpp* contains one primary function, `ParseCmd()`, and a couple of support functions (`CntInt()` and `SendErr()`). Example 10-6 shows the contents of *gs_parse.h*.

Example 10-6. GreenShield parse module include file

```
// gs_parse.h
//
// Created for "Arduino: A Technical Reference," 2016, J. M. Hughes
// Chapter 10

#ifndef GSPARSE_H
#define GSPARSE_H

void ParseCmd();
int  CvtInt(char *strval, int strt, int strlen);
void SendErr(int errtype);

#endif
```

The `ParseCmd()` function is by far one of the largest functions in the GreenShield code set. If uses a fast descending conditional tree-type parser to determine the type of an incoming command, and then extract the subfunction code and any parameters. This function will also execute any immediate commands such as enabling or disabling relays, returning relay state information, and acquiring and returning analog data to the control host or a user. Immediate command execution occurs at the endpoints of the descending tree structure.

Fabrication

It is beyond the scope of this book to provide a step-by-step walk-through for creating schematics and printed circuit boards. This is a high-level description of the steps involved to move from schematic to PCB layout to finished PCB. For the low-level details, I would refer you to the texts listed in Appendix D. A Google search for "Cad-Soft Eagle" will come back with numerous tutorials. I recommend the tutorials from SparkFun, Adafruit, and, of course, CadSoft (*http://www.cadsoftusa.com*).

The Eagle version of the Greenshield schematic is shown in Figure 10-13. Notice the block labeled "ARDUINO_R3_SHIELD." This is from the SparkFun parts library for Eagle, and it's intended specifically for creating shields. It is much more convenient than working out the placement of the pin header pads manually later on during the PCB layout phase.

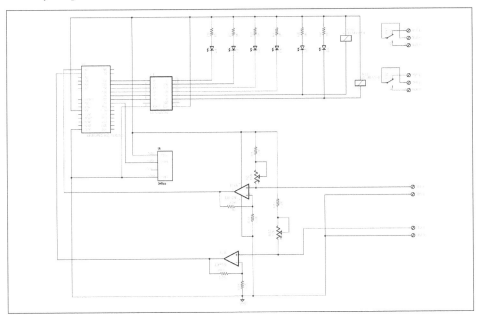

Figure 10-13. GreenShield schematic (Eagle version)

The Eagle schematic editor, like all such tools, takes some time to get used to. It's not always intuitive or obvious. Normally I use a different tool to create publication-quality line art, including schematics, but to create a PCB it's useful to have a schematic editor and PCB layout tool that can share data. You can compare Figure 10-13 with Figure 10-9 to see the difference. The Eagle tool, and the Fritzing design tool used in the Switchinator project, can keep schematics and layouts synchronized, whereas a standalone graphics tool for line art and illustrations cannot.

When creating a schematic, make sure you have selected the correct part. For example, the symbol for a resistor is the same no matter if it's an 0805 SMD (surface-mount) part or a 1/4 watt through-hole component. Sometimes it can be a challenge to find the right part in the schematic editor's parts library. With Eagle you can use wildcard characters to search the library. When looking for the ULN2003AD part for the GreenShield I entered *2003* and found what I was looking for under the *uln-udn* category. Sometimes it is necessary to go online and search for a part that someone may have already created. SparkFun, Adafruit, and others have created large libraries of parts that are available to download for free.

Once the schematic is complete the PCB can be generated. Initially the PCB layout is just a jumble of parts and thin connection lines ("air wires," as they are sometimes called) that form a wiring "rat's nest." The rat's nest shows the point-to-point connections as defined in the netlist (the network list) created from the schematic.

The first step is to move all the parts into the PCB region, and then arrange them. The main objectives when arranging the parts are to locate connectors in the desired locations, group parts by function, and minimize the number of occurrences of crossed rat's nest lines by rotating parts as necessary. Once this initial step is complete it's much easier to start creating the traces that will connect the parts.

You can elect to route each trace (or track, as they are sometimes called) by hand, using the rat's nest wires as a guide, or if you are feeling lucky you can let an autorouter take a shot at it. Eagle does have an autorouter, but I generally don't use it. Autorouting is typically an iterative process of trying to get a good layout, ripping it up, moving and rotating parts, and then trying it again. After a while it becomes obvious with some designs that it is quicker to just do it manually.

In Eagle, you do not need to manually place a via. (A via transfers a trace from one side of the PCB to the other via a plated-through hole, hence the name.) If you need to transition from the top to the bottom of the PCB (or vice versa), simply change from one side to the other with the layer selection pull-down located at the lefthand side of the trace draw toolbar. Eagle will automatically place a via at that location for you and switch the trace routing to the selected side of the PCB.

The PCB layout is shown in Figure 10-14. The top (component) side is brownish-red, and the bottom (solder) side is blue. If you happen to have the printed version of this book you should see the top side traces as light gray, and the bottom traces as dark gray. The component outlines are in light gray.

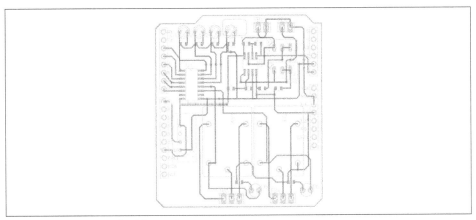

Figure 10-14. GreenShield PCB layout

The files generated by the Eagle CAM (Computer Aided Manufacturing) tool, called "Gerber" files, are used by a PCB fabricator to create the actual PCB. I used the Gerbv tool, part of the gEDA package, to load and view the Gerber files as a last step check. A screenshot of the Gerbv screen is shown in Figure 10-15.

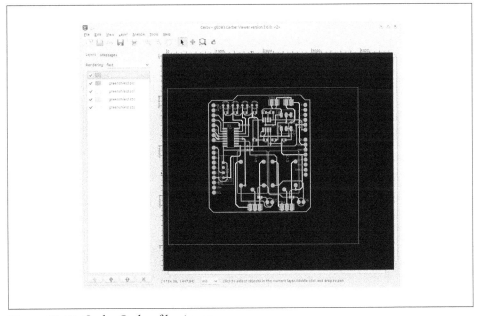

Figure 10-15. Gerbv Gerber file viewer

In order to take advantage of the low-cost service for prototype PCBs, the board outline for the GreenShield was squared to make a rectangle. It might look a little odd, but that won't affect how it works. You will not see this in the photos, but the layout was originally done using the Arduino shield outline. Converting the corners to right angles took about 5 minutes, and it happened just before the layout was sent off for fabrication. If it really mattered I could have paid a whole lot more money for an edge route and trim operation, or I could have used my own router and done it myself. I opted to just leave it as a rectangle.

It takes about 7 to 10 days to get a finished PCB back. Figure 10-16 shows the bare PCB from the fabricator.

After the parts are soldered onto the PCB it's always a good idea to spend a few minutes examining both sides for cold solder joints and shorts (called "bridges") between the pads and traces. I use a standard jeweler's loupe for this. Figure 10-17 shows what a completely assembled GreenShield looks like.

You might notice that I used stacking headers for the connections to an underlying Arduino or another shield. While it would be awkward to put another shield on the GreenShield (and I would advise against it), I wanted to have some readily accessible test and I/O points.

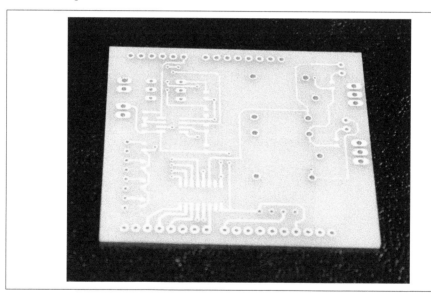

Figure 10-16. Finished bare GreenShield PCB

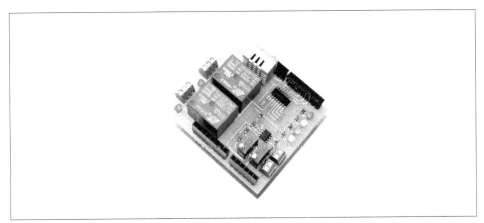

Figure 10-17. Fully populated GreenShield PCB ready to go

Final Acceptance Testing

Surface-mounted parts have their own unique set of potential problems. Before applying power to the GreenShield we should do some quick checks to make sure things are wired correctly and there are no short circuits on the PCB:

Visual inspection

Carefully examine the components on the PCB for solder bridges (solder bridging two pads or between a pad and a trace). Examine the resistors to see if any have an end that may be lifted above the pad and not connected. This can happen when using regular solder and a soldering iron (I used solder paste and a hot air SMD reflow tool). Look at the pads for the ICs (U1 and IC1) to make sure there are no solder bridges between them.

Component placement

There are nine parts that may accidentally be mounted backward. These are the two ICs, the LEDs, and the DHT22 sensor module. The op amp IC may have a dimple or dot to indicate pin 1, but some packages have a beveled edge. In addition to different lead lengths, LEDs will usually have a small flattened area next to the cathode (–) connection.

Short circuits

Using a DMM, preferably with a continuity test function (the beeper mode), check each pair of pins on the LM358 op amp (IC1) for shorts. None of the pins should short to another pin. Now check that the VCC on pin 8 of IC1 is tied to the 5V on the pin header. Also check that pin 4 is tied to one of the ground pins on the pin header.

Repeat this process for the ULN2003A driver (U1). None of the input or output pins should be shorted, pin 8 should be tied to ground, and pin 9 should be connected to the 5V supply.

Power safety

Before connecting the GreenShield to an Arduino board, use a DMM to measure the resistance between the +5V and ground pins on the pin header. You should see a value of no less than about 30K ohms, probably higher. If you get a reading of zero then there is a short somewhere, and it will need to be found and cleared before attempting to power up the GreenShield.

If the GreenShield appears to be acceptable electrically, then we can mount it on an Arduino and apply power. None of the LEDs on the GreenShield should be active initially (unless there is some software already running on the AVR on the Arduino that is controlling the digital I/O pins). Functional testing involves four basic steps:

Initial functional testing

The first part of functional testing is rerunning the same tests as were done with the prototype to test the analog inputs and the DHT22.

Upload the prototype version of the software to the Arduino and open the IDE's serial monitor window. If everything is working correctly you should see a repeating display with the temperature, humidity, and analog inputs.

Analog input testing

Connect a 470-ohm resistor across the LDR inputs. While observing the continuous output, adjust R12 until the value goes to zero. This demonstrates that this part of the circuit is working correctly. Now repeat this with R8 for the moisture sensor.

DHT22 testing

The output readings from the DHT22 should be what you would expect for the local ambient temperature and humidity. You can use a hot air source (a hair dryer, for instance) to apply warm (not hot!) air to the DHT22 and observe its response.

Software functional testing

Now load the full version of the GreenShield software. You can use the serial monitor window of the Arduino IDE for these tests. This is not an extensive suite of tests, as we've already been through some of this earlier with the prototype.

With the full GreenShield software loaded, you should see the word "OK" when it first starts. The GreenShield does not provide a prompt. Enter the command `RY:0:?` and the response should be `RY:0:0`. Now enter the command `RY:0:1`. The relay should click and the associated LED should illuminate. Using the `RY:0:?` command now will return `RY:0:1`.

We should exercise the rest of the commands as well. We can test the analog limits by shorting or opening the analog inputs. The DHT22 is self-contained, so we don't need to do a lot with that, but you can set the limits very close and use a hot air source and some steam from a pot of boiling water to verify that the temperature and humidity limits work as expected.

Testing Confession

When assembling the first GreenShield board I couldn't find a packaging tray of new Songle 5V relays I had purchased. I looked everywhere. So I grabbed a couple of blue relays from a box and soldered them onto the PCB. They were the right shape, the right color, and the right brand. Turned out they were the wrong coil voltage; 12V DC instead of 5V DC. I was baffled as to why the relays wouldn't respond to the software commands until I looked at the lettering on the tops of the two relays and discovered what I had done. In the meantime I had found the missing 5V relays (they were right where I'd left them, of course). A short while later the 12V relays were removed, the correct 5V parts were installed, and the relay control commands and the function mapping worked as expected.

The moral of the story: always examine the parts you are installing before you commit them to a PCB with solder.

Operation

The GreenShield is physically simple, but it can be functionally complex in terms of how it is integrated into its intended environment. The two main variable inputs, light level and soil moisture, must be calibrated for a specific set of conditions. The range of the light-dependent resistor and the soil moisture probe determine how the midpoints are set using the trimmer potentiometers. Not all LDRs are the same, and there will be a difference between a moisture probe that is just a pair of probes and one with a transistor on it.

There are two basic approaches for adjusting the GreenShield: (1) calibrate the GreenShield for known ranges using references of some sort, or (2) adjust the Green-Shield to a specific environment using subjective evaluation (i.e., does the soil feel wet enough?).

The calibrated approach will allow you specify ranges based on hard data, and if you know the ideal soil moisture content for, say, tomatoes, then you can use those values once you have done the calibration and worked out how the ADC readings correspond to moisture content. "Prototype testing" on page 383 described the basic procedures involved in calibrating the GreenShield for soil moisture and light levels, but you could also use expensive lab equipment as reference sources.

The subjective approach is much easier, and since the primary intent is to avoid killing off your plants it's probably just as effective after a bit of tweaking. You can adjust the trimmer potentiometers to suit the high and low ends of a subjective evaluation of what is acceptable.

I would suggest some experimentation to see which approach works best for your intended application. I would also suggest taking the time to log data from the Green-Shield and build up some profiles that you can study to achieve the best responses for your situation.

Next Steps

The GreenShield is rather minimal, to be sure, but it has a lot of potential. You can use it with a Bluetooth shield or even an Ethernet shield, and remotely monitor the soil conditions of your favorite potted plant, some tomatoes or orchids in a greenhouse, a small outdoor vegetable plot, or a garden in a Mars colony. Connect a standard 24VAC sprinkler system valve to one of the relays and you can automate the watering. You could use the other relay to enable a ventilation fan in a greenhouse to cool it down, or connect it to a heater to keep the interior nice and warm during the winter. And, of course, you could always add another relay or two with an external relay module like those described in Chapter 9.

A microSD shield could also be used if you wanted to strap the GreenShield to a tree in the forest and do some long-term data logging (a solar panel to keep it running would be a good idea, as well). Add a WiFi or GSM shield, and you could scatter GreenShields around a good-sized farm to keep an eye on soil conditions.

 The Greenshield software currently does not save the set-point values if power is lost to the Arduino. One way to get around this is to use the EEPROM in the AVR IC. Refer to Chapter 7 for an overview of the EEPROM library.

Custom Arduino-Compatible Designs

Building an Arduino hardware–compatible PCB is straightforward. In fact, the Arduino folks even provide the schematics and PCB layout files for download. They don't provide the top or bottom silkscreen masks, however, as these are copyrighted by Arduino.cc and are not covered under an open license. You will need to create your own graphics for the board.

Although much of what you need to build a copy of an Arduino or a shield is readily available, it really isn't economical to clone an existing Arduino design (either a shield or an MCU board). Unless you own or have access to a PCB production facility and a need for hundreds, or even thousands, of a particular board type, odds are you can find something already fabricated for much less per unit than it would cost you to build it yourself.

The custom approach does make sense if the existing available PCB form factors won't work for you. Perhaps you want to integrate an Arduino into a larger system, or put an AVR MCU into a uniquely constrained location such as an unmanned aircraft or a robot. If that's the case, then you might want to consider creating a software-compatible PCB. Your board doesn't have to look like an Arduino, and there's no reason it needs to be compatible with existing shields, unless of course you want to use it with a shield from some other source.

As discussed in Chapter 1, a device can be Arduino software compatible without being hardware compatible. All that is needed is a suitable AVR processor, the bootloader firmware, and the Arduino runtime libraries. Even the bootloader firmware is optional. In this section we will design and build an AVR-based DC power controller suitable for use with heavy-duty relays, high-power LEDs, and AC or DC motors.

Programming a Custom Design

If you want to use the Arduino bootloader with a brand-new AVR microcontroller, then you will need to install the bootloader in the chip's built-in flash memory. Once this is done, you can then treat your custom board like any other Arduino. The tools and procedures involved in uploading executable code to an AVR MCU are covered in "Uploading AVR Executable Code" on page 147 in Chapter 6. As an alternative to installing the bootloader firmware yourself, many sources sell AVR devices with the Arduino bootloader already installed. Check out Adafruit (*https://www.adafruit.com*) and SparkFun (*https://www.sparkfun.com*) for ATmega328 ICs with the Arduino bootloader firmware preinstalled. Entering "ATmega328 with Arduino bootloader" into the search box on Amazon.com or eBay returns numerous listings.

Once the bootloader is in place, you can use a USB-to-serial adapter or even a standard serial interface with a suitable RS-232 module like the one shown in Figure 10-26 or a USB-to-serial breakout like the SparkFun device shown in Figure 6-8. For some projects, such as the Switchinator in the next section, the serial interface is not an option for programming the MCU if it is already in use, or if the bootloader is not used. This means you will need to use an ISP (ICSP) programmer such as the Atmel-ICE or the USBtinyISP, both of which are discussed in Chapter 6.

The Switchinator

The device described here, which I am calling the Switchinator (for lack of anything better), is a remote-controlled, 14-channel DC switch with 4 analog input channels. This initial version uses an RS-232 interface and a simple command-response protocol. A possible change for a future version would be the use of an RS-485 interface instead of RS-232.

Definition and Planning

The Switchinator is a standalone PCB that can use a plain ATmega328 or ATmega328p MCU without the Arduino bootloader. It can also use an MCU with the bootloader firmware installed with a serial programmer or a USB-to-serial converter. I have elected to use the Arduino IDE for compiling, and an Adafruit USBtinyISP for the programming. Refer to Chapter 6 for more on programming the MCU.

Core hardware:

- ATmega328
- 16 MHz crystal clock source
- ICSP programming interface
- Integrated 5V DC power supply (9 to 12V DC input)

Inputs and outputs:

- 4 analog inputs
- 14 discrete digital outputs

Control interface:

- RS-232 host interface
- Command-response protocol, control host driven
- Analog readings available on demand
- Output override by control host

The Switchinator provides 14 discrete digital outputs and 4 analog inputs, and the SPI interface is available through an ICSP pin array. The discrete digital outputs and the analog inputs are terminated at the edge of the PCB using screw terminal blocks. An RS-232 interface is used for communication between the board and a control host computer via the D0 and D1 pins on the MCU.

The PCB is a 100 percent through-hole design. This simplifies the assembly at the expense of a larger PCB and a potentially more challenging PCB layout. The PCB size will not exceed a rectangular size of 100 mm by 140 mm. Mounting holes are located in each corner of the PCB.

The discrete digital outputs of the Switchinator can be used for driving relays, such as the type used on the GreenShield, or controlling up to three unipolar stepper motors using a basic circuit like the one shown in Figure 10-18.

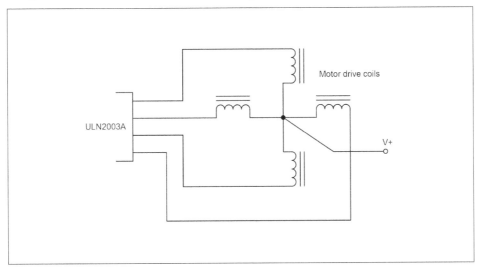

Figure 10-18. ULN2003A connected as stepper driver

The digital outputs of the Switchinator can also drive high-current LEDs, solenoids, or DC motors, as shown in Figure 10-19.

A basic prototype based on a solderless breadboard will be used to develop the initial version of the software. The final version will be completed on the finished hardware.

For the final hardware design I will use Fritzing for schematic capture and PCB layout.

Design

The Switchinator is a single PCB, 124 × 96 mm in size with four corner mounting holes. The final board size is larger than would otherwise be possible with a surface-mount design (it is also larger than the size limits set by the free version of the Eagle tool used for the GreenShield).

There is no enclosure or power supply, which greatly simplifies the design. The Switchinator is entirely self-contained, requiring only an external DC power source for operation.

Functionality

The Switchinator is a digital output device with some analog input capability. Its primary purpose is to switch DC loads, both inductive and noninductive. An

ATmega328 AVR MCU is used to decode command inputs and return status data to the control host.

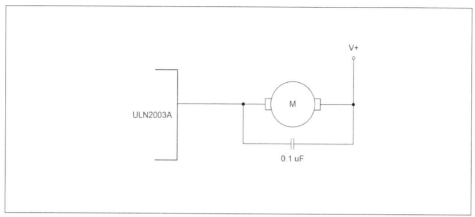

Figure 10-19. ULN2003A DC motor driver

An ATmega328 is used primarily as a command decoder to interface the I/O on the PCB to a host system. Although it has been programmed to act as an I/O device, it could also be programmed to perform autonomous functions based on the analog inputs. By using a linear temperature sensor, such as the LM35, the Switchinator could easily be reprogrammed to serve as a controller for an environmental test chamber or an epoxy curing chamber.

Circuit-wise the Switchinator is comprised of three main sections: MCU, digital I/O, and power supply. Figure 10-20 shows a block diagram of the Switchinator.

The discrete digital outputs are driven by ULN2003A ICs. Four analog inputs are also provided, and the analog reference and AVCC voltages may be supplied either on-board from the internal power supply or externally. Jumpers are used to select the analog voltage sources.

A simple, human-compatible command-response protocol is used to monitor and control the Switchinator.

Hardware

The Switchinator will have 14 digital outputs, each connected to the Darlington outputs of a pair of ULN2003A driver ICs. The ULN2003A drivers have seven channels per IC, and each ULN2003A channel can handle up to 300 mA or more in some cases.

A Microchip MCP23017 I2C digital I/O expander will be used to drive the ULN2003A parts, mainly to avoid using up all of the available digital I/O pins on the AVR MCU. Two of the digital I/O pins on the AVR are used for the serial interface, and two of the analog pins are used for the I2C interface to the MCP23017 IC. The unconnected digital I/O pins are not used, but are available for future expansion.

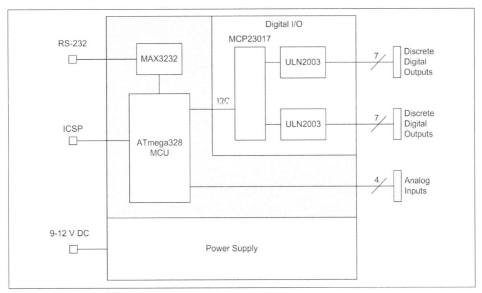

Figure 10-20. Switchinator block diagram

An RS-232 interface is implemented using a MAX232 TTL-to-RS232 transceiver IC. The serial interface will be used to communicate with a host system acting as a master controller. The master controller may be a PC, an Arduino, or some other type of programmable controller with an RS-232 interface. A DB-9 connector is used for the serial interface. This is not a full implementation of RS-232, just the RxD and TxD signals. The Switchinator does not have a USB connector.

The digital outputs and the analog inputs are terminated using 3.5 mm (0.138 inch, 138 mil) pitch screw-type terminal blocks. In addition, two jumpers allow for externally supplied analog V+ and analog reference voltage inputs via a terminal block. A standard PCB-mount DC-barrel type connector is used for DC power. The power input can range from 6 to 12V DC (9V is optimal). A 7805 in a TO-220 package is used for voltage regulation to 5V on the PCB. Figure 10-21 shows the schematic created by Fritzing.

 The schematic notation used in Figure 10-21 illustrates what happens when the parts in a tool's library don't all follow the same conventions for size and spacing. This is a common issue with open source tools because not every contributed part may have followed the same rules. That doesn't mean it won't work; it just looks odd.

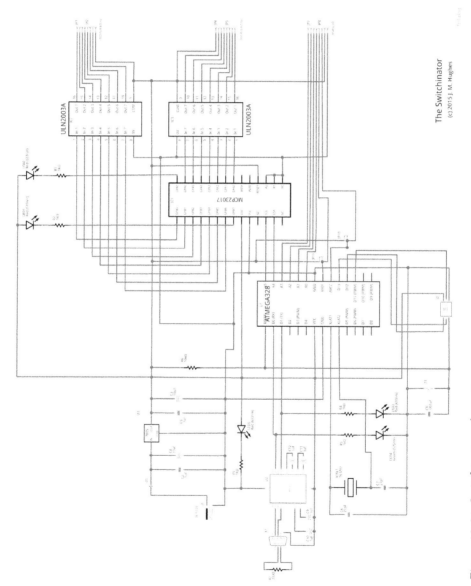

Figure 10-21. Switchinator schematic

The MCP23017 Digital I/O Expander

The MCP23017 with an I2C interface and its sibling, the MCP23S17 with an SPI interface, are register-controlled I/O devices for routing binary signals between a master device and up to 16 discrete digital inputs or outputs. The discussion of the internal workings of the MCP23017 also applies to the MCP23S17.

These devices are used in I/O expander shields such as the ones described in Chapter 8, and provide a simple means to extend the I/O capabilities of an MCU. Using the addressing pins it is possible to connect up to 8 MCP23017 ICs to an AVR MCU, resulting in 128 channels of discrete digital I/O.

The behavior of the MCP232017 is defined by the contents of a set of control and data registers. The master can read or set register values at any time via either the I2C or the SPI interface. Figure 10-22 shows a block diagram of the MCP23017. The MCP23S17 is identical except for the incorporation of an SPI interface instead of the I2C interface.

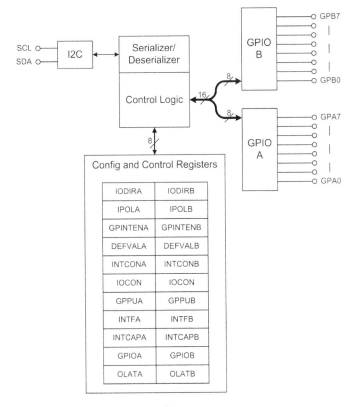

Figure 10-22. The MCP23017 I/O expander

The MCP23017 has 22 internal registers arranged as 11 pairs with an A register for the GPIO port labeled *A*, and a B register for the *B* port. Port direction (input or output) is set with the appropriate IODIRA or IODIRB register. Port polarity (active high or active low) is set with the IPOLA or IPOLB registers. GPINTENA and GPINTENB enable interrupt generation. GPPUA and GPPUB enable internal pull-ups. The signal at the port A or port B pins is read from the GPIOA or GPIOB registers. The OLATA and OLATB registers return the states of the internal output latches, which are set when a port is in output mode and data is written into the GPIOA or GPIOB registers. Table 10-9 lists the full register set. Note that the register addresses are in hex (Addr = address in MCP23017 register space), and POR/RST translates as power-on-reset or external reset.

Table 10-9. MCP23017 control registers (IOCON.BANK = 0)

Register	Addr	Bit 7	Bit 6	Bit 5	Bit 4	Bit 3	Bit 2	Bit 1	Bit 0	POR/RST val
IODIRA	00	I07	I06	I05	I04	I03	I02	I01	I00	1111 1111
IODIRB	01	I07	I06	I05	I04	I03	I02	I01	I00	1111 1111
IPOLA	02	IP7	IP6	IP5	IP4	IP3	IP2	IP1	IP0	0000 0000
IPOLB	03	IP7	IP6	IP5	IP4	IP3	IP2	IP1	IP0	0000 0000
GPINTENA	04	GPINT7	GPINT6	GPINT5	GPINT4	GPINT3	GPINT2	GPINT1	GPINT0	0000 0000
GPINTENB	05	GPINT7	GPINT6	GPINT5	GPINT4	GPINT3	GPINT2	GPINT1	GPINT0	0000 0000
DEFVALA	06	DEF7	DEF6	DEF5	DEF4	DEF3	DEF2	DEF1	DEF0	0000 0000
DEFVALB	07	DEF7	DEF6	DEF5	DEF4	DEF3	DEF2	DEF1	DEF0	0000 0000
INTCONA	08	IOC7	IOC6	IOC5	IOC4	IOC3	IOC2	IOC1	IOC0	0000 0000
INTCONB	09	IOC7	IOC6	IOC5	IOC4	IOC3	IOC2	IOC1	IOC0	0000 0000
IOCON	0A	BANK	MIRROR	SEQOP	DISSLW	HAEN	ODR	INTPOL	—	0000 0000
IOCON	0B	BANK	MIRROR	SEQOP	DISSLW	HAEN	ODR	INTPOL	—	0000 0000
GPPUA	0C	PU7	PU6	PU5	PU4	PU3	PU2	PU1	PU0	0000 0000

Register	Addr	Bit 7	Bit 6	Bit 5	Bit 4	Bit 3	Bit 2	Bit 1	Bit 0	POR/RST val
GPPUB	0D	PU7	PU6	PU5	PU4	PU3	PU2	PU1	PU0	0000 0000
INTFA	0E	INT7	INT6	INT5	INT4	INT3	INT2	INT1	INT0	0000 0000
INTFB	0F	INT7	INT6	INT5	INT4	INT3	INT2	INT1	INT0	0000 0000
INTCAPA	10	ICP7	ICP6	ICP5	ICP4	ICP3	ICP2	ICP1	ICP0	0000 0000
INTCAPB	11	ICP7	ICP6	ICP5	ICP4	ICP3	ICP2	ICP1	ICP0	0000 0000
GPIOA	12	GP7	GP6	GP5	GP4	GP3	GP2	GP1	GP0	0000 0000
GPIOB	13	GP7	GP6	GP5	GP4	GP3	GP2	GP1	GP0	0000 0000
OLATA	14	OL7	OL6	OL5	OL4	OL3	OL2	OL1	OL0	0000 0000
OLATB	15	OL7	OL6	OL5	OL4	OL3	OL2	OL1	OL0	0000 0000

The MCP23017 has other functionality, all of which may be useful for certain applications. For simple digital I/O applications all that really needs to happen is reading or writing the GPIO register. See the Microchip datasheet for the MCP23017/MCP23S17 (*http://bit.ly/micro-mcp*) for additional details about the registers.

Another aspect of the MCP23017 is how to communicate with it. A two-phase protocol is used to access the internal registers. The first step is to send a control byte and the address of the register to access. The second step is to either read from or write to the register. The most recently selected register will remain active until a new register is selected with a control byte–register address pair of bytes.

The control byte is the 7-bit address of the MCP23017 with a read/write (R/W) bit. In a system with a single IC this should always be 0 for the A0, A1, and A2 bits with the sixth bit predefined as 1, which results in byte values of 0x20 if this is a write operation, and 0x21 if we are reading data (the read/write bit is the least significant bit in the control byte).

There are several libraries available for interaction with the MCP23017 and MCP23S17 devices, and writing a custom library is not difficult. The snippet of code in Example 10-7 shows how to address the I/O direction register and the GPIOA register as an input using the I2C interface and the Wire library.

Example 10-7. Accessing the IODIRA and GPIOA registers in the MCP23017

```
// Send control and register address bytes
Wire.beginTransmission(0x20);
Wire.write(0x00);
// Write data to IODIRA register
Wire.write(0xFF);
// End session
Wire.endTransmission();

// Send control and register address bytes
Wire.beginTransmission(0x20);
Wire.write(0x12);   // GPIOA
Wire.endTransmission();
// Read data from port A
Wire.requestFrom(0x20, 1); // set R/W bit to 1 (read)
portval = Wire.read();  // still points to 0x12
```

Since the GP I/O ports are configured as inputs by default at power-up or after a reset (see Table 10-9), the first step could have been omitted. To configure the port as an output the IODIRA bits would be set to 0. This is shown in the code snippet in Example 10-8.

Example 10-8. Writing data to the output port

```
// Send control and register address bytes
Wire.beginTransmission(0x20);
Wire.write(0x00);
// Write data to IODIRA register
Wire.write(0x0);
// End
Wire.endTransmission();

// Send control and register address bytes
Wire.beginTransmission(0x20);
Wire.write(0x12);   // GPIOA
// Write data to port A
Wire.write(portval);
Wire.endTransmission();
```

The SPI interface in the MCP23S17 behaves in a similar fashion, and even the address pins can be used if desired. All other behaviors are identical to the MCP23017. The operations in the MCP23017 happen as fast as the I/O, either I2C or SPI.

I used a 16 MHz crystal in the Switchinator for the MCU clock, mainly because I have a bunch of them. For the Switchinator the internal RC oscillator in the AVR would probably do just fine. While the crystal will allow the MCU to run at a specific rate, it also adds some complications with the internal *fuse bits* the AVR MCU employs for internal configuration. Reading and setting these is discussed in "Setting the AVR MCU Fuse Bits for a 16 MHz Crystal" on page 418.

The four analog inputs are brought out to a pair of four-position terminal blocks. The remaining positions are used for V+, ground, and optional analog reference and AVCC inputs. Jumpers JP10 and JP11 are used to select the external voltage sources by removing them from the PCB.

Two of the MCU's digital pins (D0 and D1) are used for the RS-232 serial interface. Three of the digital pins (D11, D12, and D13) are used for the ICSP programming interface. The remaining digital pins are unassigned. Four of the analog inputs (A0, A1, A2, and A3) are used as general-purpose analog inputs. Pins A4 and A5 are used for the I2C interface between the MCU and the MCP23017 IC. Table 10-10 lists the pin assignments for the Switchinator.

Table 10-10. Switchinator MCU pin usage

Pin	Function	Pin	Function	Pin	Function
D0	RxD for RS-232	D7	Not used	A0	General-purpose analog input
D1	TxD for RS-232	D8	Not used	A1	General-purpose analog input
D2	Not used	D9	Not used	A2	General-purpose analog input
D3	Not used	D10	Not used	A3	General-purpose analog input
D4	Not used	D11	ICSP MOSI	A4	I2C SCL
D5	Not used	D12	ICSP MISO	A5	I2C SDA

Based on the schematic presented in Figure 10-21 we can generate a detailed parts list as shown in Table 10-11.

Table 10-11. Switchinator parts list

Quantity	Description	Quantity	Description
2	Capacitor, .1 uF	4	Red LED
2	Capacitor, 10 uf	1	Green LED
2	Capacitor, 27 pF	5	1K ohm resistor, 5%, 1/4 W
1	Capacitor, .001 uF	1	10K ohm resistor, 5%, 1/4 W
4	Capacitor, 1 uF	1	330 ohm resistor, 5%, 1/4 W
1	1N4001 diode	1	Switch, tactile pushbutton
1	MCP23017 I2C I/O expander	1	ATmega328 MCU, 28-pin DIP
2	ULN2003A output driver	1	MAX232 RS232 transceiver, 16-pin DIP
1	Power jack, 5.5 mm barrel	1	7805 voltage regulator, TO-220
1	AVR ISP connector package, 2 × 3 pins	1	DB9 connector, PCB mount, male
6	Terminal block, 3.5 mm pitch	1	16 MHz crystal
2	2-pin jumper header	1	Custom PCB

Software

As with the GreenShield project, the most complex part of the Switchinator is actually the software. Figure 10-23 shows a block diagram of the software.

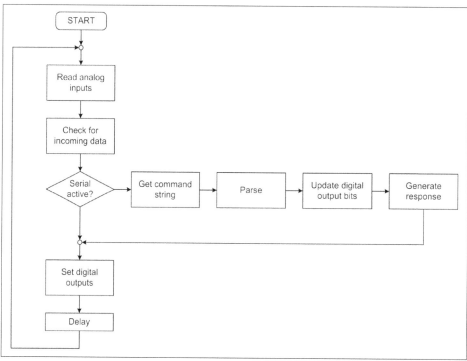

Figure 10-23. Switchinator software block diagram

When the Switchinator is waiting for command input it will send a single ">" character followed by a space. Responses are preceded with a "<" character and a space, followed by the response data. The use of the ">" and "<" characters is mainly for the convenience of software running on a control host system. All input and output lines are terminated with a newline (\n or 0x0A) character.

On every pass through the main loop the analog inputs are read and the values saved. These are returned to the control host when an analog reading is requested, which means they will be updated at the rate of the main loop.

Next, the software checks for incoming serial data. If there is data in the input buffer the input read function is called to read characters until a newline (a linefeed character), ASCII value 10 (0x0A or \n), is encountered. The parser then extracts the command from the string, parses out the parameters, and executes the commanded operation.

Control of the Switchinator is done using a simple command-response protocol. For every command or query sent by the control host the Switchinator will respond with a single response. The Switchinator will never initiate a communications transaction with the control host. The complete command-response protocol is shown in Table 10-12.

Table 10-12. Switchinator command-response protocol

Command	Response	Description	In/out format
A:n	A:n:val	Get analog input n in raw DN	Hex
R:nM	R:n:val	Read status of output n	Hex (0 or 1)
W:n:val	OK	Write 0 or 1 to output n	Hex
S:val	OK	Set all outputs to hex value	4-digit hex
G:?	G:val	Get hex value for all outputs	4-digit hex

The command set is simple, and only the W (write) command uses two parameters instead of just one. All parameter and response values are in hexadecimal notation. This allows digital port numbers to be single hex digits, while the analog input values (the A command) and the mass set and retrieve commands (S, or set all ports, and G, get all ports, respectively) use four-digit hex values. The digits 0 and 1 are nominally in hex, but they are the same in decimal notation.

Error detection is handled by examining the string returned from the Switchinator. If the A, R, or G commands return the string sent, then an error occurred. If the S or W commands encounter an error they will return the original string; if they succeed, then the string "OK" will be returned.

Octal and Hexadecimal Numbers

If you already know how to work with octal and hexadecimal notation, then feel free to skip this sidebar. But if not, then let's take a short history break and see where these software number systems came from and how they work.

Long ago, when computers filled entire rooms and generated enough heat to cook instant oatmeal for breakfast, it quickly became apparent that handling numbers in binary wasn't going to work out very well. And because the machines were inherently binary, decimal values didn't map gracefully to the bit patterns seen on the control panel lights and stored in memory.

The solution was to use a number base other than 2 or 10 to represent numeric values. It turns out that two base systems have direct mapping to binary values: the octal (base-8) system and the hexadecimal (base-16) system.

The base-8 octal number system was popular with computers that used 12-, 24-, or 36-bit words for data and addressing, since each size value can be evenly divided by 3.

An octal digit can have a value between 0 and 7, so octal digits each map to three bits. Table 10-13 shows how it works.

Table 10-13. Octal number system for a 12-bit computer

Decimal	Binary	Octal	Decimal	Binary	Octal
0	000 000 000 000	0000	9	000 000 001 001	0011
1	000 000 000 001	0001	10	000 000 001 010	0012
2	000 000 000 010	0002	20	000 000 010 100	0024
3	000 000 000 011	0003	30	000 000 011 110	0036
4	000 000 000 100	0004	40	000 000 101 000	0050
5	000 000 000 101	0005	50	000 000 110 010	0062
6	000 000 000 110	0006	100	000 001 100 100	0144
7	000 000 000 111	0007	200	000 011 001 000	0310
8	000 000 001 000	0020	511	000 111 111 111	0777

Octal numbers can still be found in modern Unix and Linux computers as the file permissions bits, but otherwise they are scarce these days. No one runs 12- or 24-bit general-purpose computers anymore, although some research groups have created 24-bit microprocessor designs for specific applications, and the DSP56303 DSP (digital signal processor) device from Freescale can be used as a 24-bit machine.

As computer architectures shifted to data and address sizes that are multiples of 4, the octal system quickly became cumbersome. The solution was the hexadecimal base-16 system, or just *hex*. In the hex number system each digit can represent a value from 0 to 15, or binary 0000 to 1111. In other words, each hex digit is a "nibble," and two hex digits represent an 8-bit "byte."

There is a problem with hex notation, however. The base-10 numbers cannot represent any value greater than 9 as a single digit. The value of 10 is written with two digits. During the 1950s different schemes were proposed and tried to address the notation problem, some of which are rather strange-looking to us today. After a few years the dust settled and we ended up with 0, 1, 2, 3, 4, 5, 6, 7, 8, 9, A, B, C, D, E, and F for the values 0 through 15. Table 10-14 shows the relationship between decimal, binary, and hex values for a selection of numbers.

In software you will sometimes see hex values written as 0x3F in C and C++, 3Fh in assembly language, $3F in Forth, or %3F in a URL. A numeric seven-segment LED display can handle hex by modifying the display of some of the digits to get symbols that look like A, b, c, d, E, and F, as shown in Figure 10-24.

Early on it quickly became apparent in both the octal and hex numbering systems that there are some special values, sometimes referred to as "magic numbers," that keep popping up. For example, in hex the value 0xFF is a byte with all bits set to 1.

0x5A5A is an alternating pattern (binary 0101 1010 0101 1010) sometimes used as filler for checking stack usage or to overwrite a disk drive to obscure previously stored data. 0x7F is 127, 0x1FF is 511, and 0x3FF is 1023. 0x3FF (1023) is the maximum value the 10-bit ADC in an ATmega328 AVR can produce, and 0x1FF is the midpoint of its range.

Table 10-14. Hexadecimal number system for a 16-bit computer

Decimal	Binary	Hex	Decimal	Binary	Hex
0	0000 0000 0000 0000	0000	10	0000 0000 0000 1010	000A
1	0000 0000 0000 0001	0001	20	0000 0000 0001 0100	0014
2	0000 0000 0000 0010	0002	30	0000 0000 0001 1110	001E
3	0000 0000 0000 0011	0003	40	0000 0000 0010 1000	0028
4	0000 0000 0000 0100	0004	50	0000 0000 0011 0010	0032
5	0000 0000 0000 0101	0005	100	0000 0000 0110 0100	0064
6	0000 0000 0000 0110	0006	200	0000 0000 1100 1000	00C8
7	0000 0000 0000 0111	0007	255	0000 0000 1111 1111	00FF
8	0000 0000 0000 1000	0008	511	0000 0001 1111 1111	01FF
9	0000 0000 0000 1001	0009	4095	0000 1111 1111 1111	0FFF

A B C D E F

Figure 10-24. Hex display with a conventional 7-segment LED display

It is possible to do math with either octal or hexadecimal numbers, and in assembly language programming there is often a need to add or subtract hex address values (for indirect jump or relative branch instructions, for instance). Doing things like multiplication and division with hex numbers is possible, but it does take some practice. Unless you want or need to do a lot of low-level assembly language programming, hex math is probably not something you need to add to your skill set, but to work with microcontrollers you should be able to translate from hex to binary (and vice versa) on a digit-by-digit basis.

The A command accepts a single digit from 0 to 3, and the analog data is the actual value returned by the ADC in the AVR MCU. It is returned as a hexadecimal value of 1 to 3 digits, with 0x3FF as the maximum possible value.

The R and W commands have the specific digital output port number in the form of a single hex digit, ranging from 0 to 0xF (15). Output channels 7 and 0xF are used for on-board LEDs; they are not brought out through the ULN2003A drivers.

Notice in Table 10-12 that to control or get the status of more than one output the S and G commands are used with a hexadecimal value. With this scheme we can enable (set to an *on* state) noncontiguous outputs, or read back the state of all the outputs.

If, for example, we wanted to set outputs 5, 6, 12, and 13 to an on state, then we would send the value 3060h, which translates to binary as follows:

```
Bit   | 15  11    7    3  0
---------+----+----+----+---
Binary|  0011 0000 0110 0000
Hex   |    3    0    6    0
```

Remember that the channel numbering is zero-based.

Sending a hexadecimal value will cause outputs to change state to match the command value. That means that if an output is on, and the command value has a zero in that position, it will be set to off.

Use a read-modify-write operation (get bits, change data, set bits) to set or clear a specific output bit without disturbing other bits. This is how the W command works.

The last step in the loop transfers the digital output state bits to the hardware. If no changes have been made since the last update the outputs will not do anything; otherwise, bits that have been changed will appear as changes in the on or off state of the ULN2003A outputs.

Prototype

The prototype mainly focuses on the RS-232 interface and the command-response control protocol, so the prototype hardware consists mainly of an ATmega328 mounted on a solderless breadboard. An RS-232 adapter module is used as a stand-in for the MAX232 in the final board. Figure 10-25 shows the prototype fixture.

The RS-232 module is a self-contained unit with a MAX3232 IC. This part is functionally equivalent to the MAX232 used in the Switchinator. The MAX3232 is nominally a 3.3V part, although it is 5V tolerant. It is also slightly more expensive than the MAX232. Figure 10-26 shows a close-up of the RS-232 interface module.

These modules are available from multiple vendors for between $3 and $6. A Google search for "arduino rs232 module" or "arduino rs232 converter" will turn up numerous results.

Figure 10-25. Switchinator prototype fixture

Figure 10-26. RS-232 interface module

Most desktop PCs still have a single RS-232 port and a DB-9 connector on the rear panel, but if you have a late-model notebook PC you may not have a serial port available. To get around this, you can use a USB-to-RS232 adapter. These range in price from around $4 to over $30, with some specialty types costing even more. For more details about RS-232 and the connectors used with it, I would recommend my books *Real World Instrumentation with Python* and *Practical Electronics: Components and Techniques* (see Appendix D). Both contain sections dealing specifically with RS-232, DB-9 connectors, gender changers, and how to wire a DB-9 to allow RxD/TxD communications without the handshaking signals.

Setting the AVR MCU Fuse Bits for a 16 MHz Crystal

Unlike the other examples in this book, the Switchinator is not an Arduino board. It can be software compatible if the bootloader is installed, but that's not a requirement. It is basically just an AVR MCU design.

That means that the MCU isn't preconfigured as it is with an Arduino board. A fresh MCU from Atmel will run in RC clock mode using an internal oscillator at about 8 MHz. If a bootloader has been installed, then some internal switches are set to indicate that a portion of the flash memory space has been reserved. MCUs with preloaded bootloader firmware will also typically have the fuse bit set for a 16 MHz external crystal, but a brand-new part will have only the default configuration.

In order to configure the MCU to use the 16 MHz crystal as its clock source we need to set the fuse bits. For more details on the fuse bits used in the AVR MCUs refer to "Fuse Bits" on page 60 in Chapter 3. The definitive source of information is, of course, the Atmel datasheet for the AVR MCU you are using.

If you want to use the Arduino configuration for AVRDUDE (and you have a Linux system), then the following command will do the job for an ATmega328:

```
sudo avrdude -cusbtiny -p atmega328 -U lfuse:w:0xFF:m \
-U hfuse:w:0xDE:m -U efuse:w:0x05:m
```

For an ATmega328p part simply change the part parameter, like so:

```
sudo avrdude -cusbtiny -p atmega328p -U lfuse:w:0xFF:m \
-U hfuse:w:0xDE:m -U efuse:w:0x05:m
```

On a Linux system, if you have set the permissions rule for the USB I/O then you won't need to use sudo to run avrdude. For more information about the command-line options and interactive commands supported by avrdude, refer to the online manual (*http://bit.ly/avrdude-manual*).

The last step is to inform the compiler that you are now running at 16 MHz. You do this by defining F_CPU like so:

```
#define F_CPU 16000000UL
```

This line is usually added by the Arduino environment when working with a conventional Arduino board, but you may need to explicitly specify the clock speed with a custom target board.

The Windows and Mac OS versions of AVRDUDE behave the same way as with Linux. See this tutorial (*http://www.ladyada.net/learn/avr/setup-win.html*) for more information about the Windows version of AVRDUDE.

Prototype software

The prototype software is essentially the same as the final version, only without the code to set the digital outputs via the MCP23017. The main focus with the prototype is the implementation of the command-response protocol. The states of the outputs are represented as bits in a 16-bit word in the software.

"Software" on page 420 contains a detailed discussion of the software, so rather than put it here and possibly repeat parts of it later, I would recommend that you look there for the details. The software is compiled using the Arduino IDE and then uploaded to the AVR MCU using an Adafruit USBtinyISP ICSP interface device.

I used the Arduino IDE to handle compiling chores, but I disabled its internal editor in the Preferences dialog. This allowed me to use a different editor (I write commercial and scientific software for living, so I have some definite preferences when it comes to text editors—I'm not a big fan of the editor in the Arduino IDE). The board was set to "Duemilanove with ATmega328" and the programmer to "USBtinyISP."

On a Linux system the USBtinyISP doesn't use a pseudo-serial port, but instead communicates directly with the underlying USB I/O subsystem. Attempting to run the Arduino IDE with the programmer will initially result in a permissions error. You can run the Arduino IDE using sudo, but this is not a convenient way to transfer your code. To get around this you need to add an access rule for the udev handler.

Create a file in */etc/udev/rules.d* called *tinyusb.rules*, and add the following string to it:

```
SUBSYSTEM=="usb", ATTR{product}=="USBtiny", ATTR{idProduct}=="0c9f", \
ATTRS{idVendor}=="1781", MODE="0660", GROUP="dialout"
```

I used vi and sudo to accomplish this:

```
sudo vi tinyusb.rules
```

Then restart the udev subsystem to make the new rule take effect:

```
sudo restart udev
```

You could also use some other type of programming device (even another Arduino, as described in Chapter 6). I have an Atmel-ICE, but ended up not using it under Linux because I don't have the latest version of AVRDUDE and I was too lazy to build it and fiddle with the configuration files. It works with Atmel's AVR Studio software, so if you're using Windows you might want to take that route. On my Linux system the little gadget from Adafruit works just fine.

Using the upload icon in my version of the IDE causes it to try to start a transfer using the USB-to-serial method built into Arduino boards (and supported by a USB-to-serial converter like the one mentioned earlier). This won't work with the USBtinyISP, so instead I use the File→Upload Using Programmer option.

One thing I noticed is that the AVRDUDE transfer software is slow getting started, but runs quickly once it has established a solid communications link with the AVR MCU. You can see this both in the Arduino IDE and by running AVRDUDE from the command line. Don't panic if it looks like things are hung; they aren't. If there is a problem AVRDUDE will eventually time out and tell you what went wrong.

Prototype testing

Testing the prototype is straightforward, and this section describes tests to specifically exercise the command parser. The ability to successfully interact with the software will demonstrate that the RS-232 interface portion of the code is working correctly. It is assumed that the external RS-232 transceiver works as intended.

First, the output (OUT), status (ST), and input (AN) commands are tested. The OUT:A:1 and OUT:A:0 commands are used to set the outputs either all on or all off. The state of each output is held in memory so there is no need for output hardware at this point.

With the ST:n:? command the n parameter is a single digit from 0 to 13 (D in hex). Note that the OUT:n:0 and OUT:n:1 command forms also use a single digit. If the software is working correctly it should be possible to set all outputs to off (0), and then selectively enable and disable any of the outputs from 0 to 13 without altering any of the other outputs.

The analog input (AN) command is tested by applying a variable voltage source (0 to 5V only) to A0 through A3 and requesting the value. As the input voltage is changed the returned data should change as well. The input value is returned as a three-digit hexadecimal value. The two most significant bits are always 0 (the AVR MCU only has a 10-bit ADC).

The SP:val and GP:? commands use a four-digit hexadecimal value, as described earlier. Testing will involve setting all odd outputs on and all even-numbered outputs off, then checking the states of each using the ST command. Then the odd-numbered outputs will be set to off and the even outputs set to on, and the states of the output bits will again be checked using the ST command.

Software

Although it is possible, and often advisable, to create a simplified version of the software for use with the prototype, in the case of the Switchinator it begins life as a multifile code set. The main file is, of course, *Switchinator.ino*. The other files in the set contain the global definitions, the global variables, the command parser, the response generator, and the I/O control code. The I/O module, *sw_io.cpp*, is not necessary in the prototype version of the software.

Source code organization

The Switchinator source code is comprised of eight files, or modules, described in Table 10-15. The primary module, *Switchinator.ino*, contains the setup() and loop() functions. It also references the other modules using #include statements.

Table 10-15. Switchinator source code modules

Module	Function
Switchinator.ino	Primary module containing setup() and loop()
sw_defs.h	Constant definitions (#define statements)
sw_gv.cpp	Global variables
sw_gv.h	Include file
sw_io.cpp	Hardware I/O functions
sw_io.h	Include file
sw_parse.cpp	Command parsing
sw_parse.h	Include file

Full listings of all the Switchinator source files are available on GitHub.

Software description

As with any Arduino program, the action begins in the main sketch file. The contents of the *Switchinator.ino* main file are listed in Example 10-9.

Example 10-9. Switchinator.ino

```
//------------------------------------------------------------------
// Switchinator.ino
//
// Created for "Arduino: A Technical Reference," 2016, J. M. Hughes
// Chapter 10
//------------------------------------------------------------------

#include <stdint.h>
#include <Wire.h>

// the MCP23017 library
#include <IOexp.h>

// Include all source modules
#include "sw_parse.h"
#include "sw_gv.h"
#include "sw_defs.h"
#include "sw_io.h"

// define the clock rate
```

```
#define F_CPU 16000000UL

bool waitinput = false;
bool usedelay  = false;

void setup()
{
    Serial.begin(9600);

    // Startup banner and start flag string
    Serial.println();
    Serial.println("SWITCHINATOR V1.0");
    Serial.println("READY");
    Serial.println("####");      // Start flag
}


void loop()
{
    // Read analog inputs and update GV array
    ScanAnalog();

    // Emit only one input prompt
    if (!waitinput) {
        waitinput = true;
        Serial.print(INPPCH);
    }

    // Check for incoming command
    if (GetCommand()) {
        waitinput = false;       // Reset input prompt output flag
    }
    else {
        ClearBuff(0);
        ResetBuffLen();
        // if no input then enable loop delay
        usedelay = true;
    }

    // Parse command (if received)
    if (DecodeCommand()) {
        // Return response to control host (or user)
        Serial.println();
        Serial.print(OUTPCH);
        Serial.println(gv_cmdinstr);
    }

    // Update the digital outputs using state bits
    SetDigBits();

    // delay briefly if no input detected
    if (usedelay) {
```

```
        usedelay = false;
        delay(50);
    }
}
```

The setup() function contains the usual initialization statements along with the initial startup message. The loop() function continuously updates the output bits and checks for incoming commands from a control host system via the GetCommand() function in the *SW_parse.cpp* module.

Referring back to Figure 10-23 the block "Get command string" refers to the GetCommand() function in the *sw_parse.cpp* module, the "Parse" and "Generate response" function blocks are contained in the DecodeCommand() function, and "Update digital output bits" is the SetDigBits() function in the *sw_io.cpp* module. Figure 10-27 shows how the output bit state buffer, gv_statebits, is used to hold an internal representation of the output bits. All bit modification functions operate on gv_statebits, not on the actual digital outputs.

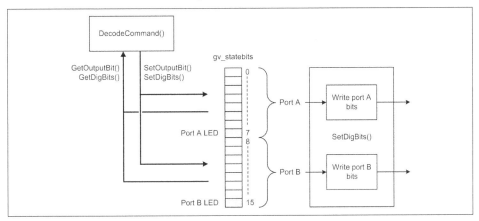

Figure 10-27. Switchinator's virtual bit buffer in operation

Note that writing 0x00 to either port A or port B will also enable the associated port LED. Writing an 0xFF to either port will disable the LED. The output bits are updated about once every 50 ms if no serial input is present to be parsed and decoded.

Fabrication

For the hardware design I used the Fritzing design tool. This is an integrated virtual breadboard, schematic capture, and PCB layout tool that is easy to install and relatively easy to use. Fritzing has a large and vibrant user base and lots of preloaded parts in its library, and best of all, it's free. Some Linux distributions may have an older version of Fritzing in their repository, but you can download the latest version from *http://fritzing.org/home*. I used version 0.8.5 running on a Kubuntu 14.04 LTS

Linux system. There are also versions of Fritzing available for Windows and Mac OS X.

Compared to other tools I've worked with over the years, I was pleasantly surprised at how easy Fritzing's user interface was to use. To be honest, I wasn't all that impressed by the autorouter, but then I've seen high-end autorouters struggle with things I thought would be easy. And autorouting software is very challenging to program, so I wasn't expecting a miracle. I ended up doing the layout manually (and it shows, I'm sure). The DRC (design rules check) worked well, and while the schematic editor had some unique quirks, it too was completely usable. My main complaint with Fritzing is the parts library. It seems that not everyone is on the same page when it comes to dimensions for schematic symbols, so parts from one contributed library might be really small, while the stock parts supplied with Fritzing are actually nice, large symbols. You will see this when you look at the schematic in Figure 10-21. The upshot is that unless all the parts in a tool's library—both the symbols and the layout footprints—adhere to the same dimensioning constraints, it is difficult to get perfectly orthogonal lines or traces without doing some serious fiddling with the grid sizing and snap functions.

Once the schematic (shown in Figure 10-21) is complete, the PCB layout work can begin. The first time the PCB layout appears it will have no traces, just the rat's nest lines to indicate which pins are connected on the parts. In Figure 10-28 I've already placed the components where I think they should go, and Fritzing is showing some of the rat's nest lines. Clicking and holding on a pin on any part will highlight all the other places it should be connected in bright yellow.

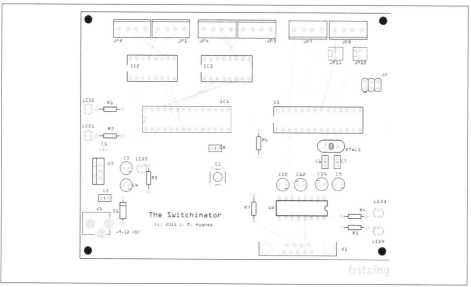

Figure 10-28. PCB after parts placement, but before routing

The components are grouped by function: the power supply is in the lower-left part of the PCB; the MCP23017 and the two ULN2003A parts are in the upper-left region; the MCU, crystal, ICSP connector, and analog inputs are located at the upper right; and the serial interface is located at the bottom right of the layout.

You might notice that I reversed the analog input terminal block symbols to simplify the wiring. They will still be mounted correctly, and since a prototype PCB like this doesn't have a top layer silkscreen it won't make any difference.

Some of the trace routing is rather tortuous, due mainly to the use of through-hole parts. Vias are used to shift the traces between the top and bottom sides of the PCB in order to route without collisions. Figure 10-29 shows the final version of the PCB that went out for fabrication.

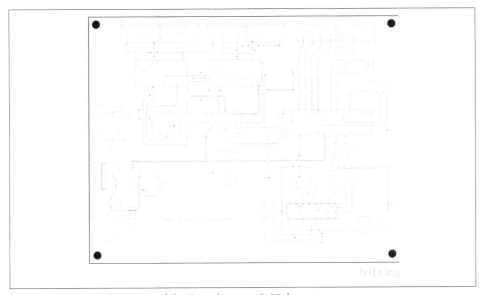

Figure 10-29. Final version of the Switchinator PCB layout

Before sending out the Gerber files, I checked the design with the Gerbv tool (which was introduced earlier with the GreenShield). A screenshot is shown in Figure 10-30.

Fortunately, the fabrication house (Advanced Circuits) discovered two pads that were almost, but not quite, connected. It would have been easy to fix on the workbench, but it would have been a challenge to find the problems. Good thing someone was paying attention and checked the layouts.

The Switchinator is a through-hole PCB, so assembly is straightforward. Figure 10-31 shows the bare PCB before any parts are installed.

The first parts to solder into place are the DC jack and the 7805 regulator, along with the associated power supply components, consisting of D1, C2, C3, C4, and C5. Note that there is no C1 (it vanished in an earlier revision). To test the power supply we also need to install R5 and LED5. Once these parts are installed we can connect a 9 to 12V DC wall supply via the DC power jack and verify that the regulator is producing 5 volts DC at the anode of LED5 (which should be glowing).

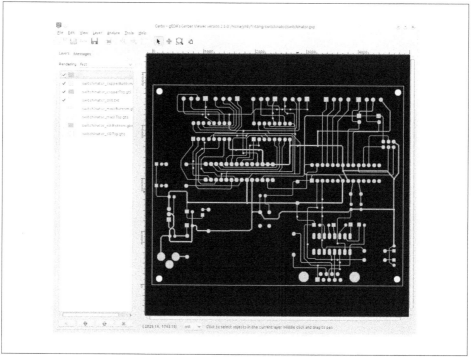

Figure 10-30. Gerbv PCB layout display

Next we will install the RS-232 connector, labeled X1 in the schematic; U2 (the MAX232 IC); and C9, C10, C11, and C12. Using the prototype breadboard we can attach the Rx and Tx signals from pins 2 and 3 of the MCU to the U1 pins 2 and 3 on the PCB. Then we run the software already on the AVR MCU and verify that our integrated RS-232 is working correctly. Remember to tie the ground and 5V DC power from the PCB to the breadboard (and disconnect the breadboard's power supply, of course).

With the power supply and the RS-232 interface installed and running, we can now install U1 (the MCU), IC1, IC2, and IC3. LED1, LED2, R1, and R2 can also be soldered in place, along with the six terminal blocks. Next, the crystal, C6, C7, C8, R3, R4, R6, LED3, and LED4 are installed. The reset switch, S1, and the ICSP connector are the last things to mount on the PCB.

Figure 10-31. The bare Switchinator PCB

Figure 10-32 shows the finished PCB with everything installed and ready to go. The maximum current that the ULN2003A drivers can supply is determined by the power supply and by the ratings for the ICs themselves. The datasheet states that the ULN2003A is capable of supplying around 300 mA per channel, or about 2.1A per IC. By comparison, the ATmega328 and the MCP23017 draw very little current, so the major concern will always be the ULN2003A devices. A power source capable of supplying at least 5A should be more than sufficient if you want to use the Switchinator for something like a small CNC tool or an LED display controller.

Acceptance Testing

Final testing of the finished Switchinator is largely a repeat of the testing done with the prototype. The big difference now is that there are two ULN2003A drivers and an integrated RS-232 interface based on a MAX232 IC. The power supply also needs to be tested, as well as the analog inputs.

Figure 10-32. The finished Switchinator

Next Steps

Not all of the discrete digital I/O pins on the MCU have been used; 10 pins (including 6 capable of PWM output) are available. I didn't bring these out due to space constraints on the PCB, but it should be possible to modify the PCB for a couple of six-position pin socket headers. It will, however, require some clever use of vias to get the traces routed through the resulting traffic jam around the MCU.

The SPI interface is available on pins D11, D12, and D13, and these are wired to the ICSP connector. If you select an unused pin for the SS line you can connect an SPI module to the Switchinator. The analog pins can be used as digital I/O pins by referencing them as pins D14 through D19.

You may have also noticed that there is no fuse, and no input protection for the analog inputs. For an example of analog input protection, see Chapter 11 and the input circuit used for the signal generator.

Resources

This chapter has covered a lot of ground, and taken a few leaps of faith. Here are some resources that will help shed more light on the topics only touched on briefly in the text:

Reference texts
> There are numerous texts available that cover all aspects of electronics. These are some that I am particularly fond of that relate directly to material covered in this chapter (refer to Appendix D for ISBN numbers and even more recommended texts):
>
> - Jan Axelson, *Making Printed Circuit Boards*
> - Paul Horowitz and Winfield Hill, *The Art of Electronics,* 2nd Edition
> - J. M. Hughes, *Practical Electronics: Components and Techniques*
> - J. M. Hughes, *Real World Instrumentation with Python*
> - Simon Monk, *Fritzing for Inventors*
> - Matthew Scarpino, *Designing Circuit Boards with EAGLE*

Schematic capture and PCB layout
> Fritzing and Eagle aren't the only electronics CAD tools available, but they are commonly encountered when downloading schematics or board layouts created and posted online by others. Fritzing is free and open source, and there are many sources for parts definitions beyond those that are supplied with it. It is easy to learn and easy to use, and for Arduino projects it is a good choice.

The free version of Eagle has the features found in the commercial versions of the tool, so it presents a clear upgrade path when you want to explore the world of professional CAD/CAM tools for electronics. Just remember that the free Eagle has some limitations, and it's not intended to be used in any situation where you plan to make money from your work. For that you will need to purchase a commercial license.

Some other CAD tool options are the open source Linux tools like the gEDA suite and KiCad, both of which have features comparable with commercial products. For more information on the tools mentioned here, see their websites:

- Eagle (*http://www.cadsoftusa.com/download-eagle*)
- Fritzing (*http://www.fritzing.org*)
- gEDA (*http://www.geda-project.org*)
- KiCad (*http://kicad-pcb.org*)

PCB fabricators

There are many PCB fabrication firms that offer low prices and quick turnaround times. Check around on the Web, and if you happen to live in a large metropolitan area, also be sure to take a look at what services are available locally. I've listed Advanced Circuits mainly because I am most familiar with them, and I've never had a problem with their work. The folks at Fritzing.org offer a PCB fabrication service, too (you can access it from within the Fritzing tool):

- Advanced Circuits (*http://www.4pcb.com/bare-bones-pcbs*)
- Fritzing Fab (*http://fab.fritzing.org/fritzing-fab*)

Component sources

Throughout this book I've referred to many different sources for everything from single components to modules and complete Arduino boards. The following companies are some starting points to consider (for more pointers, see Appendix C):

- Adafruit (*http://www.adafruit.com*)
- All Electronics (*http://www.allelectronics.com*)
- DigiKey Electronics (*http://www.digikey.com*)
- Mouser Electronics (*http://www.mouser.com*)
- Newark/Element14 (*http://www.newark.com*)
- SparkFun (*http://www.sparkfun.com*)

CHAPTER 11

Project: A Programmable Signal Generator

Almost every electronics lab, be it large or small, needs signal sources. In some cases these might be simple sine wave generators, and some situations might call for more complex instruments like a function generator. The signal generator described in this chapter, shown in Figure 11-1, is capable of both sine and square wave outputs up to 40 MHz.

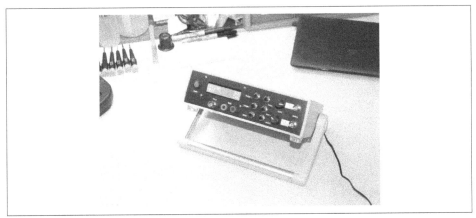

Figure 11-1. Homemade DDS signal generator

Of course, you could also purchase a signal generator. Prices vary, depending on the frequency range, features, and capabilities. You can find signal and function generator kits for anywhere from about $6 to around $50 for a device like the one shown in Figure 11-2. At the other end of the scale are commercial high-end instruments costing hundreds or thousands of dollars, or even more (for example, $72,000 for a *used* 20 GHz multifunction arbitrary waveform generator).

Figure 11-2. FG085 function generator kit (assembled)

While the kits are fine for generating waveforms, they may not have all the features or capabilities you want or need. They are, after all, someone else's idea, and that person may have a different notion of what makes a signal generator useful. By building it yourself you can get exactly what you want, and you can modify or expand it over time as your needs change.

A primary consideration is how the signal is generated. In other words, is the microcontroller doing all the work, or is it offloading the job of generating waveforms to a dedicated IC in the circuit? In the case of the FG085 function generator shown in Figure 11-2, an ATmega168 is used as the primary controller for the instrument, a CP2101 is used for the USB interface, and the signal is generated by an ATmega48 and a DAC comprised of a resistor array. There is nothing wrong with this approach, and it does allow the generator to do more than just generate sine and square waveforms; it can also produce arbitrary waveforms from preloaded data patterns (which can be very useful). The downside of using a microcontroller to do the signal generation is that it does put a limit on the highest possible frequency the device can generate. The limit is around 200 KHz for this particular function generator. This isn't a failing on the part of the FG085, or any device that uses a microcontroller in this manner—it's simply a fact.

A microcontroller can only run so fast, and that top clock rate sets a hard upper limit on how quickly it can change its outputs to create a cyclic signal. For many applications 200 KHz is fine, particularly in the world of embedded sensing and control. As I've pointed out before, things don't usually change very rapidly in the real world, at least on the microsecond time scales of microcontrollers.

If we want to go beyond about 200 KHz, we'll need to take a different route. Fortunately all the parts we need to build a signal generator capable of both sine and square wave output up to 40 MHz are readily available as Arduino-compatible modules.

You can read more about the FG085 on the JYE Tech website (*http://bit.ly/jye-fg085*). As a disclaimer, I'm not specifically endorsing this product, but I do own one and it lives on my workbench with the other pieces of test gear. I also have other signal and function generators, some fancy and some not, and each has a role to play during the development and testing of a new design.

In all fairness, I should point out now that the signal generator instrument described in this chapter is going to cost more than the $50 device shown in Figure 11-2. The total cost will be tallied when the final parts list is created in "Cost Breakdown" on page 468. You will need to decide for yourself if the cost is justified by the degree of control you will have over the design and its operation, and by the packaging method I have chosen. For me it was worth the cost, but for you it might not be.

Remember that since the main emphasis of this book is on the Arduino hardware and related modules, sensors, and components, the software shown is intended only to highlight key points, not to present complete ready-to-run examples. The full software listings for the examples and projects can be found on GitHub (*https://www.github.com/ardnut*).

Project Objectives

For this project the definition and planning phases (described in Chapter 10) are done in a single step. Physically the project isn't all that complicated, so we can compress these development steps to save effort and keep things moving along. The most complex part is the software, which is often the case when working with modular microcontroller hardware components.

The goal of this project is to build a signal generator suitable for use as a bench test instrument. While it is perfectly feasible to use the digital I/O of an Arduino to generate square waves or pulses, control a DAC of some sort, or use the PWM outputs to create a waveform simulating a sine wave, the output frequency range is limited by the speed of the microcontroller. There is, however, another way to use an Arduino to build a signal generator, and that involves a dedicated signal generator IC, the AD9850.

The AD9850 is a direct digital synthesis (DDS) chip that can output both sine and square waves. It can be programmed to generate output from 0 to 40 MHz. A readily available AD9850 module is described in Chapter 9 and shown in Figure 9-57. Because the AD9850 is handling the signal generation chores for the Arduino, we can use the free CPU cycles for other functions at the same time. Display update, detecting a gate input, and monitoring operation control inputs are just some of things the Arduino can do along with controlling the DDS IC.

This also means that the signal generator doesn't need interrupts to handle the control switches. Because the AD9850 is always running (except when the gate mode is active), the MCU can take the time to poll the pushbutton switches and it won't interfere with the signal output.

Definition and Planning

The goal of this project is to create a portable test instrument suitable for use on a workbench or in a similar environment. It will be powered from a wall-plug power supply, and the enclosure will have enough interior space to add batteries at a later date if so desired.

Signal outputs:

- Sine wave output (always active), 0 to 40 MHz, 0 to 1V P-P
- Square wave output (always active), 0 to 40 MHz, 0 to 5V P-P

Functional control inputs:

- External gate control input
- Control voltage input for VCO operation

User interface and controls:

- Two-line LCD display
- Frequency select inputs (pushbuttons)
- Signal output level controls
- Gate input jacks
- CV input BNC connector
- Power on/off

The whole project will be built into a plastic enclosure with a carry handle, like the one shown in Figure 11-3. The device will be powered by a wall-plug power supply (a so-called "wall-wart") with between 9 and 12V DC output.

On the front panel the signal generator will incorporate a two-line LCD display to show output frequency and status, various input controls, control voltage (CV) and gate inputs, and both sine wave (1V P-P) and square wave (5V P-P) outputs. The rear panel will have a DC barrel-type connector for the external power supply.

Figure 11-3. Portable instrument enclosure

The preliminary parts list for this project is given in Table 11-1. This parts list will be refined as we go along, but this gives us an idea of what will be used. An initial parts list will usually go through many changes between concept and final form, some minor and some rather drastic. It's all part of the design refinement process.

Table 11-1. Initial parts list

Quantity	Description	Quantity	Description
1	Arduino Uno	3	Female BNC connectors
1	AD9850 DDS module	2	Banana jacks (for gate input)
1	Prototype shield	1	Wall-plug power supply
1	Two-line LCD display	1	Plastic enclosure

Design

Now that we have some design objectives and a preliminary parts list, we can start to refine the design. We'll start with the intended functionality to define what exactly the device will do, and the controls and I/O needed to perform the intended functions.

After we have a clear idea of what the signal generator will do we can turn our attention to the enclosure, since that is where the parts will be mounted, and it will ultimately determine what we will be able to use for a display, control inputs, and I/O connectors. The objective here is to balance the need for a robust and compact enclosure with cost and fabrication considerations.

Because the project uses mostly prebuilt modules there is no significant circuit design or PCB layout involved. The various modules and signal connectors will need to be wired, so there will be some soldering, but we'll deal with that in the assembly step.

For the signal generator I elected to use pushbutton control inputs rather than a rotary encoder as in some of the designs for DDS signal generators found on various websites like Instructables (*http://www.instructables.com*). The reason is because the rotary encoder, while convenient for rapidly setting a value, really only does one thing: measure the amount and direction of rotation. Pushbuttons, on the other hand, can do many different things depending on how they are interpreted by software in context with other controls and the state of the devices they are controlling.

Functionality

The primary purpose of this instrument is to generate a signal at a specific frequency from 0 to 40 MHz. The AD9850 incorporates a built in comparator, and this is used to create a simultaneous square wave. The sine wave is a 1V P-P (peak-to-peak) signal, and the square wave is 5V P-P. The output frequency is continuously variable from 0 to 40 MHz under the control of an Arduino Uno. The output frequency may be externally varied by the application of a control voltage (CV), and it may be gated (or triggered) by an external signal (either active high or active low). Figure 11-4 shows a block diagram of the instrument with all the components identified in Table 11-1.

The block diagram in Figure 11-4 shows three major components. The Arduino handles all the control inputs—those from the user as well as the gate and CV inputs. The LCD displays the current state of the instrument, and the DDS module generates the sine and square wave outputs.

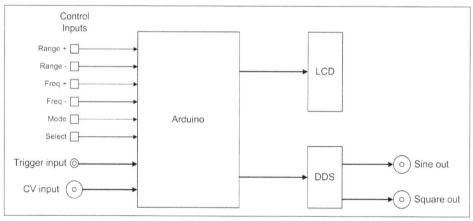

Figure 11-4. Signal generator block diagram

The block diagram doesn't provide low-level details such as pin numbers or polarities. That is not its purpose. A block diagram graphically shows how things relate to one another in a functional sense. It also serves as a check to see if the ambitions of the design have exceeded the capabilities of the components.

I've arranged Figure 11-4 so that the outputs are on the right, and the control inputs are on the left. The large circle symbol is a BNC-type connector for shielded coaxial cables, and the small circle symbol represents a banana-type jack.

A 2-line 16-character-wide LCD display will show the current frequency and the state of the gate and CV inputs. A set of pushbutton controls will be used to adjust the frequency and the operation of the external control inputs. Two potentiometers will be used to adjust the output level of the sine and square wave signals.

Enclosure

For this project I've selected the Bud Industries IP-6130 enclosure, which is shown in Figure 11-3. This enclosure includes a carrying handle that also doubles as a stand when it's on the bench, which I find very useful at times. You can download the datasheet from Mouser Electronics (*http://bit.ly/mouser-ip-6130*). The enclosure costs about $25, but I consider that to be a reasonable price to pay for something that will last a long time and looks professional. The basic dimensions are shown in Figure 11-5. For more details refer to the product datasheet.

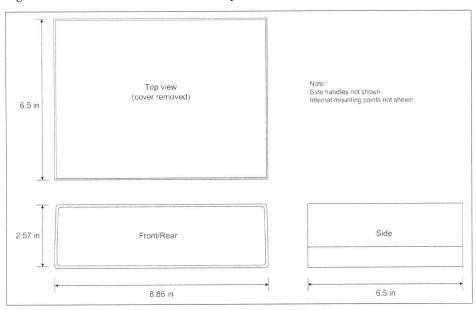

Figure 11-5. Portable instrument enclosure dimensions

One thing to notice is that the front and rear panels are not perfect rectangles; they are narrower across the top edge than the bottom to match the slope of the sides of the enclosure. This isn't a problem, but it is worth noting because the front and rear panels will only mount correctly in one orientation. It wouldn't be a total disaster to cut the holes in the panels only to find out they are upside down, but it would be slightly embarrassing (at least for me). The screws that hold the top and bottom halves together are meant to be covered by adhesive rubber bumpers on the bottom of the case.

Also notice that the front and rear panels are not of sufficient height to mount an Arduino and a stack of shields directly inside the panels. The IP-6131 model is taller (3.54 inch/8.99 cm), but it would have a whole lot of empty vertical space inside. So, I decided to mount the LCD, the input/output connectors, and the controls on the front panel, place the Arduino and the DDS generator on the floor (bottom) of the enclosure, and mount the jack for the external power pack on the rear panel. This leaves enough room inside for a built-in power supply or batteries (if I decide to incorporate either or both of those). Figure 11-6 shows the front panel layout.

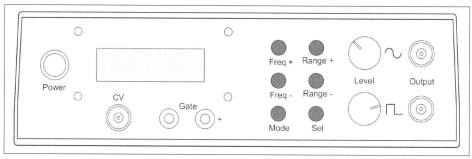

Figure 11-6. Signal generator front panel layout

Another thing to consider is the pushbutton switches used as control inputs. A module with an array of small pushbuttons with low-profile square caps is not something that can be easily found as a stock item. In most cases these are custom assemblies created for a specific product. Since we aren't going into full-scale production it doesn't make sense to spend the time and money to design and build a custom 3 × 2 switch array module, so what ends up on the front panel may not look like the drawing in Figure 11-6. But that's fine; it will still work the same regardless of what pushbutton switches are used.

Schematic

Looking at Figure 11-4, we can see that it looks like almost all the available pins on the Arduino are in use. This is indeed the case, as only the A4 and A5 pins are not assigned. With the LCD, DDS, and control switches connected all of the discrete digital I/O pins will be used, and some of the analog inputs will need to be pressed into service for digital I/O as well. This is shown in the schematic in Figure 11-7.

Table 11-2 lists the Arduino pins and assignments in the signal generator. Any further I/O expansion will need to be done using shields with an I2C interface via the A4 and A5 pins.

Table 11-2. Signal generator Arduino pin usage

Pin	Function	Pin	Function	Pin	Function
D0	DDS FQ_UP	D7	LCD E	A0	Select button
D1	DDS W_CLK	D8	Range + button	A1	Mode button
D2	LCD D4	D9	Range − button	A2	Gate input
D3	LCD D5	D10	Freq + button	A3	CV input
D4	LCD D6	D11	Freq − button	A4	SDA to I/O expansion
D5	LCD D7	D12	DDS RST	A5	SCL to I/O expansion

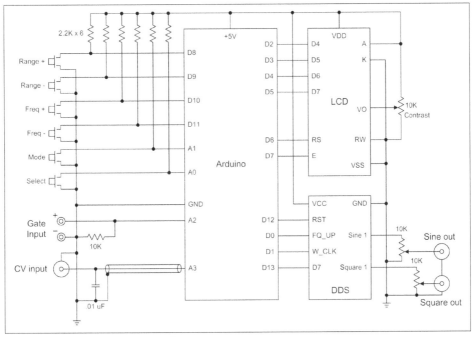

Figure 11-7. Signal generator schematic

The DDS module I'm using for this project is shown in Figure 9-57 in Chapter 9. It is available from multiple sources, including DealeXtreme (*http://bit.ly/dx-ad9850*), for about $8. Figure 11-8 shows the pinout diagram for the module.

The AD9850 supports both a parallel binary and a serial word interface. For this application the Arduino will communicate with the DDS module using the serial word interface mode. This mode uses pin D7 on the DDS module for the data.

The LCD module has 16 pin positions. These connect ground, power, data, and control signals to the LCD controller ICs, which are on the back of the PCB under blobs of black epoxy. Figure 11-9 shows the pin definitions for the LCD module.

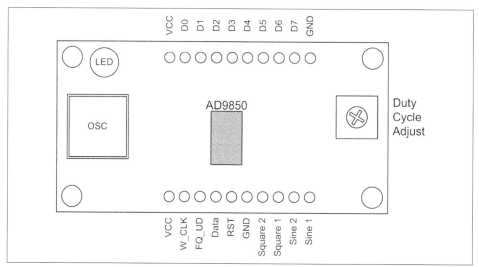

Figure 11-8. DDS AD9850 module pinout

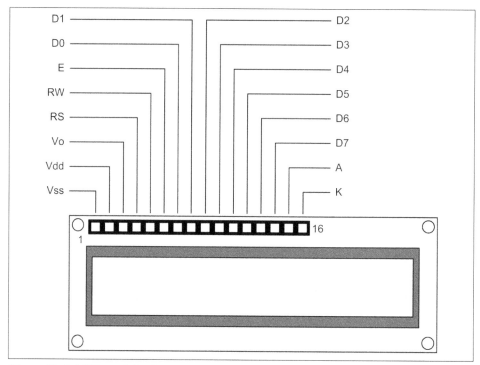

Figure 11-9. LCD module pinout

Prototype

This project uses off-the-shelf components, and a prototype built from the same shields and modules is useful for developing the software. If you are waiting for parts to arrive, a work-alike prototype can help move the software along. For this project I'm using a test setup consisting of an Arduino Uno mounted on a wood base, as shown in Figure 11-10.

Figure 11-10. Arduino prototype fixture

In the prototype I'm using an Arduino Uno, a PCB with a 5 × 4 array of miniature pushbutton switches that also includes 8 LEDs, a pair of screw terminal extenders, a prototype shield with a DDS module mounted on it, and the LCD module that will eventually end up in the final instrument. The pushbuttons will stand in for the control inputs of the signal generator. The four relays shown in Figure 11-10 are not used in this project, so they are not connected. The connections to the Arduino are identical to those shown in Figure 11-7.

Control Inputs and Modes

There are six buttons on the front panel. Table 11-3 shows how they are assigned to various functions.

Table 11-3. Signal generator control button functions

Button	Mode: Frequency	Mode: Gate	Mode: CV
Freq	Inc frequency	Gate on	CV enable
Freq –	Dec frequency	Gate off	CV disable
Range	Inc freq range	Gate	CV zero set
Range –	Dec freq range	Gate –	CV zero reset

The signal generator has three control input modes. From Table 11-3 we can see that there are actually 12 possible control inputs, depending on the mode. The Select button isn't shown because it's used as a type of "Enter" key; pressing the Select button will exit the input mode for gate or CV controls and return the instrument to the normal range and frequency mode of operation.

The waveform output is always active unless blocked by the gate mode. The frequency and range can be altered at any time that the instrument is not in the gate mode. Pressing the Freq + or Freq – button and then releasing it will cause the frequency to change by 1 Hz. When the Freq + or Freq – button is held down the value will change in increments (or decrements) of 10, 100, or 1,000 (depending on the current value) for as long as the button is pressed.

The frequency output of the generator is divided into ranges, each of which spans 10,000 Hz. So if range 1 is 0 to 9,999 Hz, then range 2 will be 10,000 to 19,999 Hz, and so on. The use of the range controls is not mandatory. The main reason for including these controls is to allow the Freq + and Freq – controls to adjust the output in manageable increments within a particular range. Otherwise, the user might be mashing the Freq + or Freq – buttons for a while to get to the desired frequency. The output frequency is updated continuously, so there is no need to press the Select button.

The range can be any value between 1 and 4,000. If the frequency is incremented past the end of a range, the range number will automatically increment. If the frequency is decremented below the minimum value of the current range, the range value will decrement by 1. The range will autoincrement or autodecrement in steps of 10, 100, or 1,000, just like the frequency control inputs.

The CV and gate settings are modified by placing the instrument into the appropriate control input mode. The Mode button selects the command input mode, and the Select button makes the current setting active and puts the instrument back into normal operation mode.

When the gate is active the generator will not produce any output until the selected gate condition is present on the gate input. The Freq + and Freq – buttons are used to enable or disable the gate. The Range + and Range – buttons select the gate sense mode: either active high or active low, respectively.

The control voltage (CV) input is an analog voltage between 0 and 5V DC, with 2.5V being the nominal zero point (negative voltages are not used). In CV mode the Freq + and Freq – buttons enable or disable the CV input. The zero point can be changed by selecting the CV mode and pressing the Range + button. The Range – button resets the CV zero point to the default of 2.5V.

Once enabled, the CV input is active until specifically disabled. An input voltage above the zero point will cause the output frequency to increase, and an input voltage below the zero point will cause the output frequency to decrease. To set the zero point, the desired voltage is applied to the CV input and the Range + button is pressed. Disabling the CV input does not alter the zero point setting.

Display Output

A big challenge with something that uses a minimal display is figuring out how to use it effectively to display information in a condensed form. The LCD used for the signal generator has two position-addressable lines with 16 characters each. Figure 11-11 shows how I elected to fit all the essential information on the display during operation.

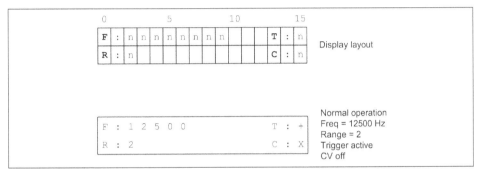

Figure 11-11. Signal generator operation display layout

When the signal generator is in any mode other than frequency output, the colon after the mode field letter will blink. So, if it's in CV mode, the colon after the C character will blink, and the colon after the G character will blink when it's in gate mode.

Pressing the Mode switch will repeatedly cycle through Gate, CV, and control inputs off (normal operation). The C and G functions may only be altered when the associated control mode is active. Once the desired value is displayed, the Select button will set the output to the displayed configuration and the instrument will resume normal operation.

The G (gate) field will display X, +, or -, depending on the state of the gate function. A + symbol means that the gate is active on a high input, and a - symbol means that the gate will respond to a low input. X means that the gate function is disabled. The gate settings will become active when the Select button is pressed.

The C (CV) field will display of X, 0, +, or -. It will show + if the CV function is active and the input control voltage is greater than the zero point, if CV is active and the control voltage is less than the zero point, and 0 if CV is active and the control voltage is equal to the zero point. The 0, +, and - symbols are updated in real time when the CV input is enabled. An X means that CV input is disabled. The CV settings will become active when the Select button is pressed.

In many ways this is a throwback to a time when many devices had small displays. Before the advent of high-resolution LCD and TFT displays, small displays like this were common, when a display was used at all. Prior to the introduction of alphanumeric LED displays this would have been done using numerals only, and interpreting some of these old displays was definitely a skill. Fortunately we can now purchase small, inexpensive alphanumeric LCD components, but they still need some creativity in the data layout, and some degree of interpretation.

DDS Module

The AD9850 DDS module is mounted on a prototype shield, as shown in Figure 11-12. This is optional, and you could just mount it to the floor of the enclosure and use wires to connect it. I would recommend the shield, though, because it provides a robust place for the module and it allows for the use of screw terminals. This is a much neater way to make connections than with solder, and they are far more reliable than the plug-in jumper wires.

The entire DDS shield consists of the parts listed in Table 11-4.

Table 11-4. DDS prototype shield parts list

Quantity	Description
1	Prototype shield, Adafruit #51 or equivalent
1	8-position 0.1 inch (2.54 mm) screw terminal block
1	6-position pin socket header
1	AD9850 DDS module

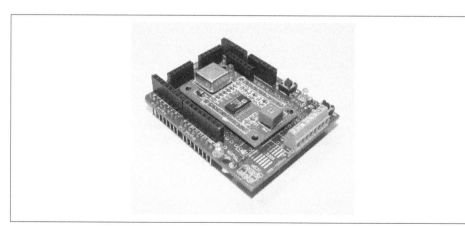

Figure 11-12. DDS module prototype shield

The inputs to the DDS module are connected to the six-position socket header, and the outputs are wired to the eight-position terminal block. Power and ground are supplied by the prototype shield. The LEDs aren't currently used, but one could be connected to power and the other to the D7 input (D13 on the Arduino). Figure 11-13 shows how the various signals and lines from the DDS modules are brought out on the prototype shield.

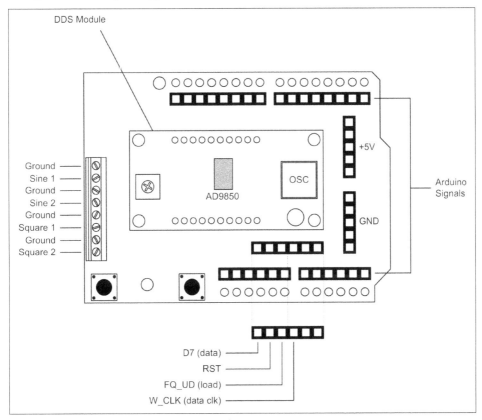

Figure 11-13. DDS prototype shield signals

The PCB terminal block and the six-position socket header are wired on the under-side of the prototype shield. I used 28-gauge wire, but anything in the 24- to 32-gauge range will work. I didn't use shielded cable for connections between the module pins and the terminal block, but it might be something to consider, particularly at high frequencies.

I would have preferred not to use a socket header for the DDS control inputs, but given the lack of space on the prototype shield there wasn't a lot of leftover room. In the final assembly I used some existing pin jumpers and snipped off the ends that would go into the screw terminals for the Arduino control signals. In the final version these wires will be secured to the socket header with some clear silicon rubber.

Software

The software for the signal generator is structured in the conventional way for an Arduino application, with a `setup()` and a `loop()`. The `setup()` function handles I/O pin modes and other configuration chores, and the `loop()` function does the control input and signal output management. Where the signal generator departs from a conventional sketch is in its organization. It is comprised of multiple source files—six in all—with associated include (i.e., "header") files.

Complexity and Interface Bloat

While this project may be conceptually simple, there is a surprisingly high degree of complexity in the software. Why? Because of the control interface. Software developers and engineers who work with GUI-oriented software on a regular basis are very familiar with this, as it's common for up to 70% or more of the code to be dedicated to just handling the GUI. This is also one of the reasons why the command-line utilities found on a Unix or Linux system tend to be functionally powerful yet very compact; there is no GUI baggage to deal with, just the command-line interface.

Microcontroller applications also experience "interface bloat" whenever code must be implemented to allow the system to interact with a human user. Humans tend to be slow, sloppy, and not very good at remembering and using compact command codes or interpreting cryptic responses. So, some part (sometimes a large part) of the MCU software must be dedicated to creating a user interface that a normal human can interact with effectively.

In some designs this is dealt with by running the user interface on a separate host system where things like limited memory and slow CPU clock speeds are not an issue, and then using a compact and efficient machine-oriented interface to pass commands, parameters, and status data between the host and the MCU. With this project that was always a possible approach, but the downside is that it limits the portability of the MCU device because it will always be tethered to the host system with the user interface. For some applications that is not a major consideration, such as with remote sensing and control devices in permanent locations connected to a central control computer. However, the signal generator we're building here is intended to be a portable test instrument, so it needs to have at least a basic user interface incorporated into its design.

The challenge then becomes devising a way to get the most functionality possible from a limited number of control inputs, putting the maximum amount of useful information onto a severely constrained display, and doing all of that without interfering with the core operations of the instrument or consuming so much of the available flash memory that the software is unable to load onto the MCU.

The software to control the DDS actually does nothing more involved than reading the gate or CV inputs and then writing the appropriate control data to the DDS IC. The tricky part is how to map the control inputs so that we can control how those actions take place, and when. As with any microcontroller-driven device, it is the software that gives it the desired functionality. Without software it's just a pile of plastic, wires, circuit boards, and some silicon.

Source Code Organization

The signal generator code is contained in multiple source files. When the Arduino IDE opens the main file, *sig_gen.ino*, it will also open the other files in the same directory. The secondary files are placed in tabs in the IDE, as shown in Figure 11-14.

Figure 11-14. The Arduino IDE after loading the sig_gen.ino sketch

The signal generator code is structured such that the global variables are in a separate compilation module, *sig_gen_gv.cpp*. The LCD, control inputs, control voltage (CV), and gate functions are also in separate modules. Table 11-5 lists the source modules and their respective functions.

Table 11-5. Signal generator source code modules

Module	Function
sig_gen.ino	Primary module containing `setup()` and `loop()`
sig_gen.h	Constant definitions (`#define` statements)
sig_gen_control.cpp	Control pushbutton input processing
sig_gen_control.h	Include file
sig_gen_cv.cpp	CV input processing
sig_gen_cv.h	Include file
sig_gen_gate.cpp	Gate input processing
sig_gen_gate.h	Include file
sig_gen_gv.cpp	Global variables
sig_gen_gv.h	Include file
sig_gen_lcd.cpp	LCD functions
sig_gen_lcd.h	Include file

Notice that the LCD and DDS objects are instantiated using the C++ `new` operator in the file *sig_gen_gv.cpp*. This is possible because the "anchor" variables for the objects, `lcd` and `ddsdev`, are defined in the global variables module and exported in *sig_gen_gv.h*. In order for this to work the main module *must* contain the include statements for *LiquidCrystal.h* and *DDS.h*, and the same include statements must appear in the global variables file as well. Due to a quirk in the way that the Arduino IDE handles scoping and include statements, any reference to an external library included in a tab file must also be included in the main file. Also notice that the auxiliary modules contain the statement `#include "Arduino.h"` to allow each one to access the Arduino environment. The `new` operator is discussed in the sidebar "Instantiating Class Objects with new" on page 99 in Chapter 5.

Software Description

The code for the signal generator is somewhat lengthy, so rather than try to list it all here I'll focus on the major sections and some of the highlights, using flowcharts and code snippets. I would suggest that you download the source code from GitHub (*https://github.com/ardnut*) and follow along with this text as you examine it.

The module *sig_gen.h* is a collection of the `#define` constants used in the program. Rather than have "naked numbers" running loose in the code, a `#define` statement can be used to reference the values. This makes it easier to change something, like an LCD position or a time delay used in multiple places, without having to hunt down every instance of the naked value. It's easy to miss one or two when making changes manually, but with the #define you can be assured that they will all be the desired value. You could, of course, also use the C++ const statement, but as discussed in Chapter 5 the #define statements consume less memory.

The file *sig_gen_gv.cpp* contains the global variable declarations. The contents of *sig_gen_gv.cpp* are given in Example 11-1. Since *sig_gen_gv.cpp* is executable source code, it can contain initialization statements.

Example 11-1. Global variables

```
// sig_gen_gv.cpp

#include "Arduino.h"
#include <LiquidCrystal.h>
#include <DDS.h>
#include "sig_gen.h"

// initialize the LCD and DDS objects and map digital pins
LiquidCrystal *lcd = new LiquidCrystal(LCD_RS, LCD_E, LCD_D4, LCD_D5,
                                       LCD_D6, LCD_D7);
DDS *ddsdev = new DDS(DDS_OSC, DDS_DATA, DDS_FQ_UP, DDS_W_CLK, DDS_RESET);

// alternate form
//LiquidCrystal lcdobj(LCD_RS, LCD_E, LCD_D4, LCD_D5, LCD_D6, LCD_D7);
//DDS ddsobj(DDS_OSC, DDS_DATA, DDS_FQ_UP, DDS_W_CLK, DDS_RESET);
//LiquidCrystal *lcd = &lcdobj;
//DDS *ddsdev = &ddsobj;

// button input state values (0 or 1)
int  fpls      = 0;      // freq +
int  fmn       = 0;      // freq -
int  rpls      = 0;      // range +
int  rmn       = 0;      // range -
int  mode      = 0;      // mode
int  sel       = 0;      // select

// last button states
int  last_fpls = 0;
int  last_fmn  = 0;
int  last_rpls = 0;
int  last_rmn  = 0;

// internal control
int  mode_cnt  = 2;      // mode cycle counter
bool new_mode  = false;  // true if mode has changed
int  btncnt    = 0;      // used to detect button hold-down
bool btnhold   = false;  // true if button is held down
```

```
int dds_load_cnt = 0;          // update cycle delay counter for DDS update

// output control settings
unsigned long fval = 1000;
unsigned int rval  = 1;

// display characters
char gval = 'X';
char gpol = '+';

char cval = 'X';
char cpol = '0';

// gates to control loop actions
bool dogate = false;
bool docv = false;
bool gate_output = true;

// CV zero control
bool set_cvzero = false;
bool reset_cvzero = false;
unsigned long cv_zero  = 512;
unsigned long cv_input = 0; // latest value from CV input

int currmode = MODE_DISP;    // init starting control input mode

// CV input
unsigned int cvinval;
unsigned int cvzero;
unsigned int fvalset;
```

Note that two different methods are presented for creating global objects. The first uses the new statement, and the second uses pointer assignment. The end result is effectively the same from a functionality standpoint, but there is a slight memory usage difference.

The include file *sig_gen_gv.h* contains the export declarations. The export statement tells the compiler that these variables will be used by other modules, so placeholders are created when a module is compiled that refers to a variable declared in the *sig_gen_gv.h* file. The linker will stitch it all together when the final executable image is built.

The output of the signal generator is controlled by a set of global variables that contain the current frequency and range values, and the state of the gate and CV functions. In embedded systems this is a common approach, particularly in situations where there isn't a lot of RAM available to store large amounts of data in the stack. Rather than pass a lot of arguments to functions, the global variables are used as a type of shared memory space. The key to using global variables effectively is to apply the "write-by-one, read-by-many" rule as much as possible. In a small system with only one active program there isn't much chance of a situation where a variable is modified by two processes at the same time, but in multithreaded applications this is a very real concern.

Since we're using the conventional setup() and loop() structure favored by the Arduino IDE, the first primary function of interest is start(). Figure 11-15 shows a detailed flowchart diagram for setup().

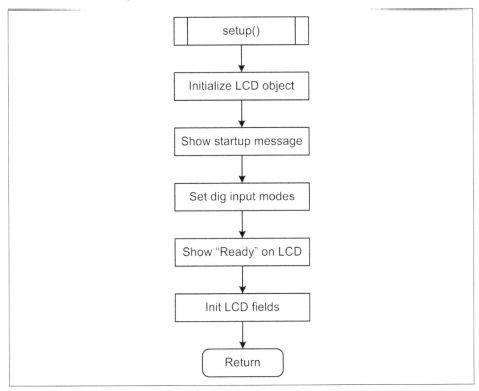

Figure 11-15. Flowchart of setup() function

The setup() function (Example 11-2) is straightforward. It initializes the LCD object, displays a startup message, initializes the digital inputs for the control buttons, displays "Ready" on the LCD, and finally writes the frequency, gate, range, and CV fields onto the LCD. These will remain unchanged for the remainder of the time that the signal generator is active.

Delay calls are used in the TitleDisp1() and TitleDisp2() functions to slow things down so that the startup text doesn't flash by too quickly for the user to read it.

Example 11-2. Signal generator setup() function

```
void setup() {
    lcd->begin(16, 2);  // Set the dimensions of the LCD

    TitleDisp1();

    pinMode(SW_FPLS, INPUT);
    pinMode(SW_FMIN, INPUT);
    pinMode(SW_RPLS, INPUT);
    pinMode(SW_RMIN, INPUT);
    pinMode(SW_MODE, INPUT);
    pinMode(SW_SLCT, INPUT);

    TitleDisp2();

    InitLCDFields();
}
```

The function InitLCDFields(), found in the file *sig_gen_lcd.cpp* and called from setup(), writes the static fields to the LCD after the power-up display is complete. It can be called multiple times from different locations in the code as necessary.

The software's main loop performs four primary steps:

1. Check for control button inputs.
2. Check for CV input (if enabled).
3. Check for gate input (if enabled).
4. Update the output frequency.

Steps 1 and 2 involve parsing command strings from a host PC and decoding the control buttons based on the current mode. This is the most complex part of the software. Steps 2 and 3 just check inputs to determine what change to the output, if any, should occur. Step 4 writes the control data to the AD9850 after performing a simple calculation. Figure 11-16 shows a high-level flowchart for loop().

The source code for the loop() function, listed in Example 11-3, looks simple because all the functionality is in the auxiliary modules. As each of the functions is called, it sets or reads the global variables.

Example 11-3. Signal generator loop() function

```
void loop() {
    // read switch states
    ReadControls();

    // handle control modes
    SetMode();
```

```
switch(currmode) {
    case MODE_DISP:
        getFreq();
        SetLCDVals();
        ShowGate();
        ShowCV();
        break;
    case MODE_GATE:
        SetGate();
        ShowGate();
        break;
    case MODE_CV:
        SetCV();
        ShowCV();
        break;
}

// check for active CV input
fval = RunCV();

// set DDS output frequency
dds_load_cnt++;
if (dds_load_cnt >= DDS_LOAD_GO) {
    dds_load_cnt = 0;

    if (RunGate()) {
        ddsdev->setFrequency(fval);
    }
    else {
        ddsdev->setFrequency(0);
    }
}

    delay(MAIN_DLY);
}
```

The control output to the DDS module is the last thing that the main loop does before it delays for MAIN_DLY milliseconds then starts over. The output frequency, fval, is modified by the CV input via RunCV() if CV is active, and it can be set on or off by the RunGate() function if that mode of operation is active. The DDS data is written after DDS_LOAD_GO number of iterations of loop() to give the module time to process the control data from the Arduino. The frequency update interval is equal to DDS_LOAD_GO times MAIN_DLY milliseconds. If you modify the value of MAIN_DLY you may also need to modify DDS_LOAD_GO.

The largest of the source files is *sig_gen_control.cpp*. It contains the code to detect control input switch activity, debounce switch inputs, determine if a button has been held down by the user, and set the input control mode.

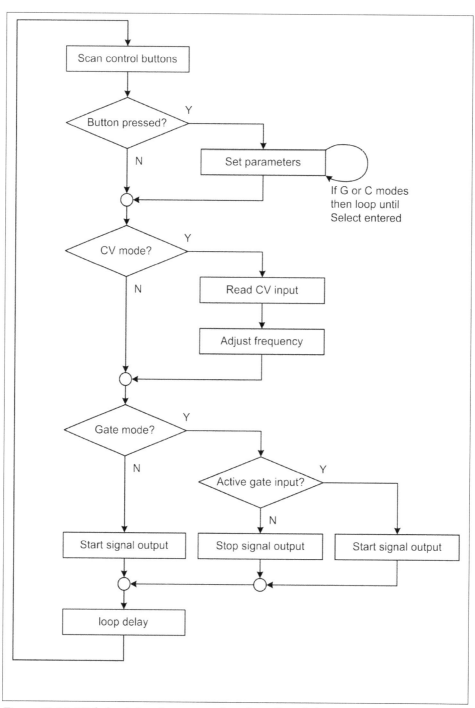

Figure 11-16. High-level signal generator flowchart (loop() function)

Switches tend to be electrically noisy. To compensate for this the signal generator incorporates a simple debounce function, as shown in Example 11-4.

Example 11-4. Debounce function

```
bool debounce(int switchid)
{
    int swval = 0;

    if (swval = digitalRead(switchid)) {
        delay(DBDLY);
        if (swval != digitalRead(switchid)) {
            return false;
        }
        else {
            return true;
        }
    }

    return false;
}
```

The idea behind debounce() is to determine if a switch has maintained the same state between two input samples. If so, then the switch is considered to be in that state; otherwise, it has glitched. The test time interval, DBDLY, is defined in *sig_gen.h* and has an initial value of 10 ms. This can be increased to improve debounce reliability, but if increased too much it will cause the control inputs to respond slowly, which may not be desirable.

The function readControls() scans through the six input pushbutton switches. If a switch has been pressed, then it will increment a count value each time loop() calls readControls(). If the count exceeds a preset value (defined as HOLD_CNT in *sig-gen.h*), then the input is flagged as being held down by setting the global variable btnhold to true. Recall that loop() executes every MAIN_DLY milliseconds.

The function getFreq() handles the frequency input control. It will automatically increment or decrement the frequency in steps of 10, 100, or 1,000 Hertz. The range value is also automatically adjusted as the frequency passes a range boundary at RSTEP Hertz.

The SetMode() function changes the current control mode of the instrument if the Mode pushbutton has been pressed and detected by readControls(). It will cycle through the three control modes (normal, gate, and CV) as long as it is held down. The behavior of the gate and CV modes may only be altered when the corresponding mode is active.

The full source code for the signal generator is available on GitHub (*https://github.com/ardnut*). It is commented and hopefully self-explanatory.

The DDS Library

A simple library for the DDS module has been included with the signal generator source code. To install the DDS library simply create a directory in your *sketchbook/libraries* directory called *DDS* and place *DDS.cpp* and *DDS.h* there. The DDS library will appear in the libraries drop down when you restart the IDE (or immediately with newer versions of the IDE). You may notice that there is no *README*, nor is there a *keywords.txt* in the DDS library subdirectory file. This class is so simple that I didn't see a need for them.

The DDS library used with the signal generator illustrates some of the custom library creation concepts presented in Chapter 5. It is a simple class that handles the necessary calculation to create the control data word used by the AD9850. The file *DDS.h*, shown in Example 11-5, contains the class definition.

Example 11-5. The DDS class

```
class DDS {
    private:
        float dds_cal;

        unsigned long dds_clock;      // ext clock, in Hz

        uint8_t dds_reset;            // DDS reset pin
        uint8_t dds_data;             // DDS data pin
        uint8_t dds_load;             // DDS data load pin
        uint8_t dds_wclk;             // DDS data load clock pin

        void pulseHigh(int pin);
        unsigned long freqData(unsigned long freq);
        void sendBit(uint8_t outbit);

    public:
        DDS(unsigned long clockfreq, uint8_t dds_data, uint8_t dds_load,
            uint8_t dds_wclock, uint8_t dds_reset);
        void setFrequency(unsigned long freqval);
        void calibrate(float calval);
};
```

After the DDS object is instantiated, you can pass in a calibration coefficient if you want to improve the accuracy of the signal generator. Refer to the datasheet for the AD9850 to see how this value is calculated. In this class it simply defaults to a value of 0.

Other than the object constructor and the `calibrate()` function, the only other method used by code outside of the class is the `setFrequency()` function. The frequency is given in Hertz. The calculated control word is pushed to the DDS IC one bit at a time.

Testing

Assuming that the prototype is wired correctly, you should be able to compile and upload the software to the Arduino Uno. Upon startup the display will show "DDS Signal Gen" and "Initializing," followed by "Ready" and then the data display fields. You can adjust the delay times to make the startup text linger longer or eliminate it completely. It's up to you.

 When the signal generator is wired as shown in Figure 11-7 you cannot use the Serial library. The library takes over the Rx and Tx pins (D0 and D1, respectively) and pulls them high. This confuses the 9850 DDS, and it will not generate any output. The Serial library is fine for debugging the control inputs, which is what I did, but it must not be instantiated when the DDS is connected. You could always shuffle the analog and digital pin assignments around to use A4 and A5 for control switch inputs and free up D0 and D1 for serial I/O, but I didn't see a need to do this for this project. The USB interface still works fine for uploading new software to the Arduino if necessary, even with the DDS module connected.

The code is preset to generate an output of 1,000 Hz when the prototype is powered on. This can be any value you like within the range of the AD9850, but I picked 1,000 as something I could easily see with any oscilloscope that happened to be handy.

 To check the DDS output you will need an oscilloscope of some sort, and the faster the better. Something like the low-cost single- or dual-channel Nano digital oscilloscopes from Seeed that look like repurposed MP3 players generally won't go above about 1 MHz. If you want to check the high frequency end of the DDS IC's range you will need something with at least a 100 MHz bandwidth. I keep my little Seeed DSO Nano handy for quick peeks at slow circuits, but I reach for something heftier when I need to work above 1 MHz.

The first thing to check is the sine wave output. The 1,000 Hz sine wave is shown in Figure 11-17, and it should be about 1V P-P. The prototype doesn't have the output level controls that the finished unit will have, so if you don't see the sine wave unplug the unit and double-check your connections.

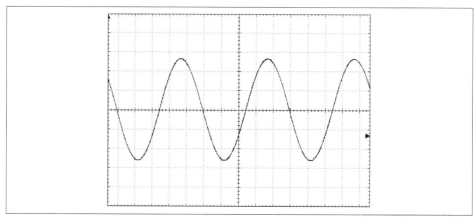

Figure 11-17. Default 1 K Hz sine wave output

The square wave output will have the same frequency as the sine wave, with an amplitude of around 5V P-P (actually, whatever V+ happens to be). The output will look like Figure 11-18.

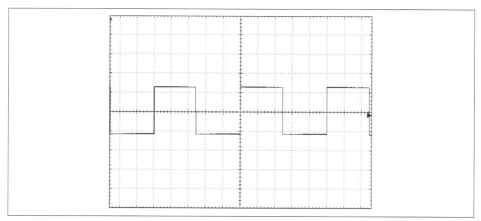

Figure 11-18. Default 1KHz square wave output

If both outputs are present, then testing can proceed to the control inputs. The frequency and range inputs can be tested by first pressing the Freq + and Freq – buttons. The frequency on the display should change by 1 Hz for each button press. Now hold the Freq + button down. The frequency should increment first by 1, then by 10, then 100, and finally 1,000. Stop at about 15,000 and check the output. Note that the range value will automatically increment when the frequency passes 9,999. Repeat this with the Freq – pushbutton. The range value should decrement as the frequency decreases.

Pressing the Range + button will cause the frequency to increment by 10,000 for each button press. Change the range to a value of 4 and observe the output. It should be near 40 KHz. Press the Range – button and observe the display. The frequency should decrease by 10,000 for each decrement in the range value.

Press the Mode button and hold it momentarily until the colon next to the G symbol blinks. Now press the Freq + and Freq – buttons to enable and disable the gate input. Press the Range + and Range – buttons to change the input sense state. It should cycle between + and –. Press the Select button to return to normal operation.

When the gate is active and there is no gate input the output should cease. Select the – (active low) input mode and observe that the output becomes active whenever a jumper is connected from the gate input (A2 on the Uno) to ground. When in the + (active high) input mode the output will be enabled whenever the jumper is connected to a positive voltage source, but will cease when the A2 input is grounded. Disable the gate input by entering the gate control mode and using the Freq – button to disable the gate operation.

To test the CV input (A3 on the Arduino) you will need either a variable power supply or a 5V power source and a 10K potentiometer, and also a DMM, shown in Figure 11-19. If your oscilloscope doesn't have a frequency display function (most modern DSO instruments do), some type of frequency counter will also be useful.

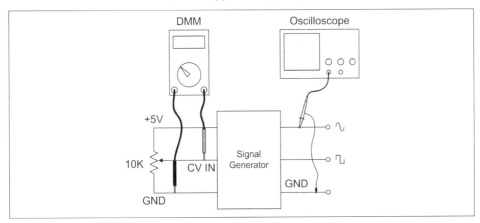

Figure 11-19. Control voltage input test setup

Enable the CV input using the Mode button, but don't set the zero point (leave it at the default midrange value of about 2.5V). Press the Select button to return the normal operation. Apply a voltage between 0 and 5V while observing the frequency output. The frequency should decrease when the CV input is less than 2.5V, and increase when it is greater than 2.5V. Also observe the polarity of the CV input by watching the + and – polarity indication. It will be next to impossible to set the CV manually at the zero point, but you may see the display briefly show a 0 as the input passes the zero value.

Now adjust the CV input voltage for about 2V on the DMM. Put the instrument into CV mode, enable the CV input, and press the Range + button to set a new zero value. The polarity symbol should now be 0, and it will stay at zero so long as the input voltage does not vary. Change the input voltage and observe that the output frequency varies up or down as the CV input goes above or below the new zero point value. When finished, use the Mode and Freq – buttons to disable the CV input function.

 Do not apply more than 5V or a negative voltage to the CV input. This prototype circuit does not have any input protection. Exceeding the input range of the ADC in the AVR MCU can destroy it. When the final unit is assembled it will have a simple input protection circuit.

This concludes the basic functional testing. If everything works correctly, then now is a good time to experiment with the signal generator to see how it responds to control inputs. This is also a good time to set the square wave duty cycle.

There is small PCB potentiometer on the DDS module's PCB on the opposite end from the silver oscillator module. Set the frequency for around 10 KHz, and using an oscilloscope adjust the square wave so that the "on" and "off" portions are equal in time. In the finished unit you might want to desolder the PCB potentiometer and replace it with a panel-mounted control on the front or rear of the signal generator.

Final Assembly

If you elect to use the prototype shield for the DDS module then I would suggest a set of screw terminal extenders, like those shown in "Adapter Shields" on page 266 in Chapter 8. These not only provide reliable connections for wires but also raise the DDS shield above the Arduino. This leaves space if you need to access the ICSP connector. You can also use a screw terminal shield, like the kit shown in "Reducing the Cost" on page 467.

Pull-up Resistor Array

Six 2.2K resistors mounted on a section of prototype perfboard are used as pull-ups for the six pushbutton switches. While the AVR MCU does have some degree of pull-up capability, the resistor array ensures that there will be a positive voltage for the switches to work with. It also means that the switch inputs are active low (0V input = on).

Electrically the pull-up module is very simple, as can be seen from the schematic shown in Figure 11-20. The six resistors are wired across the signal lines for each control input switch, and the common connection for all of the resistors is connected to +5V DC.

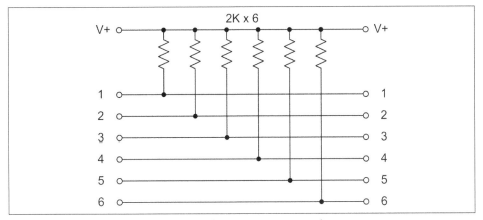

Figure 11-20. Control switch input pull-up resistor array schematic

The completed pull-up array board is shown in Figure 11-21. The use of 0.1 inch PCB terminal blocks allows for easy wiring and reliable connections. Figure 11-21 also shows the input protection module and the mounted Arduino with the DDS proto-type shield. Only the +5V and ground have been wired at this point.

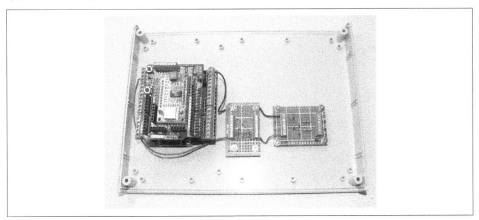

Figure 11-21. Pull-up array and input protection module

Input Protection

A small perfboard module is used to construct the simple input protection circuit shown in Figure 11-22. The idea is to prevent any voltage greater than +5V DC or less than 0V (i.e., negative) from getting into the gate or CV inputs.

The 470-ohm current-limiting resistors in series with the inputs may result in a slight decrease in the voltage that appears on the input of the AVR's ADC, but since the gate is acting as a binary input and the CV is a relative input, it doesn't really matter. The resistor values could even be increased if you are worried about either of the diodes passing too much current in an over-voltage or under-voltage situation. The completed protection module was shown in Figure 11-21. PCB terminal blocks were used here as well.

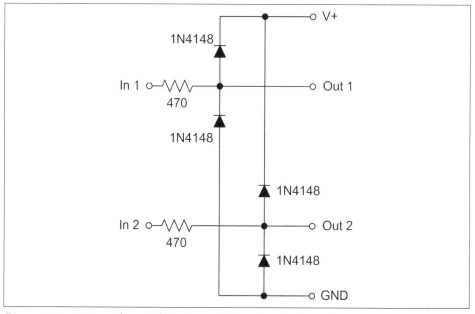

Figure 11-22. External control input protection circuit schematic

Chassis Components

The software is ready, all the parts are at hand, and it's now time to put it all into the enclosure. The most complicated part is making the holes in the front panel, and getting them right. To do this, I drill and cut from the back of the front panel. This helps to keep things neat, and if it gets scratched or cut by a wandering tool no one will ever see it. Next, the Arduino and the DDS module are mounted to the bottom of the enclosure, followed by the USB and DC power connectors on the rear panel. Last comes the task of wiring the controls, connectors, and modules to each other as appropriate.

The front and rear panels are cut and drilled first. The front panel has a rectangular cut-out for the LCD and holes for the switches and connectors. The rear panel has a panel-mounted DC power connector and a potentiometer for the LCD contrast. The USB from the Arduino is not brought out to the rear panel. The USB connector on the Arduino can be accessed after the cover of the enclosure is removed. Figure 11-23 shows the holes being drilled in the front panel using a miniature drill press, although a full-size drill press would work just as well. You could also use a hand drill, but carefully mark where the holes need to go, take it slowly, and drill small starter holes first. I transferred the dimensions from the drawing (Figure 11-24) with a machinist's rule and calipers, and then made pilot holes with a rotary tool and a small drill bit.

Figure 11-23. Drilling holes in the front panel

Figure 11-24 shows the drilling template used for the front panel of the signal generator. I specified #4 holes for mounting the LCD module, but you can use smaller-diameter holes if they are accurately placed. The rear panel has only a DC barrel-type power connector and a potentiometer for the LCD contrast, and these can be mounted as you see fit.

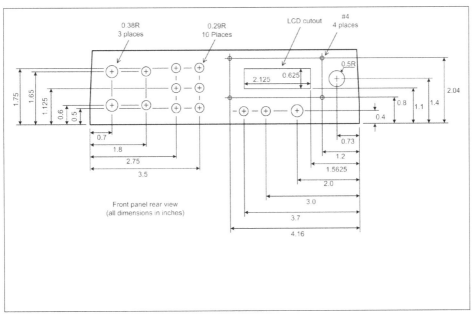

Figure 11-24. Signal generator front panel drilling layout

While it may not be necessary to go to the trouble of creating detailed drawings like the ones that have been created for this project, they do have long-term value. Design and fabrication drawings are a record of what was done, and they ensure that, should the need ever arise to build more signal generators, none of the design work needs to be repeated unless there are drastic changes.

The Arduino Uno and the DDS prototype shield are mounted on the floor of the enclosure, along with the pull-up resistor array and input protection module. The actual locations are not particularly critical, but the Arduino should be mounted close to the front panel to help keep the wiring neat. Figure 11-21 shows the internal components mounted and ready to be wired to the front panel components.

After the interior components are mounted and the holes are drilled and cut in the front panel, the next logical step is to mount the various controls and the LCD display module on the front panel. Don't forget about labeling the controls and connectors.

If you plan to use a labeling machine, you can print and apply the various labels either before or after the controls are mounted. However, if you plan to use rub-on dry-transfer lettering or paint the lettering onto the panel in some fashion (silkscreen, perhaps?), then you really should do that *before* mounting the parts. Also, when using a labeling machine make sure you have the appropriate lettering tape for the device. For example, some models don't have "white-on-black" lettering tape available. Make sure you can get the lettering tape you need, or are happy with the available options, before you commit to using the labeling machine. I used a laser printer and adhesive-backed labels to print the lettering (as white on black), and then cut out each and affixed them to the front panel.

It's surprising how quickly the enclosure can start to fill up once all the parts are in place and wires are routed between the various components. Figure 11-25 shows what the interior of the signal generator looks like just before the top cover is put in place. Note that I didn't bother to put a knob on the contrast pot mounted on the rear panel.

Don't mount the top cover just yet. Leave the instrument open until the final testing is complete. No one is perfect, and little errors can happen. If the unit is open it is a lot easier to reach in and fix a loose terminal screw or reattach a wire.

The LCD module is mounted using four nylon spacers with split-ring lock washers under the nuts. I soldered standard stacking socket headers onto the LCD PCB from the front, and then trimmed the leads on the rear by about 1/4 inch (6.5 mm) to allow socket headers to be pressed onto the pins. Wires soldered to the header pins and protected with heat shrink result in some decent homemade connectors for the LCD module. You can see the LCD connectors in Figure 11-25.

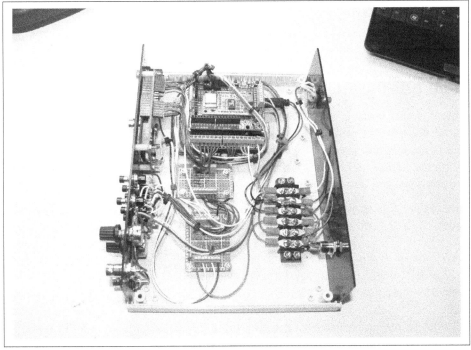

Figure 11-25. The electrically complete signal generator

I used nylon zip ties to keep the wiring neat, and the PCB terminal blocks made the connections a breeze. The only part that was a bit tricky was coming up with a way to connectorize (yes, that's a real word) the DC power input and the LCD contrast control mounted on the rear panel. I ended up settling on a six-position terminal block with #6 screws and crimped lug terminals.

DC Power

The signal generator includes a barrel-type connector on the rear panel for DC power input. In addition to the DC power jack I also purchased mating plugs. This meant that I could select a wall transformer (DC adapter) with a suitable voltage output, attach the plug, and know that it would mate correctly with the power connector. The DC voltage from the rear connector is routed to a terminal block, then to the front panel power switch, and finally to a DC connector that plugs into the Uno board. There is no fuse as the DC power is not routed anywhere where it might find a direct external path to ground, but adding one would not be difficult.

The Arduino folks recommend a 9 to 12V DC adapter for running an Arduino without the USB connected. With a 5V DC adapter there is a large drop across the internal voltage regulator on the Arduino PCB, and the 5V terminals will read around 3.5V. I modified an existing 9V DC adapter from my large box of surplus "wall-warts" (it's amazing how rapidly those things can accumulate over time). You can read more about DC adapters at the Arduino Playground (*http://bit.ly/apg-what-adapter*).

Final Testing and Closing

Now that everything is in the enclosure, the labels have been applied, and the software is loaded, it's time to give the signal generator a final shake-down. The test procedure is really just a regression test, and it is the same as the tests performed for the prototype. The primary intent is to verify that nothing has changed or failed between the prototype and the final unit.

Any further testing will require equipment that most people won't have just lying about, like a distortion analyzer or a spectrum analyzer. If you happen to know someone who does own or have access to this type of test equipment, then by all means avail yourself of it if you can. Depending on how you intend to use your new signal generator, it might be useful to know the harmonic distortion level at various frequencies (is it constant, or does it change with output frequency?), and how harmonically "pure" the sine wave output is. Other things to investigate include the rise and fall times of the square wave output, the output impedance, the hertz-to-voltage relationship for the CV input, long-term stability (i.e., does the frequency drift over extended periods of time?), and both CV and gate control input response times.

Once the final testing is complete, the last step is to mount the top cover. Don't forget to put some adhesive rubber bumpers on the bottom of the enclosure. You don't need to cover the case cover screws with them. Figure 11-26 shows the finished instrument posing on my workbench with a USB digital oscilloscope and a netbook PC.

I should mention that there is something a bit odd about the handle on the instrument in Figure 11-26. I kept the top and bottom orientation of the enclosure correct, but I accidentally flipped the front and back. I'm not going to lie and claim it was on purpose, because it was definitely a mistake on my part, but it actually works out pretty well. Normally the handle would protrude out in front of the unit when resting in a supported position. With the handle to the rear the generator can still be set in a tilted position, but now the handle doesn't waste a lot of space on the workbench. I was lucky. Mistakes usually don't work out as well as this.

Reducing the Cost

As described in this chapter, the signal generator is not a cheap project. That is due in part to the packaging selected, the connectors and controls, and additional things like the screw terminal extenders and the auxiliary pull-up array and input protection modules. But if you want to cut costs, there are a few ways to get the price down somewhat and still have something useful.

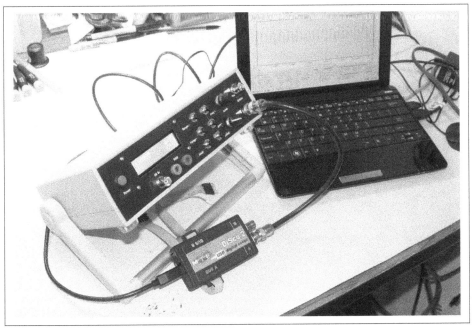

Figure 11-26. Finished signal generator ready to go to work

I've seen some Arduino clones for as little as $15. The DDS module is an essential part of the project, so there is another $10 or so. A prototype shield with screw terminals is about $16, but you could opt for a simpler prototype shield just to have a place for the DDS module for about $10. To further reduce the cost you could forgo an enclosure completely and just have a stack of shields on the Arduino. There are several LCD shields with pushbuttons available, like the one shown in "LCD displays" on page 259 in Chapter 8, and these go for around $12 each. These shields typically only have four or five pushbuttons available for programmed functions, so you may need to redesign the control interface software slightly to accommodate fewer control inputs.

If there is no need or desire for a standalone device, the LCD and pushbutton controls can be eliminated and the unit can run under the control of a host computer with just an Arduino and the DDS module. That's a different kind of device, and I didn't cover that in this chapter. For some ideas about how to implement a remote control interface, take a look at the software presented in Chapter 10 for the Green-Shield and Switchinator projects.

If you had a stack of boards consisting of an Arduino, a prototype shield to hold the DDS module, and an LCD shield it would look something like Figure 11-27.

In this setup there is an Arduino Leonardo, an Adafruit screw terminal prototype shield with the DDS module mounted on it, and a SainSmart LCD shield with push-buttons. A 0.1 inch (2.54 mm) PCB terminal block has been installed on the proto-type shield for the signals from the DDS module, but there are no connectors, no output level control potentiometers, and no enclosure. Bear in mind that without a protective enclosure there is a much higher chance that something bad could happen if a stray wire or screwdriver happened to encounter the stack of boards while it was powered on, and it will never be as robust as it would be in a solidly built enclosure.

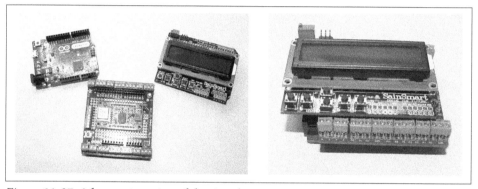

Figure 11-27. A low-cost version of the signal generator

A low-cost version of the DDS signal generator can save you about $30 over the cost of the "fancy" version described in this chapter, which means it will come in at around $65, or lower with some careful shopping. It may not be worth saving $30 to create something that isn't as rugged and portable as it could be. But, as always, the choice is yours.

Cost Breakdown

Table 11-6 lists the primary components used in the signal generator. The total cost does not include things like wire, solder, or nylon zip ties. It also doesn't include shipping charges, and in some cases there are vendors who don't offer low-cost USPS shipping, but instead insist on shipping a $4.95 part for $9 via UPS. I plan to avoid vendors like this in the future if possible, at least until they figure out how to mail things. Many Chinese vendors, on the other hand, offer free shipping, but it might take a couple of weeks to get your parts. Shop around. eBay is a good place to start looking, and if you aren't sure about eBay, you can find similar deals, often from the same vendors, on Amazon.com.

As I stated at the start of this chapter, the signal generator described here costs more to build than many of the kits that are available. But then again, you have total control over it and it will generate output at high enough frequencies to be useful for amateur radio and high-speed digital applications.

Table 11-6. Parts price list

Quantity	Item	Source	Unit price	Extended
1	Enclosure, Bud IP-6130	Mouser	25.40	25.40
1	Arduino Uno	Adafruit	24.95	24.95
1	DDS module	DealeXtreme	7.99	7.99
1	Prototype shield	Adafruit	9.95	9.95
1 set	Screw terminal adapters	Seeed Studio	7.50	7.50
1	LCD display, 16 × 2	Amazon/Uxcell	4.71	4.71
3	BNC connectors	All Electronics	1.25	3.75
2	Banana jacks	Amazon	0.67	1.34
3	Potentiometer, 10K	Amazon/Amico	1.28	3.84
1	Power switch pushbutton	All Electronics	1.35	1.35
6	Miniature pushbutton switch	All Electronics	0.60	3.60
1	DC power jack, barrel type	Parts Express	1.98	1.98

[a] The extended amount is the total cost of each line item for a given quantity.

Total parts cost = $96.33

Note: Total cost does not include wire, solder, adhesives, or the wall transformer. Prices were accurate at the time of writing but should be treated as a guide; there may be some variation.

With some careful shopping I suspect that you could get the total cost down to around $75, even for this full version. Having a large collection of parts on hand, particularly if they were purchased in bulk, can also cut the costs for things like connectors, switches, and potentiometers. Lastly, if you happen to have an old piece of test equipment that no longer works but still has a useful case, you could get creative and put the signal generator into that.

Resources

Table 11-7 lists the distributors and vendors where I purchased the parts for the signal generator (yes, I actually do keep track of all that—it's tax deductible—and you should consider keeping all your receipts and packing lists as well, if you don't already). For any given module or part there are numerous sources; these are just the ones I happened to have used at the time I made the purchases, and can give you an idea of where to start looking.

Table 11-7. Parts sources

Distributor/vendor	URL
Adafruit	*www.adafruit.com*
All Electronics	*www.allelectronics.com*
Amazon	*www.amazon.com*
DealeXtreme (DX)	*www.dx.com*
Mouser Electronics	*www.mouser.com*
Parts Express	*www.parts-express.com*
Seeed Studio	*www.seeedstudio.com*

CHAPTER 12

Project: Smart Thermostat

You may have heard of so-called "smart" thermostats (I suspect that many people have at least seen the advertisements for them). These devices are a type of programmable digital temperature controller. You might even have one installed in your house or apartment. Some allow you to change settings using Bluetooth or some other wireless connection method, along with an associated app for a smartphone or tablet. Others offer data collection capabilities with wireless download, which can be useful if you want to find out when you are using the most energy to heat or cool your home. There are also, of course, some that don't do much more than what the old-style bimetallic coil types did, except that an LCD display is used instead of a dial and some switches.

 Disclaimer: If you elect to build and use the thermostat described in this chapter, you do so **at your own risk**. While it utilizes low-voltage circuits with minimal shock hazard, there is still the risk of damage to your heating or cooling equipment from excessive power cycling or temperature settings that exceed the safe limits of the equipment (most systems have built-in safeguards, however). Use only the low-voltage control circuit for your heating and cooling equipment. DO NOT CONNECT YOUR CUSTOM THERMOSTAT TO HIGH-VOLTAGE (110V AC or greater) CIRCUITS. This includes evaporative coolers and electric heaters.

Background

There are multiple ways to improve on a classic bimetallic coil thermostat like the one shown in Figure 12-1. This type of device has been around for about a century. It uses a coil comprised of two different metals, each with a different thermal expansion coefficient, so that as the temperature changes the coil tightens or loosens slightly.

This movement is then used to bring some contacts together to open or close a circuit, or in some versions a small sealed glass tube containing two contacts and a drop of mercury is moved so that the mercury bridges the gap between the contacts and completes the circuit (the KY-017 tilt sensor module described in Chapter 9 uses this same technique). The control action is either on or off; there is nothing in between.

Figure 12-1. Electromechanical thermostat

In this chapter we'll look at what is involved in designing, building, and programming a smart thermostat using only readily available Arduino boards and common add-on components. But before we dive into that, we need to get a basic understanding of what, exactly, it is that we want to measure and control.

HVAC Overview

The main idea behind altering the temperature in a structure is to either put heat in or take heat out. Or, to put it another way, cold is just the absence of heat, and we can alter the amount of heat in a system to achieve a particular temperature. Adding heat to an environment can be accomplished with a furnace of some sort (natural gas, propane, oil, wood, or coal), electric heating elements, or solar energy. In a heat pump system the heat in the outside air is extracted and put back into the structure, which is similar to running an A/C unit in reverse. To remove heat from a structure, it is extracted using a refrigeration system that employs the condensation and evaporation cycle of a refrigerant fluid (ammonia, Freon, or any one of a number of modern coolants). The inside heat is absorbed by the refrigerant fluid, which then transfers it to the outside in a closed-cycle system. Some large-scale cooling systems use chilled water to achieve the same result, but the physical plant (the equipment and the building to house it) tends to be large and expensive to operate, so these types of systems are usually found only in university buildings, high-rise office structures, hospitals, and other large facilities. Chilled water systems aren't commonly used in residential settings outside of large apartment buildings.

In many residential settings the heating and cooling are handled by two separate systems, but in other cases the heater and air conditioner may all be in the same large cabinet situated on the roof or along the side of the house. A heat pump does both heating and cooling, depending on how it is configured at any given time, and often consists of two units, one inside the house and one outside. In any case, in a conventional thermostat there will usually be a switch to select automatic or manual fan operation, and another to select the mode (heating, cooling, or off). A dial or lever sets the target temperature, or setpoint. The entire system, which incudes the heating unit, the cooling unit, and the thermostat, can be referred to as the heating, ventilation, and air conditioning, or HVAC, system. (Actually, the term HVAC is most often heard when referring to large commercial systems, but I will use it here because it's more convenient than writing "heating and air conditioning system.") With the possible exception of an active air-exchange ventilation function, a residential HVAC system does everything the large system can do, just at a smaller scale.

An important thing to note is that most HVAC systems can and do operate as closed-loop circulation systems. In other words, they work by continuously recirculating the air in the structure. While some systems might have the ability to draw in outside air or exchange inside and outside air (the *V* in HVAC), some don't, and as a result the smell of last night's burned dinner will tend to linger for a while before it finally dissipates. This is also why many homes have a distinct smell unique to that residence. As a child I noticed that the inside of a neighbor's house always smelled like fried chicken in the summer when they ran their A/C, so I assumed that they really liked fried chicken. I never did learn the actual source of that peculiar smell.

Temperature Control Basics

Temperature control involves controlling the subsystems that will either heat or cool an environment to achieve a specific temperature (setpoint). In most cases this is an all-on or all-off type of operation; there are no intermediate heating or cooling rates in most residential HVAC systems. I say "most" because it's possible that someone, somewhere, has a variable-output resistive heater system, but I've yet to see anything like that outside of a laboratory or industrial setting.

The "all-or-nothing" nature of an HVAC system means that the actual indoor temperature will never be exactly at the setpoint, except for brief periods as it either falls or rises. For example, let's assume that we're running the air conditioner, with a target setpoint of 72 °F (22 °C). If we were to record and plot both the inside and outside temperatures over the course of a day we might get something like Figure 12-2 (this is not real data, of course, it's just for illustration purposes).

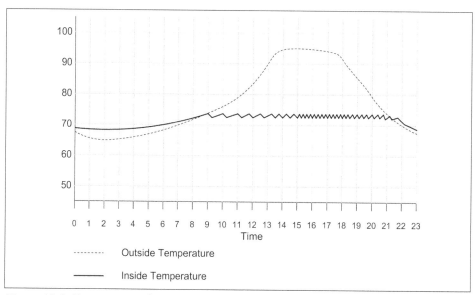

Figure 12-2. Temperature plot over the course of a day

In Figure 12-2 we can see where the air conditioner is active because the interior temperature drops. Early in the day it doesn't cycle (turn on and turn off) as often as happens later in the day. It takes time for the heat outside to work its way through the structure, and the better the insulation is, the longer it takes for the inside to start to warm up. Later in the afternoon, the air conditioner is working hard to keep the inside at the setpoint temperature. Things don't start to slow down until later in the evening as the outside temperature drops.

A controller with only two states, on or off, is called a *hysteresis controller*. It is also known as a bang-bang controller. The amount of hysteresis in the controller determines how often the cooling or heating equipment will turn on and for how long. You can think of hysteresis as the lag between when a system turns on, or turns off, based on a setpoint value somewhere in between. A physical example of this is the snap-action of a three-ring binder. It takes some effort to open the rings, and again to close them, but once in the open or closed state the rings will stay that way.

Figure 12-2 is estimated, but it is representative of the effect of a conventional thermostat on the temperature in a typical residential structure. When the inside temperature reaches about 73 °F the air conditioner is powered on, and it continues to run until the temperature drops to just below 71 °F. That means there is about 2 degrees of hysteresis in this hypothetical system. This range is referred to as the "hysteresis band" (actually, the hysteresis band in this system is too tight, and the A/C is cycling far too often later in the day).

Each on/off action of the HVAC system is referred to as a cycle. Figure 12-3 shows what happens when the difference between the on and off points becomes smaller (less range between H_{max} and H_{min}, which define the hysteresis band). The system is able to maintain the setpoint temperature better with a tighter hysteresis, but at the expense of cycling the power much more often. We would ideally want to make the hysteresis band as wide as possible, because each time the heater or A/C unit cycles brings it one step closer to wearing out and failing. It also drives up the electric bill.

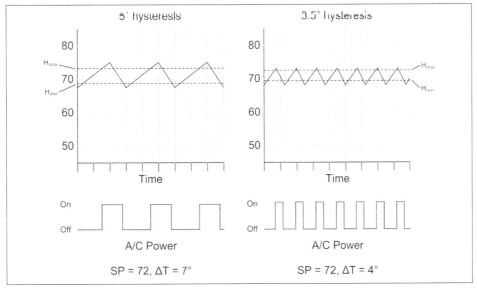

Figure 12-3. Effect of hysteresis

While the 5° hysteresis band is more energy-efficient and doesn't beat up the equipment with as much cycling as the 3.5° band, it also means that the actual temperature can swing as much as 7° between hot and cool. (Remember that these graphs are illustrative, and do not reflect real data.)

There are multiple factors that affect how often a heating or cooling system will cycle, the most obvious being how quickly heat is being removed when running the A/C, or how quickly heat is being introduced when running the heater. Less obvious, but also important, are things like humidity and interior air flow. As the humidty increases air will retain more heat, which can help when heating. An A/C unit also acts as a dehumidifier, condensing and removing moisture from the air passing over the cold evaporator coils in the system, and dry air evaporates sweat more readily, which allows us to cool off faster. Stagnant air (that is, air that's not moving unless the HVAC system is active) can lead to "hot spots" or "cold spots" in a structure, and it's possible that the thermostat will end up regulating just one part of an interior space if the rest of the air isn't passing by it.

Ideally a temperature control system should be able to sense the temperature in all parts of a structure, but in the real world this often isn't the case. There will invariably be a room that gets direct sunlight most of the day, while other areas are shaded, or perhaps a room that has poor air flow. So even though the thermostat is doing its job and working to keep the temperature in the immediate area where it is located at the desired setpoint, other parts of the structure will not be heated or cooled correctly. Just moving the air around can help avoid this, and monitoring the humidity will also help to determine when to run the fan or when to use the heating or cooling. Running the fan alone is a lot less expensive than running the heater or A/C unit.

Smart Temperature Control

The main idea behind a smart thermostat is to reduce the amount of energy wasted and achieve a more even temperature distribution within the structure. This might involve altering the cycle time (i.e., adjusting the hysteresis), only running the system when someone is in the house, altering the setpoint based on the time of day and/or day of the week, and taking advantage of the built-in fan in the HVAC system.

Some common features of modern commercial digital thermostats include adjustable cycle times, automatic heating/cooling mode changeover, and day-by-day programmability. We'll briefly review these capabilities to see which ones make sense for a small project like the one described here.

Adjustable cycle times can be used to reduce frequent on/off heating or cooling cycling. This can occur when the temperature is at or near the setpoint and varies just enough to cause the heater or air conditioning (A/C) to turn on. What it does, in effect, is increase the hysteresis in the system, so that it takes longer to cycle when the temperature is hovering around the setpoint, but the amount of hysteresis is reduced when the temperature rapidly changes one way or the other over a short period of time.

Auto-changeover allows the system to switch between cooling and heating as necessary. If, for example, you are experiencing warm days and cold nights, then the thermostat will automatically switch from cooling to heating at night to maintain a relatively constant average indoor temperature. The Arduino thermostat will support auto-changeover and adjustable cycle times as well.

Profile scheduling is a big feature with smart thermostats. Most programmable digital thermostats have the ability to schedule heating or cooling for specific days of the week, so if your house is empty during weekdays you can trim some expense from your electric bill by not running the heater or A/C when there is no one home to appreciate it. By the same token, some models allow you to automatically "dial down" the heating or cooling at night, while the home's occupants are asleep. Our controller will have the ability to create profiles (also sometimes called *programs*) for day, night, and weekend operation.

A digital thermostat can be packaged in a slick-looking enclosure with a high-tech glowing faceplate and a numeric display behind frosted plastic, but that's just marketing razzle-dazzle. It can also be housed in a commonly available plastic enclosure. It may not look as flashy and high-tech as the commercial units, but that has no bearing on how well it works, or how easily someone can use it. You can save money and time by using a low-budget enclosure for this project, and what I decided to use is definitely low-budget.

Project Objectives

The primary objective of this project is to create a replacement for a conventional residential thermostat. There are numerous descriptions for Arduino-based thermostat projects available online, and this project is similar to many of them. There are only so many ways to arrange an Arduino, a temperature sensor, a simple display, and a relay or two. What makes this project unique is the incorporation of a humidity sensor and the ability to use the fan alone to move air around and help shift cool or warm air to where it's needed without lighting up the heater or powering up the A/C unit.

Our Arduino thermostat will also have the ability to use the interior temperature and humidity data to determine if it needs to adjust the cycle time. There is also the option to add an additional sensor to read the outside temperature and humidity. Finally, it will use relays to connect to existing 24VAC control wiring used by most residential HVAC systems in the US, so there's no serious shock hazard and no real risk to your heating or cooling equipment (at least not electrically—it is still possible to cycle too fast and damage the compressor or heater ignition components).

This project is very straightforward, and requires a minimal amount of parts and very little soldering. Actually, the two main challenges are coming up with a good enclosure and programming the device.

The thermostat will be designed to replace a conventional low-voltage four-wire residential thermostat. It is not limited to four-wire systems, however. Controlling just a heater or an A/C unit alone is simply a matter of not using all the available control outputs and making some minor modifications to the software.

Definition and Planning

Based on what we've covered so far we can identify the basic features we want to incorporate into our design. These involve the functions already provided by a conventional four-wire thermostat, plus some additional capabilities based on humidity:

- Real-time clock
- Inside humidity sensor

- Inside temperature sensor
- Automatic heating or cooling operation
- Automatic fan control
- Seven-day scheduling

The basic version of the HVAC controller is similar in many ways to the units available at big-box home improvement stores (and even some of the small-box local hardware stores), from HVAC supply houses, and from various vendors online.

 This project intentionally avoids dealing directly with the high-voltage AC control circuits in an HVAC system. It is intended for low-voltage (24VAC) systems only. High-voltage AC can damage your HVAC equipment, burn down your house, and kill you (not necessarily in that order). If you need that type of system, consider a commercial controller with a UL and CSA safety rating, and hire a licensed electrician or HVAC technician to install it.

The controller will use an Arduino Nano mounted on a screw terminal prototype shield. A real-time clock (RTC) module will be used to keep track of the date and time. The Arduino PCB stack mounts to the lid of the enclosure. A quad relay module with be mounted inside the enclosure on the bottom panel. All the wiring will enter the enclosure through a hole in the bottom.

Design

The Arduino thermostat is intended to be a drop-in replacement for a standard four-wire thermostat. It is not intended for systems with multistage heating or cooling, or a heat pump system. It is best suited for older homes with a conventional thermostat, like the one shown in Figure 12-1.

Functionality

The Arduino thermostat has three basic functions: heat, cool, and fan. Figure 12-4 shows a block diagram with the primary components. The secret to success lies in how those basic functions are used to achieve the most efficient operation.

A rotary encoder will be used for various functions such as setting temperatures, stepping through days of the week, and so on. The LCD shield also has a set of push-buttons, but these will not be used.

Many older HVAC systems have only four wires, while some newer systems also have auxiliary power wiring. The Arduino thermostat can use this AC source if it is available if a small power supply is also mounted in the enclosure. For this version of the thermostat I will use an external wall power supply, or wall-wart. Ideally it would be better to draw power from the HVAC wiring, but many older systems don't have a spare 24VAC line available.

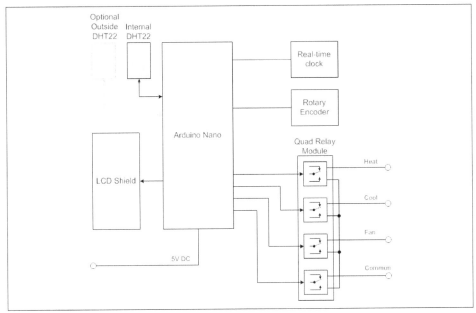

Figure 12-4. Arduino thermostat block diagram

Figure 12-5 shows how a typical older thermostat might be wired internally. The actual internal details will, of course, vary from one type to another, but the basic idea is the same.

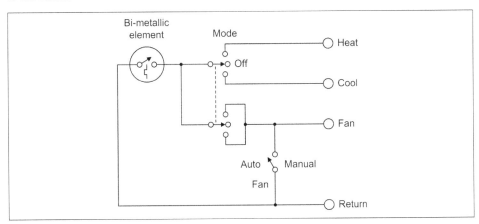

Figure 12-5. Typical old-style thermostat internal circuit

A key thing to note about this type of old-style thermostat is that the fan is wired so that it will always come on when either heating or cooling is enabled. It is not a good idea to run the A/C or the heater without the fan. In some systems the heater fan is separate from the A/C fan, and it will come on only when the internal temperature in the heater has reached a specific level.

Enclosure

For an enclosure, I've selected a plastic electrical junction box with a detachable cover suitable for wall mounting, as shown in Figure 12-6. After some finishing the box will be painted a neutral color to improve its appearance.

Figure 12-6. Arduino thermostat enclosure

Yes, the enclosure is really ugly. I won't deny that. But for this project the main concern was wall mounting and having enough internal space for the components. The side mounting tabs will be removed and the front cover will be sanded and polished, which will help improve its appearance.

The controls are simple, with just an LCD display and a rotary encoder. It will all fit comfortably on the front cover panel. The planned layout of the front panel is shown in Figure 12-8.

The relay module will be mounted to the bottom of the enclosure, and the existing wires for the thermostat will enter through a hole in the bottom of the case. The RTC module will be mounted to the inside of the enclosure, and the DHT22 will be attached to the bottom so it will be exposed to the ambient environment. I elected to mount the DHT22 this way rather than put it inside because I didn't want to drill an array of holes in the enclosure for air flow. An alternative would be to make a square hole just large enough for the DHT22 so it can mount inside but still have access to the outside air. Four screws will hold the thermostat to the wall.

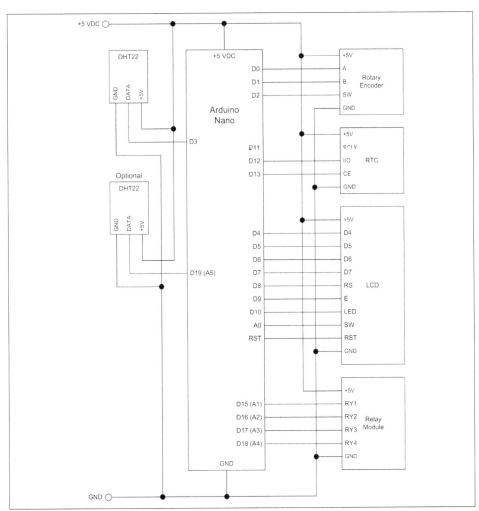

Figure 12-7. Arduino thermostat schematic

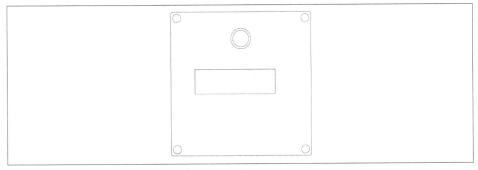

Figure 12-8. Arduino thermostat front panel layout

Schematic

There is no significant difference between the prototype and the final unit, so there is only one version of the schematic. The prototype is electrically identical to the final thermostat, but it uses an Arduino Uno rather than a Nano board (the AVR MCU is the same for both), mainly because that is what is mounted in the prototype fixture.

Notice that every single pin on the MCU is in use. The Arduino programming interface, via D0 and D1, is shared with the rotary encoder. So long as the rotary encoder is not in use this will not create a conflict.

You may also notice that the analog inputs have been pressed into service as digital I/O pins. This is completely normal, and if you refer back to Chapter 2 and Chapter 3 you can see how the ATmega328 port used for analog input (the C port) is also a standard discrete digital I/O port. By referring to the pins as D14 through D19 we can access these like any of the other digital I/O pins. Of course, the analog inputs are now unavailable, but the thermostat doesn't use any analog input (except for the A0 pin, which is used by the LCD shield), so it's not a problem.

Just because the Arduino convention is to assign functionality like analog input to specific pins doesn't mean the pins can't be used for something else. That's just how the folks at Arduino.cc decided to name them. Since an Arduino board is a breakout for the MCU with nothing between the board's pins and the MCU IC itself, what really determines what a pin can or cannot do is the MCU, not the labels on the PCB.

Software

This is a software-intensive project, but most of the software is involved with the user interface and the active profile. The actual control logic for thermostat operation is not all that complex.

Remember that since the main emphasis of this book is on the Arduino hardware and related modules, sensors, and components, the software shown is intended only to highlight key points, not to present complete ready-to-run examples. The full software listings for the examples and projects can be found on GitHub (*https://www.github.com/ardnut*).

On each iteration through the main loop the software will check for user input from the rotary encoder; update the current temperature and humidity data; and determine if the display should switch between screens, move to various settings, or

---

change values. Interrupt handlers are used to capture the inputs from the rotary encoder. A high-level version of the software block diagram is shown in Figure 12-9.

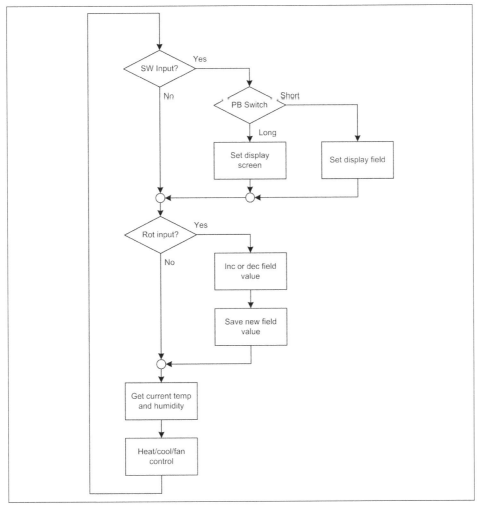

Figure 12-9. Arduino thermostat software block diagram

The main function blocks in Figure 12-9 give an indication of what to expect in terms of actual source code. In addition to the *Thermostat.ino* main file, there will be source modules for handling user inputs using interrupts, processing schedules (daytime, evening, or weekend), and updating the display.

The first thing the loop checks is the switch (SW) on the rotary encoder. This is the signal that indicates that the user wants to do something. The pushbutton switch in the rotary encoder is used to move between screens and between the fields in a screen. The type of action is determined by how long the switch is held down.

The interrupt handlers for the rotary encoder switch and the A/B inputs set flags in the global variables module. The main loop examines these to determine what display update action to take, if any. The interrupt handlers are not shown in Figure 12-9. They will be covered in "Software" on page 493.

The block labeled "Heat/cool/fan control" is the actual control logic of the thermostat. As stated earlier, this is a discrete-state hysteresis controller, also known as a "bang-bang" controller. Figure 12-10 shows the relationship between temperature, time, and hysteresis range in a timeline format for an A/C system. For heating, the operation is simply the inverse.

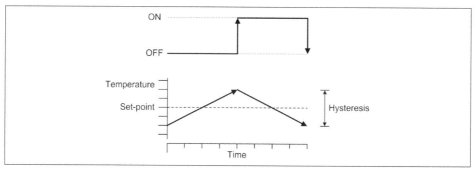

Figure 12-10. Time, temperature, and hysteresis range

Another way of representing this is shown in Figure 12-11. This graph shows hysteresis as a function of temperature. The message of both types of graphs is the same: a bang-bang control system (such as a thermostat) coupled with an on/off actuator (the heater or A/C unit) can never precisely hold a specific temperature. The system will always be at some point between the high and low ends of the hysteresis range.

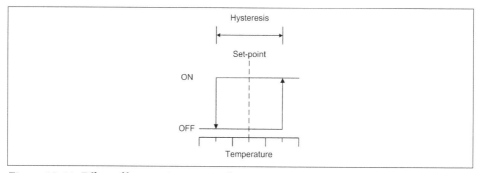

Figure 12-11. Effect of hysteresis on control system response

In a real system the rate at which the temperature increases or decreases will depend on the rate at which heat is being put into or taken out of the system, and how efficiently the heating or cooling equipment can add or extract heat. The upshot is that the heating and cooling times will almost never be the same, although both Figure 12-10 and Figure 12-11 might seem to imply that they are.

The thermostat's control logic is shown in Figure 12-12. Notice that the heating and cooling are simply inverse operations. In an old-style electromechanical thermostat the H (hysteresis) term is established using a set screw on the bimetallic coil assembly. In software we can set this dynamically if we wish.

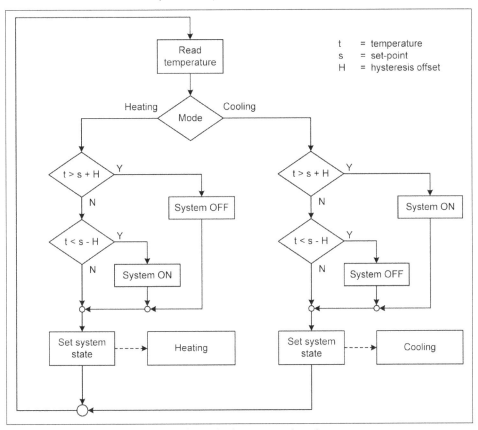

Figure 12-12. Thermostat control logic for heating and cooling

User Input/Output

The control input consists of a single rotary encoder. This might seem odd, but the encoder also has an integrated switch that is engaged when a user presses on the knob. The switch is used to move between display configurations (or screens, as I'm calling them). Each screen contains fields that display data or settings, and the rotary encoder is also used to move between the fields in each screen. The length of time the pushbutton switch in the rotary encoder is held down determines which action the software will take. A short press is a selection, and a long press commands the software to move to the next screen.

The display is the same type of 16 × 2 LCD used with the signal generator in Chapter 11. In this project it is premounted on a shield PCB. This simplifies the internal wiring. The LCD backlight is not enabled unless the control inputs are active. The small pushbutton switches on the LCD shield are not used, but they could be if you are willing to drill more holes through the front panel and find some type of extensions that a user could press. I elected to just ignore them.

Normally two primary screens will be displayed that alternate when the control inputs are not active. The first shows the current temperature readings, the setpoint, the operation mode, and the active profile (if any). The second screen shows the current date and time. Figure 12-13 shows the screens during normal operation. The symbols ^, v, and – are used to indicate if a reading is increasing, decreasing, or holding steady, respectively, relative to the previous reading.

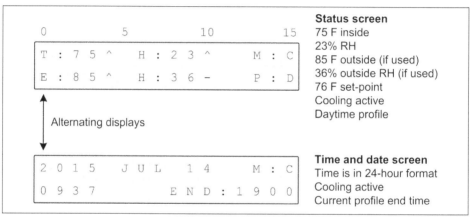

Figure 12-13. Normal operation screens

The fields shown in Figure 12-13 are defined in Table 12-1. In the second screen the "END" field shows the time at which the current profile will end.

Table 12-1. Normal screen field definitions

Field	Purpose
T	Current inside temperature, in degrees (F or C)
E	Exterior temperature (if enabled)
H	Humidity
M	Mode: A (auto), H (heat), C (cool), F (fan), or X (off)
P	Profile: D (day), N (night), W (weekend), or X (none)

The screens shown in Figure 12-13 are purely informational with no modifiable fields. The temperature units may be either Celsius or Fahrenheit. The default is Fahrenheit. I elected to use 24-hour format for the time. It can be modified to accommodate 12-hour format (AM/PM) without too much effort.

The settings screen, shown in Figure 12-14, allows the user to enable or display the profile, and if the profile is disabled the user may manually adjust the temperature setpoint, temperature scale, hysteresis, and operation mode.

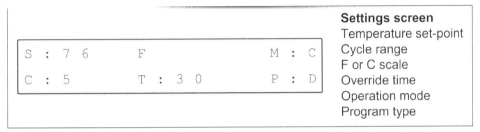

Figure 12-14. Settings screen

The settings screen is the primary control point for the thermostat. It is intentionally arranged to appear only when specifically invoked by a user. The fields used in the settings screen are listed in Table 12-2.

Table 12-2. Settings screen field definitions

Field	Purpose
S	Temperature setpoint
C	Cycle range (hysteresis, in degrees)
F	Either F or C for temp scale
O	Override time
M	Mode: A (auto), H (heat), C (cool), F (fan), or X (off)
P	Profile: D (day), N (night), W (weekend), or X (none)

The main form of preset control for the thermostat is the user-defined profiles. Profiles are defined with the screen shown in Figure 12-15.

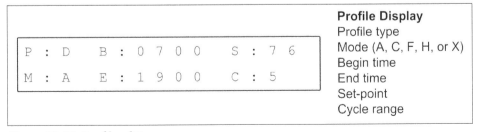

Figure 12-15. Profile editing screen

There are profiles for daytime, nighttime, and weekends; these are labeled D, N, and W, respectively. Rotating the encoder when the profile editing screen is first selected will step through the three profiles. Each profile has a start time and end time. In the case of the weekend (W) profile, the default start time is Friday at midnight, and the end time is Sunday at midnight. If the end time of a profile overlaps the start time of another, then the end time of the preceding profile always takes precedence. The fields used in the profile editing screen are listed in Table 12-3.

Table 12-3. Profile editing screen field definitions

Field	Purpose
P	Profile: D, N, W
B	Begin time
E	End time
S	Temperature setpoint
C	Cycle range (hysteresis)
M	Mode: A (auto), H (heat), C (cool), F (fan), or X (off)

Navigating through the screens and fields does require some learning, but by using the length of the rotary encoder switch press and a tree structure the user can navigate the primary screens and the fields without getting lost. Figure 12-16 shows how the screens and fields respond to either a long or a short button press.

Figure 12-17 shows all four screens in one view. This might make it easier to visualize how the screens are arranged.

A long button press is defined as 1 second or longer. A short button press is 500 ms or shorter. The software uses an interrupt and a counter to monitor the button hold-down time.

Control Output

The control output consists of a quad relay module with four independent SPDT relays. It incorporates drivers for the relays and PCB terminal blocks, so connecting it to the existing thermostat wiring is straightforward. It can be seen in Figure 12-18. The common line is routed through the fourth (lower right in the diagram) relay.

In the block diagram in Figure 12-4 you will notice that the common line from the existing HVAC system is also routed through one of the relays. This is a fail-safe measure. If the thermostat loses power the relay with the common line drops out, and none of the other functions of the HVAC can be enabled. All the external HVAC control voltage lines, including the common return, are connected to the C and NO (common and normally open) terminals of the relays.

Prototype

The prototype for this project is comprised of an Arduino Uno, a pair of terminal block extensions, and an LCD shield. These are essentially the same components that will be packaged in an enclosure later on (the Nano and the Uno use the same ATmega328 MCU).

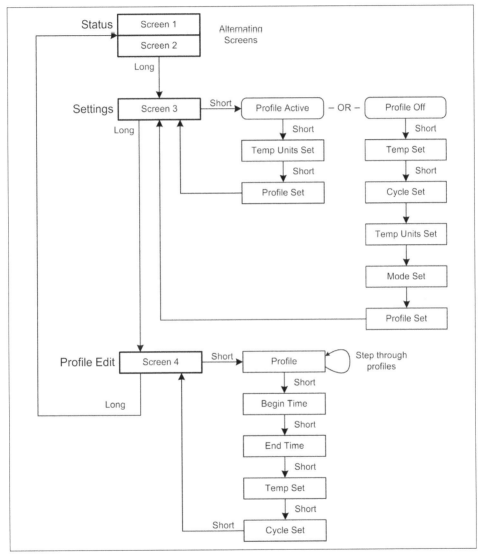

Figure 12-16. Thermostat screen and field navigation

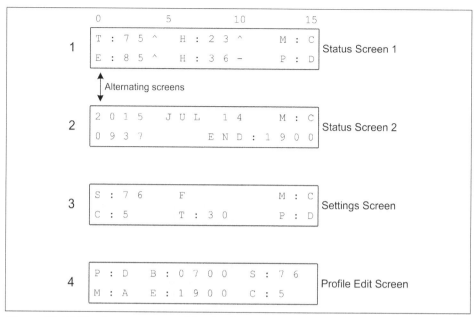

Figure 12-17. The four screens used by the thermostat

I'm reusing the Arduino-on-a-board that was used in Chapter 11 for the signal generator, but I could also have used the Duinokit introduced in Chapter 10. I elected to give the board another mission because it's small enough to mount on the wall next to the existing thermostat during testing. The prototype is shown in Figure 12-18.

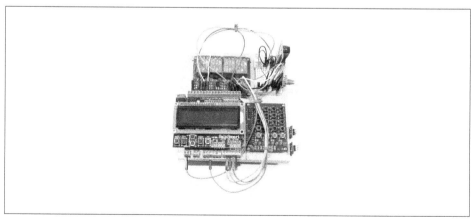

Figure 12-18. Arduino thermostat prototype

The prototype uses a slightly different, but electrically equivalent, set of parts to the final unit. The LCD shield is the same as the final unit will use, but the Arduino is an Uno whereas the final unit will use a Nano. In Figure 12-18 you can see the rotary encoder, the RTC module, and the DHT22 sensor on the small solderless breadboard module. The parts list for the prototype is given in Table 12-4.

Table 12-4. Prototype thermostat parts list

Quantity	Description
1	Arduino Uno (or equivalent)
1	LCD 16 × 2 display shield
1	DS1302 real-time clock module
1 set	Screw terminal adapters
1	Quad relay module
1	DHT22 temperature/humidity sensor
1	KEYES KY-040 rotary encoder module

DHT22 Sensor

For temperature and humidity sensing we will use the DHT22 device, which is the same device used with the GreenShield in Chapter 10. In the final version the DHT22 sensor will be connected to the Arduino Nano with leads attached to screw terminals. The pinout of the DHT22 is shown in Figure 12-19.

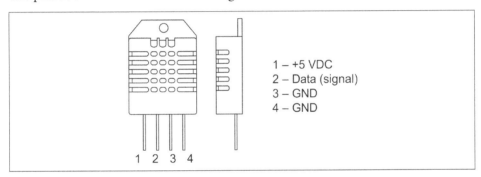

1 – +5 VDC
2 – Data (signal)
3 – GND
4 – GND

Figure 12-19. DHT22 sensor pinout

A 1K ohm pull-up will be used on the data line, as recommended in the datasheet (available from Adafruit (*http://www.adafruit.com*), SparkFun (*http://www.spark fun.com*), and other locations). In the prototype, the DHT22 will be connected using a small solderless breadboard block on the wood base, as can be seen in Figure 12-18.

Rotary Encoder

The primary (and only) control input is a KEYES KY-040 module (see Chapter 9 for information about KEYES modules) with a Bonatech rotary encoder, similar to an Alps EC11 part. The module is shown in Figure 12-20, and the pinout is shown in Figure 12-21.

Figure 12-20. Rotary encoder module

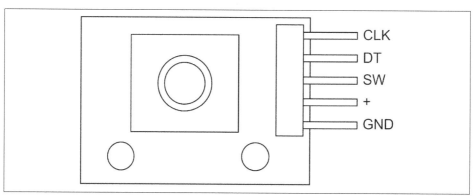

Figure 12-21. Rotary encoder module pin functions

The rotary encoder module circuit is simple, as shown in Figure 12-22. The encoder contains an internal switch that is engaged when the shaft is pressed in, and the three signal lines use 10K pull-up resistors. The pins labeled CLK and DT correspond to the A and B signals (respectively) found in most software examples and descriptions of simple quadrature rotary encoders.

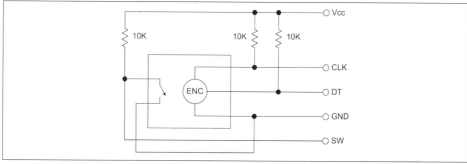

Figure 12-22. Rotary encoder module schematic

The rotary encoder's integrated switch is used to change between display screens, and also for various user-modifiable items on each screen. It and the rotary encoder itself are the only two control inputs. The internal switch is normally open.

Real-Time Clock Module

The thermostat needs to be able to keep time in order to determine which thermal profile to use. A real-time clock (RTC) module based on a DS1302 IC is used for this purpose. The RTC module is really nothing more than a carrier for the DS1302 IC, a crystal, and a battery. Figure 12-23 shows a schematic of the RTC module. The pinout is self-explanatory.

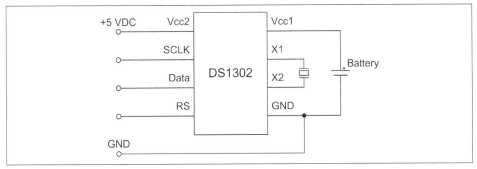

Figure 12-23. RTC module schematic

A library for the DS1302 is available from the Arduino Playground (*http://bit.ly/apg-ds1302rtc*). The thermostat will use this rather than create a new library from scratch.

LCD Shield

The LCD shield, which can be seen in Figure 12-18 and is described in Chapter 8, is a common and readily available shield. It is the largest consumer of digital I/O pins on the Nano. In this application the shield was selected because it incorporates a potentiometer for contrast and a control transistor for the display backlight. It also connects directly to the underlying Arduino, thus eliminating wiring like that used in the signal generator in Chapter 11.

Software

The thermostat software is straightforward, and like most embedded software it is designed to run continuously with a primary loop. Unlike the other examples presented so far, the thermostat utilizes interrupts to handle a rotary encoder. Implementing an interrupt handler with an Arduino isn't particularly difficult, and interrupts are by far the best way to deal with asynchronous events from the real world.

Source Code Organization

As with the other projects in this book, the source code is organized as a set of source files (see Table 12-5). These include the primary module with the `setup()` and `loop()` functions, a module for global variables, and the logic for controlling the HVAC functions in accordance with user-defined heating and/or cooling "programs." The display module borrows from work already done for the signal generator in Chapter 11, and the interface module, *tstat_iface.cpp*, also includes the interrupt handler for the rotary encoder.

Table 12-5. Thermostat source code modules

Module	Function
Thermostat.ino	Primary module containing `setup()` and `loop()`
tstat.h	Constant definitions (`#define` statements)
tstat_ctrl.cpp	HVAC control logic
tstat_ctrl.h	Include file
tstat_gv.cpp	Global variables
tstat_gv.h	Include file
tstat_iface.cpp	Control input processing
tstat_iface.h	Include file
tstat_lcd.cpp	LCD functions
tstat_lcd.h	Include file
tstat_util.cpp	Utility functions
tstat_util.h	Include file

Software Description

The software can be divided into three functional sections: user interface, control logic, and display management. The user interface functions, contained in the source module *tstat_iface.cpp*, provide the functionality to read the rotary encoder and navigate the display screens. *tstat_lcd.cpp* contains the LCD display handling functions. The source module *tstat_ctrl.cpp* provides the functionality for temperature and humidity sensing via the DHT22 sensor, the controller logic, and the user-defined programs.

Various libraries are used with this project for the DS1302 RTC, the rotary encoder, time and date functions, and the AVR's timer 1 peripheral. These are listed in Table 12-6. This is one of the nice things about working with the Arduino: if you need a library to use a particular module, sensor, or shield, chances are someone, somewhere, has taken the time to create it so that you don't have to.

Table 12-6. External libraries used with the thermostat

Name	Function	Author
Time	Time functions	Michael Margolis
DS1302RTC	RTC class	Timur Maksimov
ClickEncoder	Rotary encoder class	Peter Dannegger
TimerOne	Timer 1 class	Lex Talionis

These libraries can be found in the Arduino Playground (*http://playground.ardu ino.cc*) and on GitHub (*http://www.github.com*). Be sure to read the included documentation and the source code to gain a better understanding of what the code is doing and how to configure it if necessary. The DHT22 library is the same one that was used with the GreenShield in Chapter 10.

The file *tstat.h* contains the global definitions used by the other modules. These follow the pin assignments shown in Figure 12-7, illustrated in Example 12-1.

Example 12-1. tstat.h I/O definitions

```
#define ROTENC_A     0
#define ROTENC_B     1
#define ROTENC_SW    2

#define LCD_D4       4         // Predefined by the LCD shield
#define LCD_D5       5         //
#define LCD_D6       6         //
#define LCD_D7       7         //
#define LCD_RS       8         //
#define LCD_E        9         //
#define LCD_LED      10        //
#define LCD_SW       A0        //

#define RTC_SCLK     11
#define RTC_IO       12
#define RTC_CE       13

#define RY1          15        // A1
#define RY2          16        // A2
#define RY3          17        // A3
#define RY4          18        // A4

#define DHT1         3         // Internal DHT22
#define DHT2         19        // A5, external DHT22
```

The function setup() in *Thermostat.ino*, shown in Example 12-2, initializes the LCD, displays some startup messages, checks the RTC module, and displays the first screen.

Example 12-2. setup() function

```
void setup()
{
    lcd->begin(16, 2);  // Set the dimensions of the LCD

    TitleDisp("Initializing...", "", 1000);

    lcd->clear();

    if (rtc->haltRTC())
        lcd->print("Clock stopped!");
    else
        lcd->print("Clock working.");

    lcd->setCursor(0,1);
    if (rtc->writeEN())
        lcd->print("Write allowed.");
    else
        lcd->print("Write protected.");

    delay (2000);

    // Setup time library
    lcd->clear();
    lcd->print("RTC Sync");
    setSyncProvider(rtc->get); // The function to get the time from the RTC
    lcd->setCursor(0,1);
    if (timeStatus() == timeSet)
        lcd->print(" Ok!");
    else
        lcd->print(" FAIL!");

    delay (1000);

    TitleDisp("Initialization", "complete", 1000);

    curr_screen = 0;
    Screen1();
    disptime = millis();
}
```

The loop() function in *Thermostat.ino* isn't complicated, as you can see from Example 12-3. Most of the work takes place when the rotary encoder is in use and when the software is retrieving temperature and humidity data and executing the control functions in *tstat_ctrl.cpp*.

Example 12-3. Thermostat main loop function

```
void loop()
{
    // Get current date and time from RTC
    RTCUpdate();

    if (input_active) {
        HandleInput();
    }
    else {
        // Toggle between screen 1 and screen 2
        if ((millis() - disptime) > MAX_DISP_TIME) {
            if (curr_screen) {
                Screen1();
                curr_screen = 0;
                disptime = millis();
            }
            else {
                Screen2();
                curr_screen = 1;
                disptime = millis();
            }
        }
    }

    GetTemps();
    SystemControl();
}
```

When the encoder is not in use the display will alternate between screens 1 and 2, which show the current conditions and operating state and the date and time, respectively. When the encoder is used the input_active flag is set to true, and it will remain true until the user returns to screen 1 by pressing the encoder knob to engage the pushbutton switch. Notice that the system control is still active while the input screens are in use.

Testing

Testing is a three-step process, and this section applies to both the prototype and the final unit. The first step involves setting the date and time in the RTC. This is done by changing to the time and date setup screen using the encoder pushbutton. After returning to the main displays, the correct time and date should appear. The next step involves setting the target temperature, the hysteresis range (the C parameter in the display), and the operation mode. I started by verifying that I could manually enable the system fan, and then entered a temperature setpoint and set the mode. You will want to watch the display and verify that the system stops heating or cooling when it has reached a temperature equal to the setpoint plus or minus one-half of the cycle (hysteresis) range.

Lastly, you should step through the profiles and set each one up the way you think is best for your situation. Remember that a conflict between the end time of one profile and the start of another is resolved by honoring the end time of a preceding profile, not the start time of a subsequent profile.

You should keep an eye on the operation of the thermostat over a period of a couple of weeks. It will probably require some fine-tuning to get the times and setpoints established for the most efficient operation. If you have built a GreenShield you can use it to log the temperature and humidity to capture performance data for the thermostat. A simple Python script will work nicely to query the GreenShield at regular intervals to collect data.

Final Version

The final version of the thermostat differs from the prototype in a few ways physically, but it is otherwise identical in terms of both function and signal connections. The three primary differences are the use of Nano board, a bare LCD module rather than an LCD shield, and an enclosure to mount it all into. The parts list for the final version of the thermostat is given in Table 12-7.

Table 12-7. Final version thermostat parts list

Quantity	Description
1	Arduino Nano (or equivalent)
1	Screw terminal prototype board
1	16 × 2 LCD display module
1	Quad relay module
1	DHT22 temperature/humidity sensor
1	KEYES KY-040 rotary encoder module
1	Real-time clock module
1	Plastic enclosure

Assembly

The final assembly mainly consists of mounting the components in and on the enclosure. But before that, the enclosure itself went through some changes. The embossed lettering was sanded off the top cover, the side mounting tabs were removed, and holes were cut and drilled for the display and the rotary encoder. Holes were also drilled into the bottom of the enclosure for mounting screws and the existing heater and

A/C control wires. The case was painted a nice neutral shade of ivory after all the sanding, drilling, and cutting was done.

I used a screw terminal prototype shield to hold the Nano board, and wired the Nano's pins to the appropriate terminals and pin headers. Figure 12-24 shows the topside view of the Nano's prototype shield, along with the LCD shield, the rotary encoder, and the RTC module. The RTC module will be mounted to the inside of the enclosure.

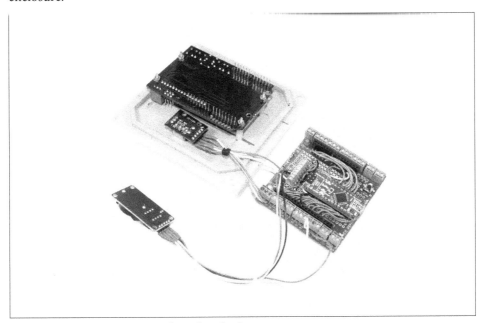

Figure 12-24. Nano prototype board and other top cover components

As I elected to keep all the wiring on the top of the prototype shield, the bottom of the shield is nothing but soldered pads. I used the smallest gauge of insulated stranded wire that I had available (28-gauge 7-strand) without resorting to wire-wrap wire. Solid 30 AWG wire with Kynar insulation is great for wire-wrap construction, but it can be less than ideal for something like this.

The big advantage of the screw terminal prototype shield is the ease with which wires to or from other components may be connected to the terminals. In the case of the thermostat this includes the DHT22 sensor, the real-time clock module, the relay module, and the rotary encoder.

Figure 12-25 shows what the enclosure looked like after its makeover. The LCD shield and the rotary encoder are already mounted. What isn't shown is the DHT22 sensor, which is mounted on the outside of the bottom of the enclosure. You might also notice that the cover screws aren't installed.

The small hole to the upper left of the LCD window is an access hole for the trim potentiometer on the LCD shield to set the display contrast. The LCD window will get some edge trim as the very last step, and a clear plastic cover might also be a good idea. Yes, it looks like something from an industrial plant, but as I said before, how it looks has no bearing on how it works. The gold knob is there (if you can see the image in color) because that was the only knob I had available at the time that didn't have a pointer molded into it.

Figure 12-25. Almost finished thermostat enclosure

The quad relay module is mounted to the bottom of the enclosure. The RTC module is mounted to an interior side of the box. Figure 11-25 shows how things are arranged in the enclosure. The existing HVAC wiring is brought in through an access hole in the bottom of the enclosure. Note that the hole is offset vertically from center to allow clearance for the relay module.

It's a tight fit, to be sure, but it does all fit. There are still several steps necessary before closing it up and turning it loose:

1. Locate the mounting screw locations on the wall and drill holes.
2. Pull the existing HVAC control wires through the bottom hole and mount the enclosure to the wall.
3. Connect the HVAC wires to the relay module.
4. Connect the power source and check the display.
5. Attach the cover to the enclosure.

Assuming that the software was already loaded onto the Nano, the thermostat should now be operational. There is no power switch, so it will start functioning as soon as the power is connected. The display backlight should remain active for 15 seconds after power is applied, so you may need to turn the encoder knob slightly if it has timed out and turned off.

Figure 12-26. Thermostat internal components

The enclosure comes with a rubber gasket, which you can use if you think it is necessary. I used it simply because it allows for a snug fit of the cover and gives a bit more space inside for the components, but I'm not worried about rain getting into the unit.

Testing and Operation

Testing consists of using the encoder to step through the display screens and verify that the long and short switch sensing is working correctly. You will need to set the parameters again, since loading the software doesn't load the EEPROM, just the flash program memory. If the installed unit appears to operate correctly then you can start setting it up for your particular environment and preferences.

I would suggest something simple to begin with. Just controlling the heating or cooling is a good place to start. After you are satisfied that the unit is working as it should, then you might want to try setting up daytime/nighttime profiles and see how that works. As you gain more experience with the unit you can incorporate the humidity and cycle time settings. A device like this requires some amount of "tuning" to get it set so it will integrate effectively with your environment.

You may recall from Figure 12-4 that there is an input for a second DHT22 sensor. I elected to leave it out for the first iteration of this unit so that I could focus on the basic functionality. Incorporating the input from a second DHT22 is trivial electrically, but it introduces a new level of complexity into the software.

The reason is that with a second sensor the thermostat must not only monitor the inside environment, but now must also factor in the outside environment—things like internal versus external temperature difference, the rate of change at any given moment in time, and the estimated heat loss or gain through walls and windows. With enough effort and clever programming, the thermostat could attempt to balance the various factors and arrive at an optimal heating or cooling profile. This may sound really interesting, but it's not trivial.

As with all the software in this book, the source code for the thermostat is available on GitHub (*https://www.github.com/ardnut*). I will occasionally post updates for the thermostat and the other example projects.

Cost Breakdown

One of the objectives of this project was to use off-the-shelf boards and modules as much as possible. This not only saved time and effort, but also helped hold down the cost (custom PCBs are not inexpensive). Build-wise, this project required much less effort than the signal generator in Chapter 11. Table 12-8 lists the primary components.

Table 12-8. Thermostat parts price list

Quantity	Item	Source	Unit price	Extended
1	Arduino Nano	SainSmart	$14.00	$14.00
1	Screw terminal prototype shield	Adafruit	$15.00	$15.00
1	RTC module	DealeXtreme	$2.00	$2.00
1	Rotary encoder module	Various	$6.00	$6.00
1	LCD display shield	SainSmart	$10.00	$10.00
2	DHT22 sensors	Adafruit	$10.00	$20.00
1	Quad relay module	SainSmart	$7.00	$7.00
1	Electrical enclosure	Home Depot	$7.00	$7.00

[a] The extended amount is the total cost of each line item for a given quantity.

Total parts cost = $81.00

Note that KEYES modules, like the KY-040 rotary encoder, are often found in kits of modules. They can be purchased as single items from vendors on both eBay and Amazon, but picking up one of the kits is a better bargain.

As with the other example projects in this book, you can probably do better price-wise than what is shown in Table 12-8 by shopping around and doing some bargain hunting. I've seen Nano-type boards going for as little as $5, and DHT22 sensors in the $7 to $8 range.

Next Steps

This project is an example of the limitations that you can encounter when trying to create complex devices using only off-the-shelf shields and modules. If I were to create a Version 2 of the thermostat I would build a custom PCB that would contain everything on one board. Like the GreenShield from Chapter 10 it would have on-board relays, and a power supply would be incorporated as well.

This design is also a prime candidate for an I/O multiplexer like the MPC23017 used in the Switchinator PCB in Chapter 10. The LCD display can be controlled via an MPC23017. Adafruit sells a version of the display shield with an MPC23S17, which is the SPI version of the IC.

Additionally, I would use surface-mount parts for everything except perhaps the relays, the DHT22, and the terminal blocks. I would think that the relays could be downsized a bit, perhaps even to surface-mount parts, since they don't really need to handle a lot of current through the contacts. With some careful design and parts selection the size of the entire PCB could be significantly reduced.

Some nice-to-have features that simply won't fit in the current design include a Bluetooth or WiFi interface, a microSD flash card carrier for long-term data logging, and perhaps some additional LEDs for the front panel. As it stands, the Arduino thermostat is almost an Internet of Things (IoT) device, but it will require some redesign and enhancement to get it all the way there. Of all the things touted as IoT devices, a thermostat might make the most sense. I'm not convinced that letting the coffee maker or the microwave oven talk to the refrigerator is a good idea.

Resources

Table 12-9 lists the distributors and vendors where I purchased the parts for the thermostat. As with all of the example projects in this book, there are multiple sources for most of the components. These are simply the ones I used at the time I made the purchases.

Table 12-9. Parts sources

Distributor/vendor	URL
Adafruit	www.adafruit.com
DealeXtreme (DX)	www.dx.com
DFRobot	www.dfrobot.com
SainSmart	www.sainsmart.com

I found the enclosure in the electrical section of my local Home Depot store, but you can find similar electrical enclosures at any electrical supply house or well-stocked hardware store.

CHAPTER 13

Model Rocket Launcher: A Design Study

This chapter is a departure from the previous three chapters. Instead of using an example project (or projects, in the case of Chapter 10) to illustrate how the concepts, tools, and components described in earlier chapters are used in real applications, this chapter illustrates how to perform a design study. We won't actually build anything, but we will define some possible approaches, identify suitable tools and components, and evaluate design trade-offs. This type of engineering activity applies to more than just model rocket launchers; it can be applied to an Arduino design project with any degree of complexity.

Remember that since the main emphasis of this book is on the Arduino hardware and related modules, sensors, and components, the software shown is intended only to highlight key points, not to present complete ready-to-run examples. The full software listings for the examples and projects can be found on GitHub (*https://www.github.com/ardnut*).

Overview

An Arduino-controlled rocket launcher is an interesting application for several reasons: first, it is very extensible and you can take it as far as you want (and can afford) within the limits imposed by the Arduino hardware; second, physically the launcher can be anything from a small plastic box with a switch and an LED to a console with digital readouts and a keyswitch or two; third, the launcher can support a wide range of functions beyond simply applying current to an igniter. It could have WiFi or Bluetooth connectivity, the ability to continuously check igniter continuity, synthetic speech generation for the countdown and launch status reports, range safety inputs, and even the ability to disconnect DC power from a rocket prior to launch to keep the internal batteries fully charged. How much (or how little) effort you want to put into it will depend on your objectives, abilities, and finances.

By the end of this chapter you should be able to make informed decisions about what hardware to buy, how difficult the software will be to write and test, and how to anticipate the level of effort a given design will entail. Using the material presented in earlier chapters you should be able to create a parts list and get an idea for how much the design will cost and how difficult it will be to build. You should also be able to create a preliminary block diagram and software flowchart for your design. Once all these things are in place, all that remains is gathering up the parts, assembling it, and then creating and loading the software to make it go.

 Before attempting to build something like this, make sure you understand the safety considerations that are involved with something like a rocket launcher. The National Association of Rocketry (NAR) and the Tripoli Rocketry Association have published guidelines for safely handling and launching model rockets, and many states have basic rules in place regarding where and when you can launch, and what qualifies as a model rocket.

The Design Cycle

When we design anything with even a moderate degree of complexity, it is rare that we don't find ourselves going back to an earlier step in the process to make modifications or seek alternatives. Sometimes that might even mean tossing out the original design and starting over. This is not uncommon. Design is an iterative process, and as it progresses things become apparent that weren't obvious at the outset. It is almost impossible to anticipate everything in advance.

The design process follows the general outline presented in Chapter 10. We begin with a set of objectives. The objectives are refined to create functional requirements. From the functional requirements we can identify the types of hardware and software that will fulfill the requirements. Then we look at the details of the selected components and techniques to evaluate their suitability. Figure 13-1 shows how the design cycle works.

At the end of each step we will stop and consider what we've done up to that point. If there is problem, that is where we want to evaluate how to resolve it. Depending on the type and severity of the problem, we could elect to step back and revise a previous step, or we might decide to drop something that isn't essential and go forward with reduced expectations. We might also find issues during the activity for a particular step. Say, for example, that we discover that a particular shield, sensor, or module simply isn't available, and we can't justify building it ourselves. In that case we need to stop, step back, and see what drove that selection in the first place and if it can be revised to accommodate the current situation.

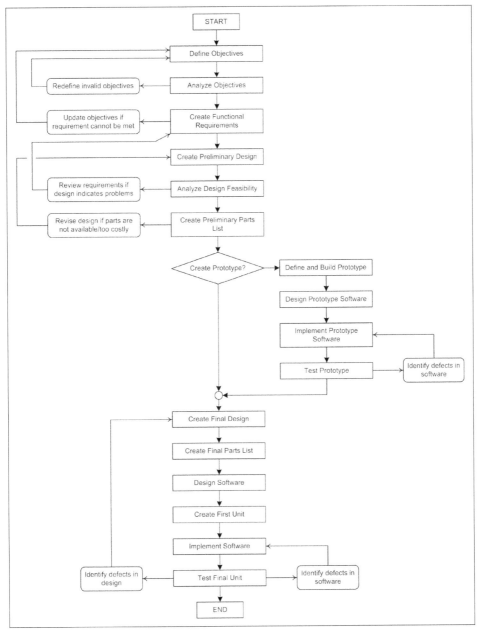

Figure 13-1. The design cycle

In some cases it may not be necessary to build a real prototype. This is often the case when a design consists primarily of ready-made modules, as is the case with many Arduino-based designs. I like to build prototypes with prototyping platforms I have created or purchased, such as the Duinokit and solderless breadboard used in Chapter 10, or the Arduino-on-a-board used in Chapters 11 and 12. It allows me to connect things without resorting to soldering wires and drilling holes, and still get the same functionality that the finished unit will have. You may find that you can skip the prototyping step for some design projects, but be aware that if a problem does pop up in the final design it may be expensive and time-consuming to deal with it.

 In a real-world situation the cost of dealing with a defect in the design or implementation, either hardware or software, increases as the project progresses through the steps shown in Figure 13-1. In other words, it is much cheaper to fix an invalid functional requirement than it is to rewrite defective software or deal with a hardware design flaw. By the same token, it is easier and cheaper to deal with a problem in a prototype than it is to address a problem in a design that is about to go into production. This is why building prototypes is so popular.

Objectives

The first step for a design activity is to decide what objectives the design is intended to meet. We may not have all the details starting out, but we should be able to decide on the fundamental characteristics, even if it's just a small set initially. The list of objectives can grow over time, sometimes dramatically. In fact, it is often a challenge to keep the design objectives in line with reality, particularly when the person defining the objectives is not the same person who will be building the final design. Fortunately, this is not the case here. So long as we keep our enthusiasm and optimism in line with reality we should be fine.

When setting down a list of objectives I recommend organizing the list by importance. In other words, put the "must have" features at the top of the list, and the "bells and whistles" features toward the end of the list. That will allow you to trim off items from the list should the level of effort, cost, or time constraints exceed acceptable limits without losing the essential functionality of what you're trying to build.

Here is my list of prioritized objectives for an automatic rocket launch controller (aka launcher). The primary objectives are:

1. The launch controller will incorporate a countdown timer with presettable times (5, 10, 15, 30, and 60 seconds).

2. The countdown may be aborted at any time for a number of reasons (given next).

3. The launch controller must have a key-operated safety interlock (arming) switch. The launcher will not start the count or activate the igniters unless the interlock is set to the "on" position.

4. The launch countdown timer is started by pressing a pushbutton switch.

5. The launch countdown cannot proceed unless a safety switch is depressed prior to pressing the countdown start button and during the entire period of the count. This could be a handheld pendant or small box, and it is separate from the count start switch.

6. If the safety switch is released prior to launch, the countdown will abort and the controller will enter a "safe" mode (igniter circuits disabled and grounded).

7. The launcher must be capable of energizing multiple igniters at the same time. Two would be a minimum number, with up to eight if possible.

8. Each igniter should be monitored independently of any others to detect circuit continuity problems prior to launch.

9. A continuity failure of any igniter will abort the launch (the same response as a premature release of the safety switch).

10. The controller will use a flashing LED to track the countdown in seconds. Other LED indicators will show the status of the controller (armed, counting, safety abort, or igniter fault).

The secondary objectives are as follows:

1. A 7-segment LED display will display the countdown.

2. A 7-segment LED display will display the current time.

3. A 16 2 LCD will display system status data and error messages.

4. All components will be mounted in a low-profile sloped-top console chassis.

And these are the optional objectives:

1. The launch controller will have the ability to activate external devices at specific points in the countdown sequence, such as ground power disconnect, launch warning siren, and on-board electronics activation (camera start and so on).

2. The controller will have a connector for attaching a large 7-segment display for group event display purposes.

3. The controller will incorporate the ability to generate synthetic speech for count-down and status announcements.

4. A Bluetooth interface will be used to connect to a remote weather monitoring system (primarily used to detect excessive wind speed in the launch area). If this

is used then the system must also have the ability to abort on out-of-range wind speed and indicate this to the launch director (the user).

That's a rather extensive and ambitious list, to be sure. Achieving the primary objectives doesn't seem like it would be too difficult, which is what we want for primary objectives. The secondary objectives are a bit more challenging, and the optional objectives, while intriguing, could tax the stamina of the builder (not to mention costing more money).

Selecting and Defining Functional Requirements

The way I've written the objectives allows us to move directly into the functional requirements with very little translation. A requirement needs to be clear, concise, coherent, and verifiable, and in some cases the objectives are worded like requirements, so that makes it easier for us.

The big question now is, how many of the objectives do we adopt and move forward with as functional requirements? In order to answer that question we need to use some type of scoring system to determine how hard, how expensive, and how necessary each objective will be to achieve.

We'll start with the primary objectives. Table 13-1 shows a list of the primary objectives with my best estimates for difficulty, expense, and overall necessity. Each aspect is scored on a scale of 1 to 5, with 1 being the least and 5 being the most.

Table 13-1. Primary objectives ranking

Objective	Difficulty	Expense	Necessity
1.1	2	2	2
1.2	3	2	5
1.3	3	2	5
1.4	2	2	4
1.5	2	3	5
1.6	2	1	5
1.7	3	3	3
1.8	4	3	5
1.9	3	2	5
1.10	3	2	2

From this we can conclude that objective 1.1, a presettable countdown timer, could be replaced with a fixed 10- or 15-second count if need be. It is also apparent that objective 1.10, the blinking lights, might fall under the category of "bells and whistles," or secondary objectives, and could be dispensed with if necessary (although we would likely want to keep at least the countdown LED to let us know something is happening).

We can do the same thing for the secondary and optional objectives, as shown in Table 13-2.

Table 13-2. Secondary and optional objectives ranking

Objective	Difficulty	Expense	Necessity
2.1	3	3	2
2.2	3	3	2
2.3	3	3	2
2.4	3	3	2
3.1	4	4	1
3.2	4	3	1
3.3	4	4	1
3.4	4	4	1

You may notice that there is an interesting pattern emerging between Table 13-1 and Table 13-2. The primary objectives are all relatively inexpensive and easy to implement. Also, with the exceptions of 1.1 and 1.10, they are all necessary. In Table 13-2 we see that as the expense and difficulty increase, the necessity decreases. None of the optional objectives are necessary, although they would be really fun to implement.

I didn't intentionally arrange the objectives like this when I wrote them down; they just fell out that way. Honest. I suppose it might be the result of doing this sort of thing quite a bit over the years, but in any case you'll want to organize your objectives first on the basis of necessity, then on the basis of either cost or difficulty, depending on what is more important to you: money or time.

This brings us to a trade-off point. If something is necessary but it will be expensive, do we eliminate something that is not as necessary to keep the project budget sensible, or would we be willing to throw more money at it to get both objectives met? In some cases, such as choosing between the primary and optional objectives, the choice is clear-cut. The end result might not be as dazzling as it could otherwise be, but we won't break the bank building it, either. It will still launch rockets just fine. On the other hand, when choosing between two primary objectives with equal necessity and equal expense, such as objectives 1.1 and 1.10, the choice could hinge on how difficult the objectives will be. In this case it would probably be easier to install one rotary switch to select the countdown time than it would be to drill holes for LEDs, wire and install them, and then create the software to control them all.

For the sake of continuing with the design analysis, let's assume that we will keep all of the objectives so we can see how they will drive the design and component selections in later steps. Table 13-3 lists the functional requirements derived from the objectives.

Table 13-3. Rocket launcher functional requirements

Req #	Description
1.1	The launch controller will incorporate a countdown timer. At the count of zero, the igniter(s) will be energized.
1.1.1	The countdown timer will have five presettable times of 5, 10, 15, 30, and 60 seconds.
1.1.2	The countdown times will be selected using a five-position rotary switch.
1.1.3	If the countdown time is changed in mid-count the countdown will be aborted.
1.2	The countdown may be aborted at any time for a number of reasons (given below).
1.2.1	The countdown will be aborted if any igniter circuit is open that should be closed (see 1.7.1 and 1.9.1).
1.2.2	The countdown will be aborted if the count time is changed while the countdown is active.
1.2.3	The countdown will be aborted if the safety switch is released during the count.
1.3	The launch controller must have a key-operated safety interlock (arming) switch.
1.3.1	The launcher will not start the count or activate the igniters unless the interlock switch is set to the on position.
1.3.2	The countdown will abort if the interlock switch is set to off during the countdown.
1.4	The launch countdown timer is started by pressing a pushbutton switch.
1.4.1	A "launch" button will be used to start the countdown.
1.4.2	The launch button will not function unless the range safety switch is engaged.
1.5	The launcher will use a range safety switch to control the countdown and launch.
1.5.1	The launch countdown cannot proceed unless a safety switch is depressed prior to pressing the launch button.
1.5.2	The range safety switch must be manually engaged during the entire period of the countdown.
1.5.3	The range safety switch can be a handheld pendant or small box, and it must have at least 10 feet (3 m) of electrical cable.
1.6	If the countdown is aborted for any reason the launcher will enter "safe mode."
1.6.1	In safe mode all igniter lines will be de-energized and grounded.
1.6.2	Exit from safe mode will require that the launcher be disarmed using the safety interlock switch.
1.7	The launcher must be capable of energizing multiple igniters at the same time.
1.7.1	The launcher will support up to six active igniter circuits.
1.7.2	A rotary switch will be used to select the number of active igniter circuits.

Req #	Description
1.8	Each igniter should be monitored independently of any others to detect circuit continuity problems prior to launch.
1.8.1	Each igniter circuit must have continuity to verify that the igniter is connected.
1.8.2	The sense current used to determine continuity shall not exceed 250 uA on any igniter circuit.
1.9	A continuity failure of any igniter will abort the countdown timer.
1.9.1	The countdown will not start if any active igniter circuit has no continuity.
1.9.2	The countdown will abort if any active igniter circuit loses continuity during the count.
1.9.3	The launcher will use LEDs to indicate active and ready igniter circuits (see 1.1.2).
1.10	The controller will use LEDs to indicate the countdown and status.
1.10.1	A flashing LED will track the countdown in seconds, one flash per second.
1.10.2	An array of LEDs will show the active state and continuity of the igniter circuits.
1.10.3	An LED will indicate the active state of the range safety switch.
1.10.4	An LED will indicate a range safety abort.
1.10.5	An LED will indicate the interlock switch state.
1.10.6	A flashing LED will indicate when the launcher is in safe mode after an abort.
2.1	A 7-segment LED display will display the countdown.
2.1.1	A 2-digit 7-segment red LED display will be used to display the countdown.
2.1.2	The countdown will display dash characters after a successful launch.
2.1.3	In event of an abort the display will show the word "Abort."
2.2	A 7-segment LED display will display the current time.
2.2.1	A 4-digit 7-segment yellow LED display with HH:MM format will show the current time.
2.2.2	The time display will be settable using pushbuttons on the front panel of the launcher.
2.2.3	The time display will be in 24-hour format.
2.3	A 16 × 2 LCD will display system status data and error messages.
2.3.1	A 16 × 2 LCD display will show messages for readiness status and faults.
2.3.2	The LCD display will also be capable of displaying messages from external sources such as a remote weather station.
2.4	All components will be mounted in a low-profile sloped-top console chassis.
2.4.1	A low-profile metal chassis will be used to contain the launcher.
2.4.2	The chassis will be at least 14 inches wide by 10 inches (35.5 × 25.5 cm) deep, with a height of 1.5 inches at the front and 3 inches at the rear of the chassis (3.8 × 7.6 cm).
3.1	The launch controller will have the ability to activate external devices at specific points in the countdown.
3.1.1	The controller will be capable of activating external devices via relays at user-definable points in the countdown.

Req #	Description
3.1.2	The action points will be set using a USB connection to the Arduino in the launcher.
3.2	The controller will have a connector for attaching a large 7-segment display for group event display purposes.
3.2.1	The signals to the internal two-digit countdown display will be duplicated to a connector on the rear of the chassis.
3.2.2	The external display signals will be controlled by a high-current IC to provide sufficient drive current.
3.3	The controller will incorporate the ability to generate synthetic speech for countdown and status announcements.
3.3.1	The launcher will incorporate a speech synthesis shield to produce intelligible speech.
3.3.2	The speech output will be user defined.
3.4	A Bluetooth interface will be used to connect to a remote weather monitoring system.
3.4.1	The user can define an upper limit for wind speed. Exceeding the limit will cause a launch abort.
3.4.2	An LED on the front panel will indicate a wind speed launch abort.

I endeavored to maintain a one-to-one relationship to the objectives listed earlier, but there is some deviation, mainly involving the elimination of redundancy.

 This design utilizes a "range safety switch" that must be held down (active) during the countdown. This is an essential safety feature. In practice there could be two people involved: a launch director and a range safety officer. It is the range safety officer's job to abort the launch if anything is not as it should be prior to ignition (e.g., wind, or people in the launch area).

Creating the Preliminary Design

Now that the functional requirements have been defined, we can see some obvious design criteria as we read through them. We know that we are going to need two rotary switches, an LCD display, two 7-segment displays, a real-time clock of some sort, a pendant or small box with a pushbutton switch (the range safety switch), some connectors on the rear panel, some relays, and multiple LEDs of various colors.

In order to get a better idea of what will be involved in the design we can create a preliminary block diagram like the one shown in Figure 13-2. From the block diagram it becomes immediately apparent that something like an Arduino Uno isn't going to be able to handle all the required I/O. We need to consider a Mega2560 instead. We might also want to consider more than one Arduino to spread the workload around.

Figure 13-3 is a more refined version of the block diagram. Notice that it uses one Uno for the speech synthesis, another for the Bluetooth, and a Mega2560 for the primary control functions. The ATmega2560 MCU used in the Mega2560 has multiple USART interfaces, so each of the outboard Unos can have its own dedicated communications channel.

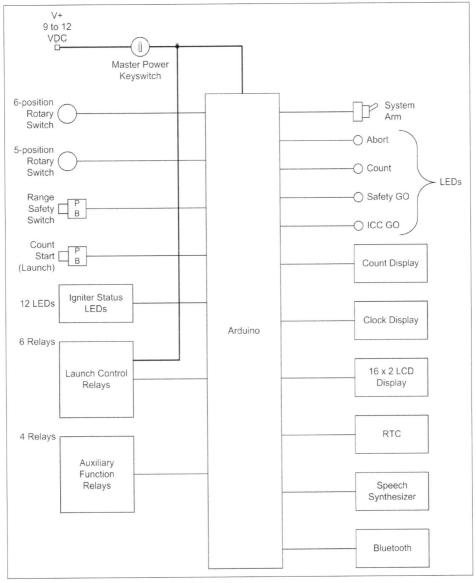

Figure 13-2. Preliminary launcher block diagram

In Figure 13-3 several of the interfaces to external modules have been refined and designated as I2C types. In addition, the digital I/O (DIO) ports have been identified. An I/O expander (similar to what was used in the Switchinator in Chapter 10) has been selected to handle the relays and the 12 igniter status LEDs.

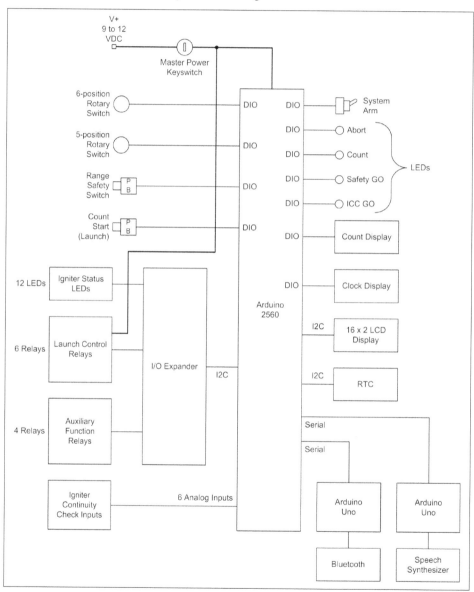

Figure 13-3. Refined launcher block diagram

Build or Buy

At some point with moderately complex projects trade-off decisions may need to be made regarding buying a ready-made hardware module versus designing and building something tailored to the specific requirements of the design. This decision can be difficult, and it involves more than just cost.

A commercial off-the-shelf (COTS) component or module may or may not have all the features you need for your design. It might have the correct functions, but the wrong connectors, or even vice versa. It might be the wrong size, or maybe just a bit off in one dimension or another. It might even require a supply voltage that's different from all the other components in the design, which could require some additional circuitry to accommodate it.

So, does it make sense to adapt the design to the COTS part, or would it be better to build a custom part that has exactly what you need? If the required change to the module is minimal, then it's probably smart to simply adapt it rather than spend the time and money on a custom design. However, if the issue is a showstopper, like physical dimensions or unneeded functions that will create conflicts, then it may make more sense to build something from a clean-sheet design and keep it consistent with the rest of the components.

If you're not sure, the decision usually comes down to weighing the cost and time required to use the COTS component versus the cost and time to build a custom item. If it's a close call, then one other criterion can be used to cast the deciding vote: how much control do you really need over the production and internal functions of the component in question? If you are building a commercial product and there's a possibility that the COTS module you want to use will be discontinued while your product is still in production, then you might want to consider building it yourself. Likewise, if you need visibility into the inner workings of the COTS component but it's a "black box" to you, then you might want to consider creating your own version of it.

Design Feasibility

So just how realistic is the design we've come up with? It is sufficient? Yes, undoubtedly. Is it overkill? Perhaps. Now is the time to step back and review the preliminary design with a critical eye and a willingness to cut out unnecessary or frivolous features without remorse or regret.

Bells and Whistles

In engineering the term "bells and whistles" is often used to describe features that don't really contribute to the functionality of a system, but just add some razzle-dazzle to the final product. Things like a color TFT display on a coffee maker. Does it really help the device make better coffee? No, of course not. Does it look cool and high-tech in your kitchen? Yes, and that's why it's there, and that's why you might be willing to pay extra for that bit of flashiness.

Consider this: modern automobiles are rolling palaces of bells and whistles. All that's really needed for a car is four wheels, an engine, some seats, and a body with windows to keep the rain, wind, and dust at bay. No one really needs a 10-speaker sound system, hands-free cell phone, navigation system, and multiposition electric seats just to drive to work or make a run to the grocery store, right? A basic car without all the gadgets and gizmos might be ugly as sin, but it will still get you to where you want or need to be. Better yet, that plain, ugly car will probably still be running while the fancy ride with all the bells and whistles is spending its time at the dealership in the repair department.

The upshot is that not only are simpler things less expensive, but they tend to be more reliable—there are fewer things to fail. So, when reviewing the feasibility of a design it is important to pay attention to more than just the *technical* feasibility. Also look out for the bells and whistles that don't really contribute to the operation of the device, but may end up contributing to early failures.

Feasibility is a relative assesement. In other words, feasible in relation to what? Some things that might not be feasible for someone working on the kitchen table could be perfectly feasible for another person with access to a complete shop. Budget limitations are another important concern, as are level of effort limitations. One person with a small budget and limited time cannot hope to achieve what someone else with a generous amount of money to spend and lots of time can do. Since I'm just one person, and I don't have lots of spare time and have only a limited amount of funds at my disposal, I have to be prudent about how much I can take on and how quickly I can expect to accomplish it. I'm sure I'm not alone in that regard.

I like to use the following five criteria when evaluating the feasibility of a design and its features:

- Does the feature or design make sense functionally?
- Does it make sense cost-wise?
- Is there any component that might be hard to get?
- Are there any unsafe aspects to the feature or the design?

- Can it be assembled, programmed, and tested by one person in a reasonable period of time?

If the answer to any of these questions is no, or even a maybe, then the component or design feature in question should be considered for removal.

The design of the rocket launcher, as represented by Figure 13-3, would indeed be a wondrous thing, but I don't think it's something I could expect to finish in a reasonable period of time. The better approach might be to keep it simple (as simple as possible, anyway) and create a design that can be expanded in the future.

Because of the way the objectives for the launcher were organized, trimming back the design is actually rather easy. The level 3 functional requirements define things that will be both time- and money-intensive. They also don't contribute anything critical to the core function of the launcher, which is of course launching model rockets. I am going to designate all the level 3 functional requirements as optional future add-on features for future implementation.

Let's call the basic launcher, which is comprised of the level 1 and level 2 functional requirements, the Phase I version. We can refer to the extended design as the Phase II version. While we're thinking about the Phase I design we can also be looking at ways to further simplify the design without sacrificing any of the required functionality.

For example, in Figure 13-3 there is an I/O expander. What is this? It could be something like a Centipede I/O expander shield, described in Chapter 8. The Centipede has 64 digital I/O lines, so one of these could easily handle the igniter status LEDs, the 4 system status LEDs, and the 2 rotary switches. We can further simplify the wiring by using SPI and I2C as much as possible. The RTC, 16 × 2 LCD, and I/O expander are all available with I2C interfaces, and 7-segment display modules for the count and time display are available with SPI interfaces. The revised block diagram for Phase I is shown Figure 13-4.

I elected to leave the launch control relays connected to discrete digital I/O pins on the Mega2560 rather than route them through the I/O expander. This is so that there is nothing between the launch relays and the Arduino controlling them. It probably won't make that much difference, but it does take a potential fault source out of the primary control path in the launcher. The same reasoning is why the count start, system arm, and range safety switches are connected directly to the Arduino. Also, by connecting the switches directly we can take advantage of the AVR MCU's pin change interrupt capability, should we want to do so.

The diagram in Figure 13-4 is what we will use going forward. At some point in the future it might be feasible to try for Phase II, but not at this time. It will still be there waiting, when it is time to make it happen.

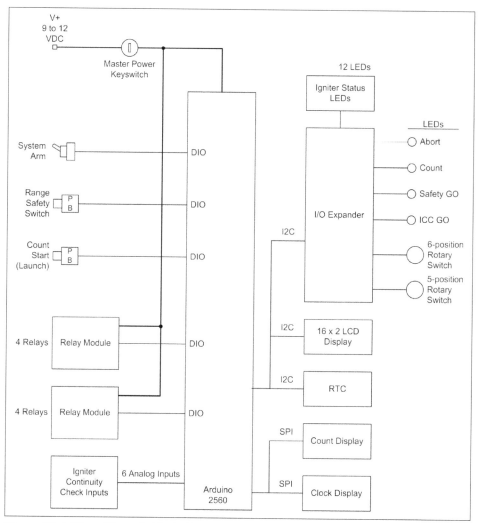

Figure 13-4. Phase I launch block diagram

Preliminary Parts List

With a preliminary design in hand for the Phase I version we have enough information to assemble a preliminary parts list. By going through the functional requirements and counting the number of times various types of controls and functions are mentioned, we can get a start on the quantities that will be needed.

The preliminary parts list doesn't include things like prototype PCBs for patching the wiring and mounting small terminal blocks (like the ones shown in Chapter 11). These requirements can be discovered and documented when it comes time to connect it all into a working unit, either as a prototype or the final version.

Table 13-4. Preliminary launcher parts list

Quantity	Description	Quantity	Description
1	Arduino Mega2560	1	2-digit 7-segment LED display
1	Mega screw terminal shield	1	4-digit 7-segment LED time display
1	LCD display shield	1	Keyed switch
1	Real-time clock module	1	Large pushbutton switch
2	Quad relay modules	1	Sloped-top metal console

Prototype

As discussed earlier, a prototype serves several valuable functions. First off, it allows you to work with the design and its member components in a flexible and easily modifiable form. If a module or sensor doesn't work out the way you thought it would, then it's a lot easier to change it in the prototype than it would be by drilling more holes in a metal chassis. Second, for designs that involve a custom PCB (or two or three), there will be a lead time between when the PCB design is sent off and when the finished boards arrive and assembly can start. If you're making a prototype, this time can be put to good use working on the software or the documentation, or both. Finally, in some cases you may have most of what you need for a prototype already on hand, and getting it to the point where it can serve as a stand-in for the final product is relatively simple. Figure 13-5 shows a simple prototype fixture for an Arduino Mega2560 built from some angle-cut sections of pine board.

Prototypes can definitely save time and hassle, but as was also mentioned earlier, sometimes you can skip over this step. In the case of the model rocket launch controller, we have a design that is not overly complex physically, there are no custom PCBs involved, and all the components are well-understood COTS parts.

You could definitely build a prototype if you wanted to do so. If you plan to incorporate the optional objectives, then a prototype would probably be a good idea. You could mount all the parts on a large board and make sure that everything works as expected before committing it to a metal enclosure. A cutting board from the kitchen would probably be about the right size. This, by the way, is where the term "breadboard" arose, in case you've ever wondered about that. Mounting Arduino boards, terminal blocks, and other assorted things to a board is part of a long tradition going back to the early days of radio at the start of the 20th century. Radio experimenters would literally use a breadboard, like those found in kitchens, as a base for terminal posts, tube sockets, and other components. Sometimes a schematic would be pasted to the board to give the builder something to follow. Many of the early radios were built mostly of wood, and the base for the components in the production radio wasn't much different from the breadboard that had been used to prototype and test the radio initially.

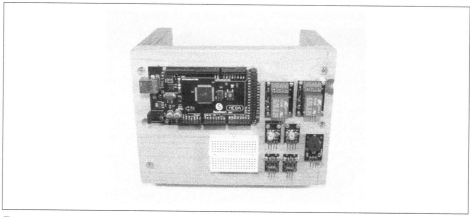

Figure 13-5. Mega2560 prototype fixture

But while the breadboard may have a storied legacy, for this design study I will assume that we will go directly to the final unit (for examples of prototypes, see the previous three chapters). For the launcher we have a good set of Phase I functional requirements and we know what primary components will be needed, so let's move on to the final hardware and software design.

Final Design

The final design is where the functional requirements are applied, along with what was learned from the prototype (if one was created), to come up with a design that describes what the finished device will look like and how it will function. Ideally it should behave the same as the prototype, and the software should be the same, or very nearly so.

Electrical

The schematic shown in Figure 13-6 covers all of the circuitry for the launcher except the igniter continuity check circuit, which is shown in Figure 13-7. Notice that the RTC, the LCD display, and the I/O expander shield all use the I2C interface. The two 7-segment displays use the SPI interface. Only the relays use discrete digital outputs.

A typical igniter for model rockets requires around 0.5A or greater for ignition. A circuit like the one shown in Figure 13-7 can be used to determine if the igniter circuit is open or closed by measuring the voltage at the point labeled "sense." The main concern is sensing the open or closed state of the igniter without applying enough current to cause it to ignite. With this circuit only about 2 mA is flowing through the igniter, and the 4.9V Zener diode will prevent the full launch voltage from getting back into the Arduino's analog inputs.

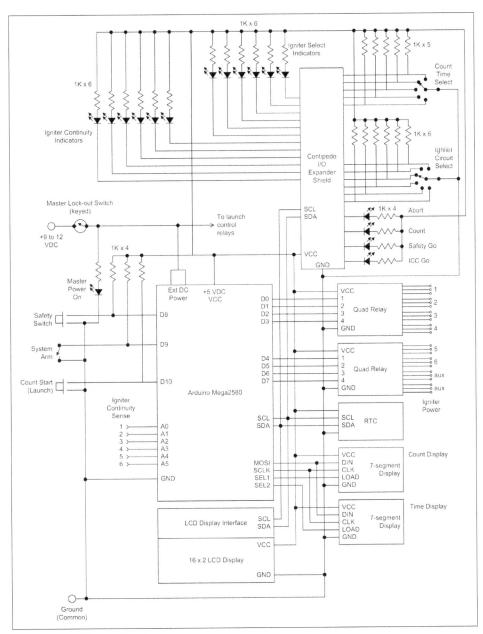

Figure 13-6. Phase I final schematic

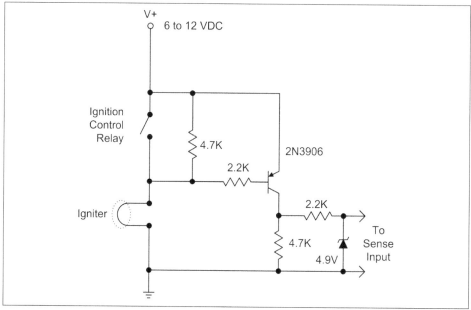

Figure 13-7. Igniter continuity check circuits

 Other igniter continuity check circuits have been devised, some more complex than others (there seems to be a minor cottage industry involved in creating launch controllers and continuity test circuits for model rockets). If you want to learn more, one source is J. R. Brohm's detailed study of igniter continuity test techniques (*http://bit.ly/brohm-igniter*).

In the schematic shown in Figure 13-6, six continuity circuits would be connected to the six analog inputs of the Mega2560 labeled "Igniter Continuity Sense." For this I would suggest that the six identical circuits be built using a piece of prototype board and some 0.1 inch (2.54 mm) terminal blocks, similar to what was done for input protection in Chapter 11.

Binding posts, like the types used for connecting speakers to high-end stereo receivers, can be used to connect the igniters to the launch controller. I would also suggest using binding posts to connect an external battery. If you want to have batteries in the launch controller enclosure, then an additional pair of binding posts can be used to connect the internal batteries to the igniter circuits. The will allow you to choose which power source will be used, depending on the number of igniter circuits that will be active. Connecting the igniter continuity sense circuits to the igniter binding posts is shown in Figure 13-8.

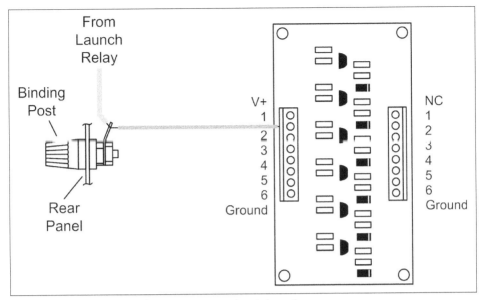

Figure 13-8. Connecting the igniter sense circuit board

The idea behind the 12 LEDs connected to the I/O expander and labeled "Igniter Continuity Indicators" and "Igniter Select Indicators" in Figure 13-6 is to show how many igniter circuits are active and what state they are in at any given time. This is determined by the six-position rotary switch S6, with the settings of 1, 2, 3, 4, 5, and 6. The corresponding continuity LEDs will glow to indicate that the selected igniters are connected and ready. After launch the continuity LEDs should be dark, and only the active selection LEDs will remain lit.

Figure 13-9 shows a different view of the wiring with an emphasis on the igniter outputs, the perfboard modules for LED and switch pull-up circuits, and the DC power routing for the various modules. Note that the individual signal lines for the I2C, SPI, and DIO and AIN functions are not shown. Instead, bus notation consisting of a slash with a number is used to indicate how many discrete signals are involved. Also not shown in the schematics are things like terminal blocks for DC power and ground.

A point of interest in Figure 13-9 is the use of one of the relays on the second relay module as an ignition safety interlock. Unless this relay is energized the igniters cannot be powered. Also note that there are four perfboard modules used to hold the igniter continuity sense circuits and pull-up resistors for various LEDs and switches. The use of 0.1 inch (2.54 mm) pitch terminal blocks makes wiring these into the system much easier than soldering, and they can be removed and replaced if necessary in the future.

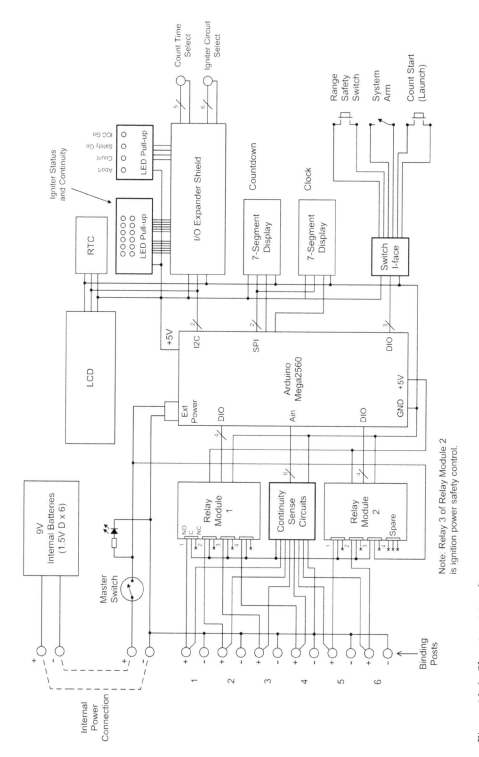

Figure 13-9. Chassis wiring diagram

With schematics in hand we can now create a detailed parts list like the one in Table 13-5.

Table 13-5. Final launcher parts list

Quantity	Description	Quantity	Description
1	Arduino Mega2560	1	Large pushbutton switch
1	Mega screw terminal shield	8	Binding posts, red
1	LCD display shield	8	Binding posts, black
1	Real-time clock module	1	3.5 mm jack (single-circuit)
2	Quad relay modules	1	Battery holder, six D cells
1	2-digit 7-segment LED display	1	Sloped-top metal console
1	4-digit 7-segment LED time display	1	Keyed switch

Physical

I would recommend that the launcher be built into a sloped metal chassis. Although it appears that this particular enclosure is no longer available, something like the one shown in the drawing in Figure 13-10 would do nicely. This is a drawing of an old chassis that has been lying around my workshop not doing much, and I stripped out the electronics that used to be inside long ago. The chassis is a two-piece design that is 14 inches (35.6 cm) wide, 10 inches (25.4 cm) deep, and 1.5 inches (3.8 cm) at the front and 2.75 inches (7 cm) at the rear. It also has a 3.5 inch (8.9 cm) flat "shelf" along the top at the rear of the cover. It is currently too ugly to photograph, so a drawing will have to suffice.

This isn't the final word on a chassis, by any means; it's just something that I happened to have on hand. You may have something equally suitable, or access to a surplus electronics outlet. If it comes down to it, a nice chassis similar to this can be purchased from multiple distributors. You can expect to pay between $30 and $50 for a new sloped-front chassis.

After deciding on the enclosure, the next step is to lay out the front panel and decide where the boards and connectors will be located. With a chassis like the one shown in Figure 13-10, I recommend mounting as much as possible to the inside of the top panel. The main reason is to avoid running wires between boards and controls on the top panel. Since the rear of the chassis is just a continuation of the top panel, you won't have to worry about disconnecting things if you need to remove the top panel piece to get at something inside.

Figure 13-11 shows a layout for the panel. This is just one way of doing it, and you might want a different arrangement. The I/O expansion shield has lots of I/O points (64 total), so there is plenty of room to grow.

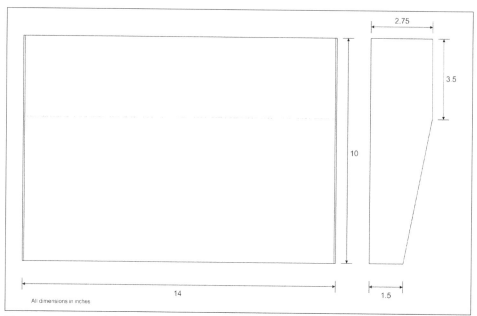

Figure 13-10. Candidate chassis for the launcher

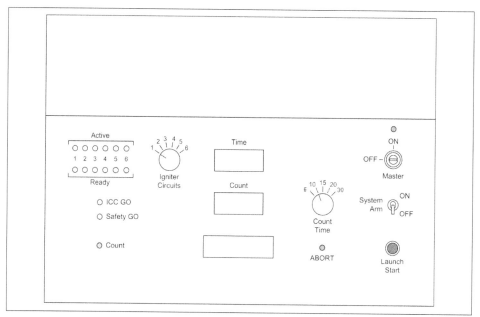

Figure 13-11. Launcher control panel layout example

Again, I recommend mounting all of the components to the inside of the cover piece, except perhaps any batteries. This makes it easy to assemble and test the launcher without also dragging along the base piece and a requisite bundle of wires to connect top and bottom components. The downside is that you will end up with screw heads on the front panel, but you can always paint over them or use flat-head screws (if the panel metal is thick enough).

If I were to actually build this (and I might, since I've already put this much work into it), I would mount the Mega2560 and the I/O shield under the flat section of the top of the chassis. This is a 3.5-inch-wide (8.9 cm) space that is more than wide enough to accommodate an Arduino (or two) without interfering with the controls and displays on the front panel.

One way to get everything arranged where you want it is to create a mockup using footprint models for the various boards and modules made from cardboard or foam-core material and some adhesive labels. Cut out shapes with the same dimensions as the actual parts, and some round adhesive stickers will work as stand-ins for switches and LEDs. Then arrange the mockup models on a piece of paper with a rectangular outline of the panel to get an optimal arrangement. I would do this as if I had X-ray vision and I was looking through the panel. That way I don't have to flip things over mentally to visualize where the LED and LCD displays will go, or where the LEDs and switches will be located. You can take measurements directly from your mockup and create a fabrication drawing like the one shown in Chapter 11. (Of course, if you happen to be adept with a CAD tool then you can just go straight to the fabrication drawing and skip the mockup step.)

You might also want to consider bringing out the USB connector on the Mega2560 to a panel-mounted B-type connector on the rear panel. This can be used to update the software, capture operational data during launch, or even provide a real-time display that can be sent back to a classroom for everyone to watch.

Software

For a design study like this, the software is largely just some suggestions and a few block diagrams. The main intent is to determine the overall level of effort in relation to the number of functions supported by the software. If it's designed and implemented in accordance with the functional requirements, then we should be able to map the software functionality directly to the requirements. Furthermore, we shouldn't have functionality that isn't defined in the requirements. In industry, and the aerospace industry in particular, that's considered is a bad thing, because software functionality without a driving requirement is functionality that won't be tested completely, if at all. Untested software is a risk.

Since this is a typical Arduino-type design, the `setup()` and `loop()` functions will be there in a main module. We will also need to have a selection of global variables to hold various state and time data. The `loop()` function will need to perform a series of continuous operations such as monitoring the system arm and launch start switches, turn on or turn off LEDs as necessary, and manage the countdown while monitoring the range safety switch during the count. Figure 13-12 shows the actions encompassed by the main loop.

At reference 1 in the diagram the loop starts by reading the rotary switches (igniter select and count time select). If the count time has changed since the last time the switch was read and the count is active (i.e., a launch is in progress), then the launcher will stop the count, disarm the system, and enter an abort state.

At reference 2 the software looks at the arm switch. If the switch was off in the previous iteration of the loop but is now on, then the system arm flag is set and the master launch power relay (see Figure 13-9) is energized. If the system arm flag is true, then we check to see if a count is in progress. If it is, then we don't read the launch start switch (reference 3); otherwise, we see if the user has pressed the launch button. If so, then we set the launch state to true (on) and start the count.

The righthand side of Figure 13-12 is only active if the system is in the launch state and a count is in progress. At reference 4 a check is made to verify that the range safety switch is depressed (on). If not, then we abort the system and place it into a safe state.

Reference 5 is where the real action occurs. When the count reaches zero the launch relays corresponding to the select ignition circuits are energized. The relays are held in an on state for 2 seconds, and then released. This is likely not the optimal way to do this. It would be better to look at the igniter status and de-energize the relays once all the igniters are open. The time delay would be a maximum permissible time before declaring a fault.

At reference 6, either the launch has been a success or the rocket is still (hopefully) sitting on the pad. In any case, the system is restored to a standby state, the relays are powered off, and the master power relay is de-energized. I would also suggest disarming the system, so that the user must set the arm switch to off and then back to on to rearm the launcher.

Reading switch inputs, setting LED states, reading the RTC, and controlling the relays can take place very rapidly. It doesn't look like interrupts will be necessary for this design. That being said, you might want to consider an interrupt for the range safety switch. This is why it was connected directly to the Mega2560 instead of through the I/O expander.

The range safety switch can be used to trigger an interrupt that will disengage the igniter power relay (relay 3, relay module 2) the instant that the switch is released. So even if the MCU takes a few tens of milliseconds to read the state of the switch and respond by putting the system into an abort state, the power will have already been cut to the rocket motor igniters.

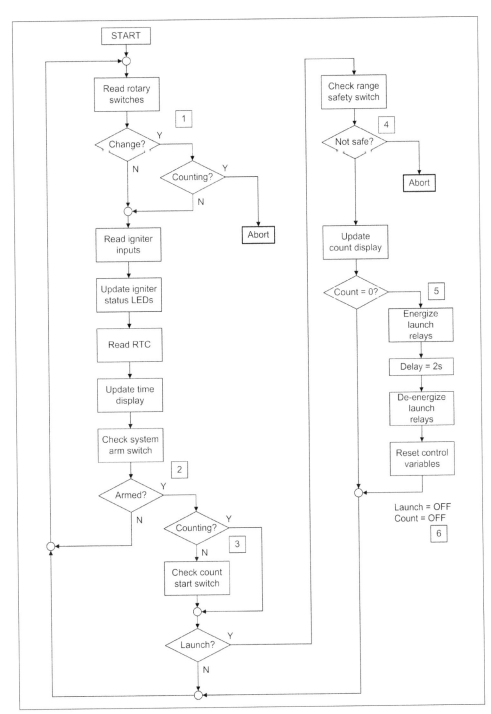

Figure 13-12. Main loop operations

The LCD can be used to display messages during the loop. It would be good to see the current state and get messages if a fault should occur. It would not be particularly useful to repeat the time or the count with the LCD. That would just add latency (i.e., execution delay) to the main loop.

When writing the software, the primary guide is the functional requirements. These are the things the software must support in order to meet the requirements, and consequently fulfill the objectives for the project. For examples of ways to arrange the software into modules, refer to Chapters Chapters 10, 11, and 12.

From Figure 13-12 we can see that there are multiple possibilities for additional modules. The igniter status read and update is one, as is the RTC and the time display update. The launch section of the loop (everything between reference 4 and reference 6) is another possible candidate for its own module. The RTC module, the LCD, the LED displays, and the I/O expander will most likely have classes defined, but there may not be much need to use classes for the main loop functions. They aren't that complex, really.

Since this is a design analysis, not a full-on design, I will leave the software here. If you want to pursue this and create your own rocket launcher, then I would suggest that you look at the examples provided in the previous three chapters. With this design analysis as a starting point, you should not have too much difficulty working out the remaining details.

We can't really define a time estimate because there's no way to know who will be writing the software. Different people work at different rates; what might take one person an hour or so to code up could take another person half a day. So, to get some idea of how much effort the software may require, I recommend that you try writing some test code with a breadboard or undedicated prototype fixture (as seen in earlier chapters) to do something with the RTC or the LCD. Once you have a general idea of how long this takes, then you can multiply that by the number of unique functions in the software design, and then multiply that number by two. That might end up being close to how long it will really take, and if the software gets done before that, then all the better.

Testing and Operation

A test plan is always a good idea. Fortunately the launcher is relatively easy to test, as it is basically just switch inputs and relay outputs. You can simulate igniter continuity by simply connecting a jumper between the pairs of igniter output binding posts, and an open circuit is just the lack of a jumper.

If you look over the functional requirements you can see that they are all verifiable. In other words, if you rotate the igniter circuit select knob you should see the active LEDs light up in succession from 1 to 6. If the binding posts are connected to simulate igniters, the "ICC GO" light should also be active. Arm the system and then press and hold the range safety switch, and you should see the "Safety GO" light come on. If you start the count and then disconnect any of the active igniters, change the number of igniter circuits or the count time in mid-count, or release the range safety switch, the system should halt the count and go into an abort state.

Be careful not to let the count reach zero if you are using jumpers on the binding posts to simulate igniters. When the relays close, full current from the battery will flow through the jumpers. This could melt the jumper wires or damage the relays, or both. A better testing approach would be to use replaceable fuses or even actual igniters mounted on a wood base. You might also be able to use small incandescent bulbs with a correct voltage rating, provided that they have a sufficiently low cold (unlit) resistance. A typical igniter is about 0.5 ohms.

Cost Analysis

The final cost of the launcher depends on where you get the parts, and if you can find bargains. A major cost item is the chassis, which can run upwards of $30. You could elect to build a console from 1 by 1 inch (2.5 by 2.5 cm) hardwood and 1/4 inch (0.6 cm) fiberboard, but that is assuming that you have the necessary woodworking tools. After purchasing the materials and adding in your time, you may find that it's more cost-effective to just buy the chassis.

Based on the final parts list in Table 13-5, the total cost for the Phase I version will likely be somewhere around $125, give or take $25. The Phase II features could easily drive the cost up to well over $300, but you could likely hold this down by doing some smart shopping and bargain hunting.

Tools and Accessories

For many Arduino projects you don't need any tools, just some jumper wires, shield and sensor PCBs, and of course an Arduino. But after graduating to more advanced projects you will find that a selection of basic tools and accessories becomes essential. A set of hand tools, a soldering iron, and a few other items are usually sufficient for all but the most complex projects. If I may be so bold, I would recommend my book *Practical Electronics: Components and Techniques* (O'Reilly) as a reference for things like screw and bolt sizes, electronic components, and PCB fabrication.

In this chapter I will describe the basic tools you might want to consider having on hand for your own projects. Everything presented here can easily fit into a medium-sized toolbox when it's not needed.

Hand Tools

A good selection of hand tools is essential. With patience and some effort you can accomplish just about any task with good hand tools. Before the introduction of electricity, hand tools were really the only way for most people to build anything, and they built a lot of amazing things. So can you, as long as you are willing to take the time to do it correctly. We won't look at techniques here, as there are other books that cover that, but I will describe some tools you might want to consider, and where to find them.

Screwdrivers

For most projects involving an Arduino all you need in the way of screwdrivers is a good set of the miniature types or a combination kit, such as the one shown in Figure A-1, and a set of larger screwdrivers. You can find various sets of miniature screwdrivers at most well-stocked hardware stores, some big-box home improvement stores, and just about any electronics store that specializes in components and tools.

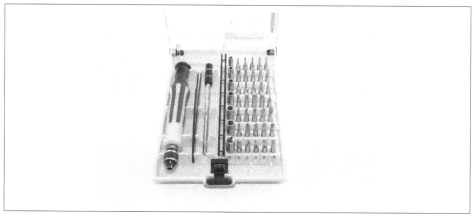

Figure A-1. A set of miniature screwdrivers

Full-size screwdrivers, like those shown in Figure A-2, can be found in numerous places, including the home repair aisle of a large grocery store. Avoid the very large tools, and look for a kit that has smaller tip sizes. You will need those, but the large tools not so much (unless you also need to do residential power wiring or work on an automobile).

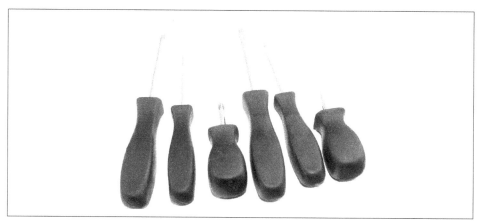

Figure A-2. A set of standard full-size screwdrivers

Pliers and Cutters

Needle-nosed pliers, diagonal cutters, and a pair of good flush cutters are essential. You might also want to consider a pair of standard pliers and perhaps even lineman's pliers, but these aren't absolutely necessary. Figure A-3 shows a selection of basic pliers and cutters that can be purchased as a set. You can also pick and choose from individual tools to suit your needs.

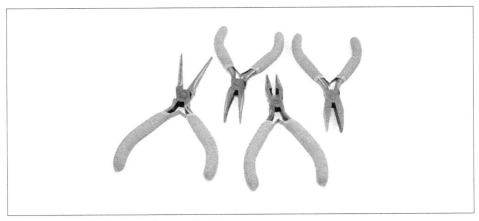

Figure A-3. A set of basic pliers and cutters

Resist the temptation to try to use the wire cutters that come with the bundled selections from some hardware and home improvement stores for doing PCB-level electronics work. Flush cutters are made specifically for trimming component leads and cutting small-gauge wire, and they do a fine job of it. Figure A-4 shows a typical flush cutter tool.

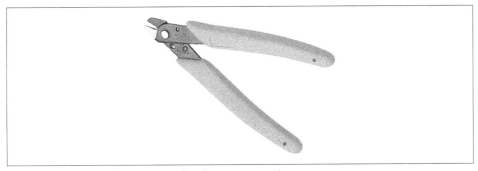

Figure A-4. Typical flush cutters for electronics work

Wire Strippers

Another essential tool is a wire stripper. Although you may be tempted to use a pair of cutters to do this, it's generally not a good idea. It's very easy to cut one or more of the fine wires that make up a strand, and a nick on a solid conductor is where it will usually break if it is flexed. I keep two types on hand, and which one I pick up depends largely on how many wires I need to strip and which tool happens to be the easiest to reach.

The simplest wire stripper consists of a pair of blades with an adjustable stop, like the tool shown in Figure A-5. The downside to this tool is that you have to adjust the stop each time you use a different gauge of wire. But if you always use the same wire, then it's really not a problem (I use #24 gauge insulated twisted strand wire for almost everything, so I seldom need to adjust my tool).

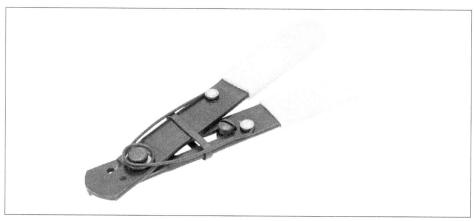

Figure A-5. Basic wire strippers

My favorite wire strippers are sold by Klein, and they not only handle different wire gauges but also pull off the cut insulation, all in one motion. Figure A-6 shows an example of this type of tool. These are surprisingly affordable, and you can purchase an additional cutter blade for even more wire gauges. The downside is that they are big and somewhat bulky, so they won't fit into tight spaces.

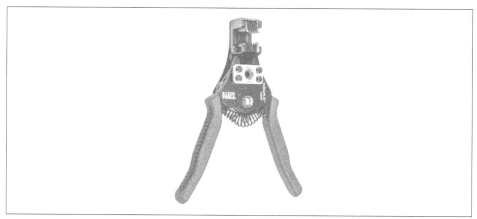

Figure A-6. Fancy wire strippers

Connector Crimping Tools

One of the main annoyances encountered when working with Arduino boards, shields, and the various available modules is connecting everything. Jumper wires with pins and sockets are fine for assembling something on the bench (or kitchen table) to see how it works, but this can present some long-term reliability issues. A better approach is to use an I/O extension shield (like those described in Chapter 9) that provides multiconductor connectors. Figure A-7 shows such a shield with cables attached.

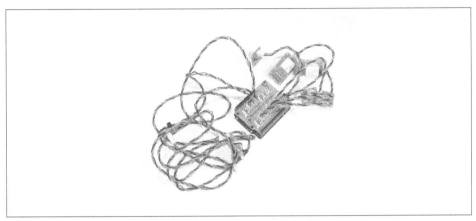

Figure A-7. I/O extension shield with connectors attached

The metallic connectors attach to wires by crimping, and that means you will need a special tool (and the correct connector bodies and inserts). Fortunately the price of these tools has dropped dramatically over the past few years. You can now buy a tool for about $30 that does the same basic job as a tool that used to cost $200. Figure A-8 shows a selection of crimping tools.

Once the contacts (either pins or sockets) have been crimped onto the wires, the next step is to insert them into a connector housing, also called a shell or a body. These are available in 0.1 inch (2.54 mm) pitch (spacing), which is a de facto industry standard and is what is commonly found on Arduino components. Figure A-9 shows 1-, 2-, 3-, and 4-position connector housings. The pin or socket connectors lock into the plastic housings and can be easily removed by gently lifting a small locking tab using a miniature screwdriver.

Some Arduino shields and modules use connectors similar to those found on the ends of telephone or network cables. These can be assembled with tools available at most big-box home improvement stores, electronics distributors, and of course from online suppliers. Figure A-10 shows a shield that uses these types of connectors.

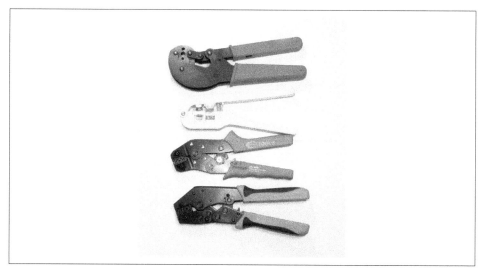

Figure A-8. Various types of low-cost crimping tools

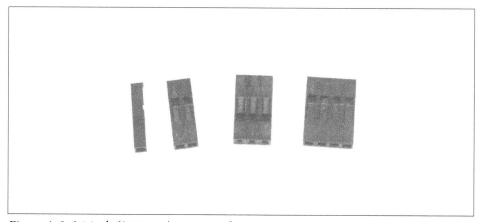

Figure A-9. 0.1 inch (2.54 mm) connector housings

Lastly, there are the so-called lug connectors used in electrical systems and vehicles, like the part shown in Figure A-11. These are readily available, but not very commonly used with Arduino projects (although they are used in the signal generator in Chapter 11). The connectors come in a variety of styles and types, and the crimping tools are available from many different sources.

Figure A-12 shows one type of tool used with lug connectors. Do not attempt to use this type of crimping tool with the small connectors used for the pins and sockets on a PCB like an Arduino—the end result will just be a smashed and useless connector.

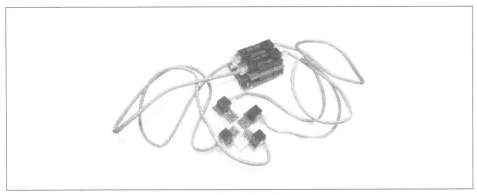

Figure A-10. I/O extension shield with RJ45 (8P8C) connectors

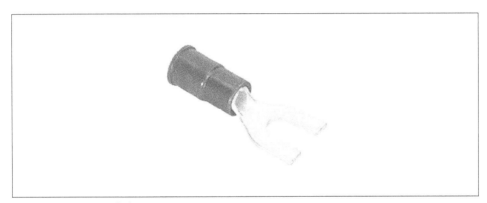

Figure A-11. A spade lug-type connector

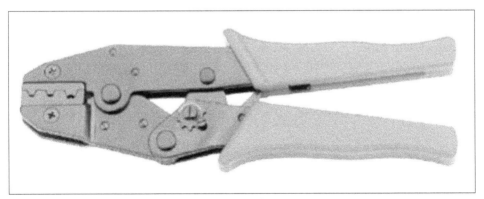

Figure A-12. A common spade lug crimping tool

Crimped connectors are easy to install, reliable (if done correctly—it can take some practice), and cheap. The downside is the initial investment in the tools. If you are willing to make that investment, then your soldering iron will spend most of its time in your toolbox and your projects will have a polished and professional look.

Saws

A couple of types of small saws are handy to have on hand when you need to trim a circuit board, cut out a small section of a plastic enclosure, or cut a section of plastic tubing. Nothing else can do those things as quickly and easily as a saw.

A jeweler's saw like the one shown in Figure A-13 is useful for doing very fine precision cuts, but it's not very good at cutting large items. The trick to using a jeweler's saw is to let the saw do the work without forcing it into the cut (this can generally be said of any saw, by the way). The thin blades won't take much in the way of stress, but they will cut through almost anything with enough care and patience.

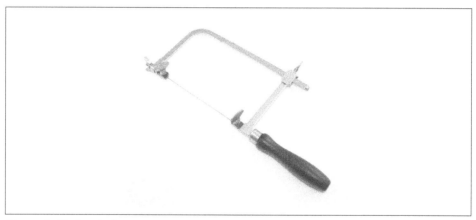

Figure A-13. Jeweler's saw

For larger jobs, particularly those involving metal, a hacksaw is the way to go. A typical generic hacksaw is shown in Figure A-14. Newer models may have a more streamlined look, but the basic idea is the same. You can also buy hacksaws that are little more than a blade with a handle at one end.

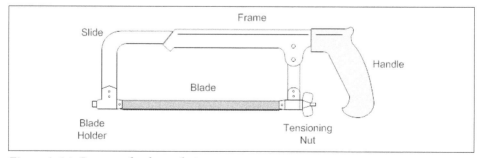

Figure A-14. Common hacksaw design

When using a hacksaw remember that the saw will only cut in one direction, either push or pull. It depends on how the blade is installed. I prefer to mount the blade so that the saw cuts when pulled, but some people like to do it the other way.

Power Tools

For many tasks some good hand tools will get the job done and, if used correctly, do it nicely as well. But other tasks might need more power than a hand tool can deliver without causing muscle cramps. Drilling and grinding are two examples.

Drills

An electric hand drill is great for many things. Drilling precise holes is generally not one of those things, but when you need just one 1/8 inch hole in a panel, and it doesn't have to be super-precise, then a hand drill is very useful. I recommend a cordless drill, like the one shown in Figure A-15, if for no other reason than that it is less of a hassle without a power cord. Although a battery-operated cordless drill might not have the same amount of torque as a drill that plugs into a wall outlet, most small projects involve plastic, thin wood or wood-like materials, and thin metal, and a cordless drill will work just fine.

Figure A-15. Cordless drill with interchangeable battery pack

Miniature Grinder

Although a grinder isn't actually an essential tool, it is a very useful and handy tool to have around. A miniature grinder, like the one shown in Figure A-16, can be used to sharpen screwdrivers, take the rough edges and burrs off of the end of a metal rod after it is cut, clean up the edges of plastic pieces, and even trim up a PCB.

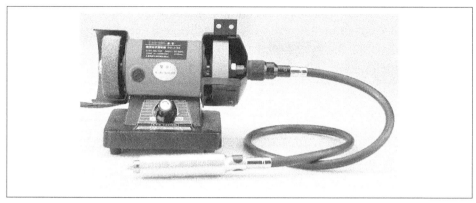

Figure A-16. Miniature grinder

This particular grinder is from Harbor Freight, and it includes a rotary tool attachment. It might not do everything a standalone rotary tool will do, but it does come in handy for lightweight jobs.

Miniature Drill Press

If you need some precisely sized holes, in precise locations, then you really need a drill press. Although a full-sized drill press can be used for jobs like this, they tend to be large things that don't easily tuck away into a closet when you don't need them. The solution is a miniature drill press like the one shown in Figure A-17.

In addition to drilling holes for switches and LEDs in a small plastic enclosure, you can also drill holes in a PCB. Accessories such as a miniature vise are available to hold the work steady while drilling.

Figure A-17. Miniature drill press

Soldering

If any one activity could be said to characterize electronics, it would have to be soldering. Soldering is not really necessary if you are using ready-made PCBs and modules with an Arduino, but if you want to integrate an Arduino into a larger system, then soldering may be required. And if you happen to purchase a shield with a packet of pin and socket connectors and empty holes on the PCB, then soldering is no longer optional.

Soldering Irons

Soldering irons come in a range of prices, from ultra-low-cost tools with no temperature control and tips of dubious quality, to soldering stations with interchangeable tips and integrated temperature control costing hundreds of dollars. Avoid the cheap tools, as they can do some serious damage to a circuit board and the components soldered onto it. Spend as much as you can afford, but at least consider something like the iron shown in Figure A-18, which sells for about $15.

Figure A-18. Inexpensive soldering iron

If you can afford it, consider a soldering station like the one shown in Figure A-19. These tools range in price from about $50 to somewhere around $300. A good soldering station is a good investment, but you really need to have some serious soldering work to do in order to justify a pricey model.

Soldering Accessories

A soldering iron or soldering station is nice to have, but without some basic accessories it won't be very useful. At a minimum you'll need some solder. Don't buy solder for electronics at the local hardware store unless it specifically states that it is for electronics work. A good electronics-grade solder will have a flux core (usually rosin), and most are on the thin side. I like to purchase solder in one-pound (454 g) spools, like the one shown in Figure A-20.

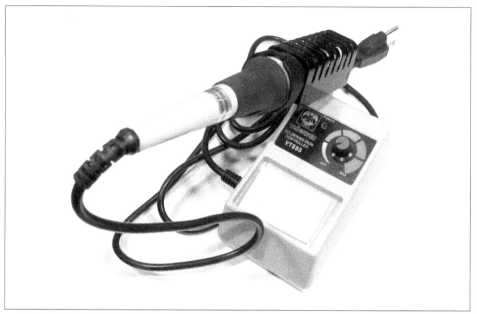

Figure A-19. Soldering station

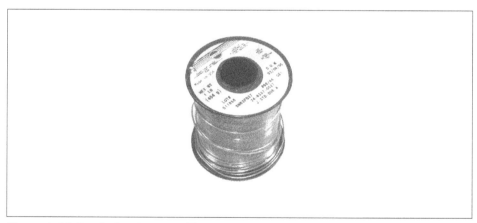

Figure A-20. Spool of rosin-core solder

Other useful accessories include solder wick (copper braid for removing solder), liquid or paste flux, and solder paste. You can learn more about the tools and accessories, and find soldering tutorials, in numerous texts and in online videos.

Tool Sources

Table A-1 lists some sources for the tools covered in this appendix. This is a very short list, as there are a large number of suppliers selling tools of all kinds.

Table A-1. Tool suppliers

Distributor/vendor	URL	Distributor/vendor	URL
Adafruit	www.adafruit.com	Maker Shed	www.makershed.com
Apex Tool Group	www.apexhandtools.com	MCM Electronics	www.mcmelectronics.com
CKB Products	www.ckbproducts.com	SainSmart	www.sainsmart.com
Circuit Specialists	www.circuitspecialists.com	SparkFun	www.sparkfun.com
Electronic Goldmine	www.goldmine-elec-products.com	Stanley	www.stanleysupplyservices.com
Harbor Freight Tools	www.harborfreight.com	Velleman	www.vellemanusa.com

Don't overlook your local used tool shop. Many cities have one or more shops that specialize in used tools, with selections that include everything from buckets full of used screwdrivers to used machine shop tools like vertical mills. Other places to look include organizations that accept donations, such as Goodwill (here in the US). They may not have much of a selection when it comes to tools, but if you have a moment to spare they can sometimes yield up some amazingly good deals.

AVR ATmega Control Registers

The register summaries in this appendix are intended as a quick reference. This appendix is not a comprehensive description of each control register. For detailed descriptions of each control register for a particular MCU type, see the Atmel documentation. Pay special attention to the notes included with the control register summaries in the Atmel documents. Each MCU has a slightly different set of things to watch out for.

In general, reserved bits (marked with a "–") should not be accessed. Registers in the range of 0x00 to 0x1F are directly bit-accessible with the SBI and CBI instructions (set I/O bit and clear I/O bit, respectively). Register addresses in parentheses are the SRAM addresses of the control registers, whereas the addresses not in parentheses reside in the 64-byte address space reserved for I/O control registers. The reserved locations can be used with the IN and OUT instructions, and the SRAM addresses must be accessed with the ST/STS/STD and LD/LDS/LDD instructions.

The information in this appendix was derived from the following Atmel technical documents, all of which are available from Atmel.com:

Document number	Title
Atmel-8271I-AVR- ATmega-Datasheet_10/2014	Atmel ATmega48A/PA/88A/PA/168A/PA/328/P
2549Q–AVR–02/2014	Atmel ATmega640/V-1280/V-1281/V-2560/V-2561/V
7766F–AVR–11/10	Atmel ATmega16U4/ATmega32U4

ATmega168/328

Address	Name	Bit 7	Bit 6	Bit 5	Bit 4	Bit 3	Bit 2	Bit 1	Bit 0
(0xFF)	Reserved	–	–	–	–	–	–	–	–
(0xFE)	Reserved	–	–	–	–	–	–	–	–
(0xFD)	Reserved	–	–	–	–	–	–	–	–
(0xFC)	Reserved	–	–	–	–	–	–	–	–
(0xFB)	Reserved	–	–	–	–	–	–	–	–
(0xFA)	Reserved	–	–	–	–	–	–	–	–
(0xF9)	Reserved	–	–	–	–	–	–	–	–
(0xF8)	Reserved	–	–	–	–	–	–	–	–
(0xF7)	Reserved	–	–	–	–	–	–	–	–
(0xF6)	Reserved	–	–	–	–	–	–	–	–
(0xF5)	Reserved	–	–	–	–	–	–	–	–
(0xF4)	Reserved	–	–	–	–	–	–	–	–
(0xF3)	Reserved	–	–	–	–	–	–	–	–
(0xF2)	Reserved	–	–	–	–	–	–	–	–
(0xF1)	Reserved	–	–	–	–	–	–	–	–
(0xF0)	Reserved	–	–	–	–	–	–	–	–
(0xEF)	Reserved	–	–	–	–	–	–	–	–
(0xEE)	Reserved	–	–	–	–	–	–	–	–
(0xED)	Reserved	–	–	–	–	–	–	–	–
(0xEC)	Reserved	–	–	–	–	–	–	–	–
(0xEB)	Reserved	–	–	–	–	–	–	–	–
(0xEA)	Reserved	–	–	–	–	–	–	–	–
(0xE9)	Reserved	–	–	–	–	–	–	–	–
(0xE8)	Reserved	–	–	–	–	–	–	–	–
(0xE7)	Reserved	–	–	–	–	–	–	–	–
(0xE6)	Reserved	–	–	–	–	–	–	–	–
(0xE5)	Reserved	–	–	–	–	–	–	–	–
(0xE4)	Reserved	–	–	–	–	–	–	–	–
(0xE3)	Reserved	–	–	–	–	–	–	–	–
(0xE2)	Reserved	–	–	–	–	–	–	–	–
(0xE1)	Reserved	–	–	–	–	–	–	–	–
(0xE0)	Reserved	–	–	–	–	–	–	–	–
(0xDF)	Reserved	–	–	–	–	–	–	–	–
(0xDE)	Reserved	–	–	–	–	–	–	–	–
(0xDD)	Reserved	–	–	–	–	–	–	–	–

Address	Name	Bit 7	Bit 6	Bit 5	Bit 4	Bit 3	Bit 2	Bit 1	Bit 0
(0xDC)	Reserved	–	–	–	–	–	–	–	–
(0xDB)	Reserved	–	–	–	–	–	–	–	–
(0xDA)	Reserved	–	–	–	–	–	–	–	–
(0xD9)	Reserved	–	–	–	–	–	–	–	–
(0xD8)	Reserved	–	–	–	–	–	–	–	–
(0xD7)	Reserved	–	–	–	–	–	–	–	–
(0xD6)	Reserved	–	–	–	–	–	–	–	–
(0xD5)	Reserved	–	–	–	–	–	–	–	–
(0xD4)	Reserved	–	–	–	–	–	–	–	–
(0xD3)	Reserved	–	–	–	–	–	–	–	–
(0xD2)	Reserved	–	–	–	–	–	–	–	–
(0xD1)	Reserved	–	–	–	–	–	–	–	–
(0xD0)	Reserved	–	–	–	–	–	–	–	–
(0xCF)	Reserved	–	–	–	–	–	–	–	–
(0xCE)	Reserved	–	–	–	–	–	–	–	–
(0xCD)	Reserved	–	–	–	–	–	–	–	–
(0xCC)	Reserved	–	–	–	–	–	–	–	–
(0xCB)	Reserved	–	–	–	–	–	–	–	–
(0xCA)	Reserved	–	–	–	–	–	–	–	–
(0xC9)	Reserved	–	–	–	–	–	–	–	–
(0xC8)	Reserved	–	–	–	–	–	–	–	–
(0xC7)	Reserved	–	–	–	–	–	–	–	–
(0xC6)	UDR0	USART I/O Data Register							
(0xC5)	UBRR0H	USART Baud Rate Register High							
(0xC4)	UBRR0L	USART Baud Rate Register Low							
(0xC3)	Reserved	–	–	–	–	–	–	–	–
(0xC2)	UCSR0C	UMSEL01	UMSEL00	UPM01	UPM00	USBS0	UCSZ01/ UDORD0	UCSZ00/ UCPHA0	UCPOL0
(0xC1)	UCSR0B	RXCIE0	TXCIE0	UDRIE0	RXEN0	TXEN0	UCSZ02	RXB80	TXB80
(0xC0)	UCSR0A	RXC0	TXC0	UDRE0	FE0	DOR0	UPE0	U2X0	MPCM0
(0xBF)	Reserved	–	–	–	–	–	–	–	–
(0xBE)	Reserved	–	–	–	–	–	–	–	–
(0xBD)	TWAMR	TWAM6	TWAM5	TWAM4	TWAM3	TWAM2	TWAM1	TWAM0	–
(0xBC)	TWCR	TWINT	TWEA	TWSTA	TWSTO	TWWC	TWEN	–	TWIE
(0xBB)	TWDR	2-wire Serial Interface Data Register							
(0xBA)	TWAR	TWA6	TWA5	TWA4	TWA3	TWA2	TWA1	TWA0	TWGCE
(0xB9)	TWSR	TWS7	TWS6	TWS5	TWS4	TWS3	–	TWPS1	TWPS0
(0xB8)	TWBR	2-wire Serial Interface Bit Rate Register							

Address	Name	Bit 7	Bit 6	Bit 5	Bit 4	Bit 3	Bit 2	Bit 1	Bit 0
(0xB7)	Reserved	–	–	–	–	–	–	–	–
(0xB6)	ASSR	–	EXCLK	AS2	TCN2UB	OCR2AUB	OCR2BUB	TCR2AUB	TCR2BUB
(0xB5)	Reserved	–	–	–	–	–	–	–	–
(0xB4)	OCR2B	Timer/Counter2 Output Compare Register B							
(0xB3)	OCR2A	Timer/Counter2 Output Compare Register A							
(0xB2)	TCNT2	Timer/Counter2 (8-bit)							
(0xB1)	TCCR2B	FOC2A	FOC2B	–	–	WGM22	CS22	CS21	CS20
(0xB0)	TCCR2A	COM2A1	COM2A0	COM2B1	COM2B0	–	–	WGM21	WGM20
(0xAF)	Reserved	–	–	–	–	–	–	–	–
(0xAE)	Reserved	–	–	–	–	–	–	–	–
(0xAD)	Reserved	–	–	–	–	–	–	–	–
(0xAC)	Reserved	–	–	–	–	–	–	–	–
(0xAB)	Reserved	–	–	–	–	–	–	–	–
(0xAA)	Reserved	–	–	–	–	–	–	–	–
(0xA9)	Reserved	–	–	–	–	–	–	–	–
(0xA8)	Reserved	–	–	–	–	–	–	–	–
(0xA7)	Reserved	–	–	–	–	–	–	–	–
(0xA6)	Reserved	–	–	–	–	–	–	–	–
(0xA5)	Reserved	–	–	–	–	–	–	–	–
(0xA4)	Reserved	–	–	–	–	–	–	–	–
(0xA3)	Reserved	–	–	–	–	–	–	–	–
(0xA2)	Reserved	–	–	–	–	–	–	–	–
(0xA1)	Reserved	–	–	–	–	–	–	–	–
(0xA0)	Reserved	–	–	–	–	–	–	–	–
(0x9F)	Reserved	–	–	–	–	–	–	–	–
(0x9E)	Reserved	–	–	–	–	–	–	–	–
(0x9D)	Reserved	–	–	–	–	–	–	–	–
(0x9C)	Reserved	–	–	–	–	–	–	–	–
(0x9B)	Reserved	–	–	–	–	–	–	–	–
(0x9A)	Reserved	–	–	–	–	–	–	–	–
(0x99)	Reserved	–	–	–	–	–	–	–	–
(0x98)	Reserved	–	–	–	–	–	–	–	–
(0x97)	Reserved	–	–	–	–	–	–	–	–
(0x96)	Reserved	–	–	–	–	–	–	–	–
(0x95)	Reserved	–	–	–	–	–	–	–	–
(0x94)	Reserved	–	–	–	–	–	–	–	–
(0x93)	Reserved	–	–	–	–	–	–	–	–

Address	Name	Bit 7	Bit 6	Bit 5	Bit 4	Bit 3	Bit 2	Bit 1	Bit 0
(0x92)	Reserved	–	–	–	–	–	–	–	–
(0x91)	Reserved	–	–	–	–	–	–	–	–
(0x90)	Reserved	–	–	–	–	–	–	–	–
(0x8F)	Reserved	–	–	–	–	–	–	–	–
(0x8E)	Reserved	–	–	–	–	–	–	–	–
(0x8D)	Reserved	–	–	–	–	–	–	–	–
(0x8C)	Reserved	–	–	–	–	–	–	–	–
(0x8B)	OCR1BH	Timer/Counter1: Output Compare Register B High Byte							
(0x8A)	OCR1BL	Timer/Counter1: Output Compare Register B Low Byte							
(0x89)	OCR1AH	Timer/Counter1: Output Compare Register A High Byte							
(0x88)	OCR1AL	Timer/Counter1: Output Compare Register A Low Byte							
(0x87)	ICR1H	Timer/Counter1: Input Capture Register High Byte							
(0x86)	ICR1L	Timer/Counter1: Input Capture Register Low Byte							
(0x85)	TCNT1H	Timer/Counter1: Counter Register High Byte							
(0x84)	TCNT1L	Timer/Counter1: Counter Register Low Byte							
(0x83)	Reserved	–	–	–	–	–	–	–	–
(0x82)	TCCR1C	FOC1A	FOC1B	–	–	–	–	–	–
(0x81)	TCCR1B	ICNC1	ICES1	–	WGM13	WGM12	CS12	CS11	CS10
(0x80)	TCCR1A	COM1A1	COM1A0	COM1B1	COM1B0	–	–	WGM11	WGM10
(0x7F)	DIDR1	–	–	–	–	–	–	AIN1D	AIN0D
(0x7E)	DIDR0	–	–	ADC5D	ADC4D	ADC3D	ADC2D	ADC1D	ADC0D
(0x7D)	Reserved	–	–	–	–	–	–	–	–
(0x7C)	ADMUX	REFS1	REFS0	ADLAR	–	MUX3	MUX2	MUX1	MUX0
(0x7B)	ADCSRB	–	ACME	–	–	–	ADTS2	ADTS1	ADTS0
(0x7A)	ADCSRA	ADEN	ADSC	ADATE	ADIF	ADIE	ADPS2	ADPS1	ADPS0
(0x79)	ADCH	ADC Data Register High Byte							
(0x78)	ADCL	ADC Data Register Low Byte							
(0x77)	Reserved	–	–	–	–	–	–	–	–
(0x76)	Reserved	–	–	–	–	–	–	–	–
(0x75)	Reserved	–	–	–	–	–	–	–	–
(0x74)	Reserved	–	–	–	–	–	–	–	–
(0x73)	Reserved	–	–	–	–	–	–	–	–
(0x72)	Reserved	–	–	–	–	–	–	–	–
(0x71)	Reserved	–	–	–	–	–	–	–	–
(0x70)	TIMSK2	–	–	–	–	–	OCIE2B	OCIE2A	TOIE2
(0x6F)	TIMSK1	–	–	ICIE1	–	–	OCIE1B	OCIE1A	TOIE1
(0x6E)	TIMSK0	–	–	–	–	–	OCIE0B	OCIE0A	TOIE0

Address	Name	Bit 7	Bit 6	Bit 5	Bit 4	Bit 3	Bit 2	Bit 1	Bit 0
(0x6D)	PCMSK2	PCINT23	PCINT22	PCINT21	PCINT20	PCINT19	PCINT18	PCINT17	PCINT16
(0x6C)	PCMSK1	–	PCINT14	PCINT13	PCINT12	PCINT11	PCINT10	PCINT9	PCINT8
(0x6B)	PCMSK0	PCINT7	PCINT6	PCINT5	PCINT4	PCINT3	PCINT2	PCINT1	PCINT0
(0x6A)	Reserved	–	–	–	–	–	–	–	–
(0x69)	EICRA	–	–	–	–	ISC11	ISC10	ISC01	ISC00
(0x68)	PCICR	–	–	–	–	–	PCIE2	PCIE1	PCIE0
(0x67)	Reserved	–	–	–	–	–	–	–	–
(0x66)	OSCCAL	Oscillator Calibration Register							
(0x65)	Reserved	–	–	–	–	–	–	–	–
(0x64)	PRR	PRTWI	PRTIM2	PRTIM0	–	PRTIM1	PRSPI	PRUSART0	PRADC
(0x63)	Reserved	–	–	–	–	–	–	–	–
(0x62)	Reserved	–	–	–	–	–	–	–	–
(0x61)	CLKPR	CLKPCE	–	–	–	CLKPS3	CLKPS2	CLKPS1	CLKPS0
(0x60)	WDTCSR	WDIF	WDIE	WDP3	WDCE	WDE	WDP2	WDP1	WDP0
0x3F (0x5F)	SREG	I	T	H	S	V	N	Z	C
0x3E (0x5E)	SPH	–	–	–	–	–	(SP10)	SP9	SP8
0x3D (0x5D)	SPL	SP7	SP6	SP5	SP4	SP3	SP2	SP1	SP0
0x3C (0x5C)	Reserved	–	–	–	–	–	–	–	–
0x3B (0x5B)	Reserved	–	–	–	–	–	–	–	–
0x3A (0x5A)	Reserved	–	–	–	–	–	–	–	–
0x39 (0x59)	Reserved	–	–	–	–	–	–	–	–
0x38 (0x58)	Reserved	–	–	–	–	–	–	–	–
0x37 (0x57)	SPMCSR	SPMIE	(RWWSB)	–	(RWWSRE)	BLBSET	PGWRT	PGERS	SELFPRGEN
0x36 (0x56)	Reserved	–	–	–	–	–	–	–	–
0x35 (0x55)	MCUCR	–	BODS	BODSE	PUD	–	–	IVSEL	IVCE
0x34 (0x54)	MCUSR	–	–	–	–	WDRF	BORF	EXTRF	PORF
0x33 (0x53)	SMCR	–	–	–	–	SM2	SM1	SM0	SE

Address	Name	Bit 7	Bit 6	Bit 5	Bit 4	Bit 3	Bit 2	Bit 1	Bit 0
0x32 (0x52)	Reserved	–	–	–	–	–	–	–	–
0x31 (0x51)	Reserved	–	–	–	–	–	–	–	–
0x30 (0x50)	ACSR	ACD	ACBG	ACO	ACI	ACIE	ACIC	ACIS1	ACIS0
0x2F (0x4F)	Reserved	–	–	–	–	–	–	–	–
0x2E (0x4E)	SPDR	SPI Data Register							
0x2D (0x4D)	SPSR	SPIF	WCOL	–	–	–	–	–	SPI2X
0x2C (0x4C)	SPCR	SPIE	SPE	DORD	MSTR	CPOL	CPHA	SPR1	SPR0
0x2B (0x4B)	GPIOR2	General Purpose I/O Register 2							
0x2A (0x4A)	GPIOR1	General Purpose I/O Register 1							
0x29 (0x49)	Reserved	–	–	–	–	–	–	–	–
0x28 (0x48)	OCR0B	Timer/Counter0 Output Compare Register B							
0x27 (0x47)	OCR0A	Timer/Counter0 Output Compare Register A							
0x26 (0x46)	TCNT0	Timer/Counter0 (8-bit)							
0x25 (0x45)	TCCR0B	FOC0A	FOC0B	–	–	WGM02	CS02	CS01	CS00
0x24 (0x44)	TCCR0A	COM0A1	COM0A0	COM0B1	COM0B0	–	–	WGM01	WGM00
0x23 (0x43)	GTCCR	TSM	–	–	–	–	–	PSRASY	PSRSYNC
0x22 (0x42)	EEARH	EEPROM Address Register High Byte							
0x21 (0x41)	EEARL	EEPROM Address Register Low Byte							
0x20 (0x40)	EEDR	EEPROM Data Register							
0x1F (0x3F)	EECR	–	–	EEPM1	EEPM0	EERIE	EEMPE	EEPE	EERE
0x1E (0x3E)	GPIOR0	General Purpose I/O Register 0							
0x1D (0x3D)	EIMSK	–	–	–	–	–	–	INT1	INT0

Address	Name	Bit 7	Bit 6	Bit 5	Bit 4	Bit 3	Bit 2	Bit 1	Bit 0
0x1C (0x3C)	EIFR	–	–	–	–	–	–	INTF1	INTF0
0x1B (0x3B)	PCIFR	–	–	–	–	–	PCIF2	PCIF1	PCIF0
0x1A (0x3A)	Reserved	–	–	–	–	–	–	–	–
0x19 (0x39)	Reserved	–	–	–	–	–	–	–	–
0x18 (0x38)	Reserved	–	–	–	–	–	–	–	–
0x17 (0x37)	TIFR2	–	–	–	–	–	OCF2B	OCF2A	TOV2
0x16 (0x36)	TIFR1	–	–	ICF1	–	–	OCF1B	OCF1A	TOV1
0x15 (0x35)	TIFR0	–	–	–	–	–	OCF0B	OCF0A	TOV0
0x14 (0x34)	Reserved	–	–	–	–	–	–	–	–
0x13 (0x33)	Reserved	–	–	–	–	–	–	–	–
0x12 (0x32)	Reserved	–	–	–	–	–	–	–	–
0x11 (0x31)	Reserved	–	–	–	–	–	–	–	–
0x10 (0x30)	Reserved	–	–	–	–	–	–	–	–
0x0F (0x2F)	Reserved	–	–	–	–	–	–	–	–
0x0E (0x2E)	Reserved	–	–	–	–	–	–	–	–
0x0D (0x2D)	Reserved	–	–	–	–	–	–	–	–
0x0C (0x2C)	Reserved	–	–	–	–	–	–	–	–
0x0B (0x2B)	PORTD	PORTD7	PORTD6	PORTD5	PORTD4	PORTD3	PORTD2	PORTD1	PORTD0
0x0A (0x2A)	DDRD	DDD7	DDD6	DDD5	DDD4	DDD3	DDD2	DDD1	DDD0
0x09 (0x29)	PIND	PIND7	PIND6	PIND5	PIND4	PIND3	PIND2	PIND1	PIND0
0x08 (0x28)	PORTC	–	PORTC6	PORTC5	PORTC4	PORTC3	PORTC2	PORTC1	PORTC0
0x07 (0x27)	DDRC	–	DDC6	DDC5	DDC4	DDC3	DDC2	DDC1	DDC0

Address	Name	Bit 7	Bit 6	Bit 5	Bit 4	Bit 3	Bit 2	Bit 1	Bit 0
0x06 (0x26)	PINC	–	PINC6	PINC5	PINC4	PINC3	PINC2	PINC1	PINC0
0x05 (0x25)	PORTB	PORTB7	PORTB6	PORTB5	PORTB4	PORTB3	PORTB2	PORTB1	PORTB0
0x04 (0x24)	DDRB	DDB7	DDB6	DDB5	DDB4	DDB3	DDB2	DDB1	DDB0
0x03 (0x23)	PINB	PINB7	PINB6	PINB5	PINB4	PINB3	PINB2	PINB1	PINB0
0x02 (0x22)	Reserved	–	–	–	–	–	–	–	–
0x01 (0x21)	Reserved	–	–	–	–	–	–	–	–
0x00 (0x20)	Reserved	–	–	–	–	–	–	–	–

ATmega1280/2560

Address	Name	Bit 7	Bit 6	Bit 5	Bit 4	Bit 3	Bit 2	Bit 1	Bit 0
(0x1FF)	Reserved	–	–	–	–	–	–	–	–
...	Reserved	–	–	–	–	–	–	–	–
(0x137)	Reserved	–	–	–	–	–	–	–	–
(0x136)	UDR3	USART3 I/O Data Register							
(0x135)	UBRR3H	–	–	–	–	USART3 Baud Rate Register High Byte			
(0x134)	UBRR3L	USART3 Baud Rate Register Low Byte							
(0x133)	Reserved	–	–	–	–	–	–	–	–
(0x132)	UCSR3C	UMSEL31	UMSEL30	UPM31	UPM30	USBS3	UCSZ31	UCSZ30	UCPOL3
(0x131)	UCSR3B	RXCIE3	TXCIE3	UDRIE3	RXEN3	TXEN3	UCSZ32	RXB83	TXB83
(0x130)	UCSR3A	RXC3	TXC3	UDRE3	FE3	DOR3	UPE3	U2X3	MPCM3
(0x12F)	Reserved	–	–	–	–	–	–	–	–
(0x12E)	Reserved	–	–	–	–	–	–	–	–
(0x12D)	OCR5CH	Timer/Counter5: Output Compare Register C High Byte							
(0x12C)	OCR5CL	Timer/Counter5: Output Compare Register C Low Byte							
(0x12B)	OCR5BH	Timer/Counter5: Output Compare Register B High Byte							
(0x12A)	OCR5BL	Timer/Counter5: Output Compare Register B Low Byte							
(0x129)	OCR5AH	Timer/Counter5: Output Compare Register A High Byte							
(0x128)	OCR5AL	Timer/Counter5: Output Compare Register A Low Byte							
(0x127)	ICR5H	Timer/Counter5: Input Capture Register High Byte							
(0x126)	ICR5L	Timer/Counter5: Input Capture Register Low Byte							
(0x125)	TCNT5H	Timer/Counter5: Counter Register High Byte							

Address	Name	Bit 7	Bit 6	Bit 5	Bit 4	Bit 3	Bit 2	Bit 1	Bit 0
(0x124)	TCNT5L	Timer/Counter5: Counter Register Low Byte							
(0x123)	Reserved	–	–	–	–	–	–	–	–
(0x122)	TCCR5C	FOC5A	FOC5B	FOC5C	–	–	–	–	–
(0x121)	TCCR5B	ICNC5	ICES5	–	WGM53	WGM52	CS52	CS51	CS50
(0x120)	TCCR5A	COM5A1	COM5A0	COM5B1	COM5B0	COM5C1	COM5C0	WGM51	WGM50
(0x11F)	Reserved	–	–	–	–	–	–	–	–
(0x11E)	Reserved	–	–	–	–	–	–	–	–
(0x11D)	Reserved	–	–	–	–	–	–	–	–
(0x11C)	Reserved	–	–	–	–	–	–	–	–
(0x11B)	Reserved	–	–	–	–	–	–	–	–
(0x11A)	Reserved	–	–	–	–	–	–	–	–
(0x119)	Reserved	–	–	–	–	–	–	–	–
(0x118)	Reserved	–	–	–	–	–	–	–	–
(0x117)	Reserved	–	–	–	–	–	–	–	–
(0x116)	Reserved	–	–	–	–	–	–	–	–
(0x115)	Reserved	–	–	–	–	–	–	–	–
(0x114)	Reserved	–	–	–	–	–	–	–	–
(0x113)	Reserved	–	–	–	–	–	–	–	–
(0x112)	Reserved	–	–	–	–	–	–	–	–
(0x111)	Reserved	–	–	–	–	–	–	–	–
(0x110)	Reserved	–	–	–	–	–	–	–	–
(0x10F)	Reserved	–	–	–	–	–	–	–	–
(0x10E)	Reserved	–	–	–	–	–	–	–	–
(0x10D)	Reserved	–	–	–	–	–	–	–	–
(0x10C)	Reserved	–	–	–	–	–	–	–	–
(0x10B)	PORTL	PORTL7	PORTL6	PORTL5	PORTL4	PORTL3	PORTL2	PORTL1	PORTL0
(0x10A)	DDRL	DDL7	DDL6	DDL5	DDL4	DDL3	DDL2	DDL1	DDL0
(0x109)	PINL	PINL7	PINL6	PINL5	PINL4	PINL3	PINL2	PINL1	PINL0
(0x108)	PORTK	PORTK7	PORTK6	PORTK5	PORTK4	PORTK3	PORTK2	PORTK1	PORTK0
(0x107)	DDRK	DDK7	DDK6	DDK5	DDK4	DDK3	DDK2	DDK1	DDK0
(0x106)	PINK	PINK7	PINK6	PINK5	PINK4	PINK3	PINK2	PINK1	PINK0
(0x105)	PORTJ	PORTJ7	PORTJ6	PORTJ5	PORTJ4	PORTJ3	PORTJ2	PORTJ1	PORTJ0
(0x104)	DDRJ	DDJ7	DDJ6	DDJ5	DDJ4	DDJ3	DDJ2	DDJ1	DDJ0
(0x103)	PINJ	PINJ7	PINJ6	PINJ5	PINJ4	PINJ3	PINJ2	PINJ1	PINJ0
(0x102)	PORTH	PORTH7	PORTH6	PORTH5	PORTH4	PORTH3	PORTH2	PORTH1	PORTH0
(0x101)	DDRH	DDH7	DDH6	DDH5	DDH4	DDH3	DDH2	DDH1	DDH0
(0x100)	PINH	PINH7	PINH6	PINH5	PINH4	PINH3	PINH2	PINH1	PINH0

Address	Name	Bit 7	Bit 6	Bit 5	Bit 4	Bit 3	Bit 2	Bit 1	Bit 0
(0xFF)	Reserved	–	–	–	–	–	–	–	–
(0xFE)	Reserved	–	–	–	–	–	–	–	–
(0xFD)	Reserved	–	–	–	–	–	–	–	–
(0xFC)	Reserved	–	–	–	–	–	–	–	–
(0xFB)	Reserved	–	–	–	–	–	–	–	–
(0xFA)	Reserved	–	–	–	–	–	–	–	–
(0xF9)	Reserved	–	–	–	–	–	–	–	–
(0xF8)	Reserved	–	–	–	–	–	–	–	–
(0xF7)	Reserved	–	–	–	–	–	–	–	–
(0xF6)	Reserved	–	–	–	–	–	–	–	–
(0xF5)	Reserved	–	–	–	–	–	–	–	–
(0xF4)	Reserved	–	–	–	–	–	–	–	–
(0xF3)	Reserved	–	–	–	–	–	–	–	–
(0xF2)	Reserved	–	–	–	–	–	–	–	–
(0xF1)	Reserved	–	–	–	–	–	–	–	–
(0xF0)	Reserved	–	–	–	–	–	–	–	–
(0xEF)	Reserved	–	–	–	–	–	–	–	–
(0xEE)	Reserved	–	–	–	–	–	–	–	–
(0xED)	Reserved	–	–	–	–	–	–	–	–
(0xEC)	Reserved	–	–	–	–	–	–	–	–
(0xEB)	Reserved	–	–	–	–	–	–	–	–
(0xEA)	Reserved	–	–	–	–	–	–	–	–
(0xE9)	Reserved	–	–	–	–	–	–	–	–
(0xE8)	Reserved	–	–	–	–	–	–	–	–
(0xE7)	Reserved	–	–	–	–	–	–	–	–
(0xE6)	Reserved	–	–	–	–	–	–	–	–
(0xE5)	Reserved	–	–	–	–	–	–	–	–
(0xE4)	Reserved	–	–	–	–	–	–	–	–
(0xE3)	Reserved	–	–	–	–	–	–	–	–
(0xE2)	Reserved	–	–	–	–	–	–	–	–
(0xE1)	Reserved	–	–	–	–	–	–	–	–
(0xE0)	Reserved	–	–	–	–	–	–	–	–
(0xDF)	Reserved	–	–	–	–	–	–	–	–
(0xDE)	Reserved	–	–	–	–	–	–	–	–
(0xDD)	Reserved	–	–	–	–	–	–	–	–
(0xDC)	Reserved	–	–	–	–	–	–	–	–
(0xDB)	Reserved	–	–	–	–	–	–	–	–

Address	Name	Bit 7	Bit 6	Bit 5	Bit 4	Bit 3	Bit 2	Bit 1	Bit 0
(0xDA)	Reserved	–	–	–	–	–	–	–	–
(0xD9)	Reserved	–	–	–	–	–	–	–	–
(0xD8)	Reserved	–	–	–	–	–	–	–	–
(0xD7)	Reserved	–	–	–	–	–	–	–	–
(0xD6)	UDR2	USART2 I/O Data Register							
(0xD5)	UBRR2H	–	–	–	–	USART2 Baud Rate Register High Byte			
(0xD4)	UBRR2L	USART2 Baud Rate Register Low Byte							
(0xD3)	Reserved	–	–	–	–	–	–	–	–
(0xD2)	UCSR2C	UMSEL21	UMSEL20	UPM21	UPM20	USBS2	UCSZ21	UCSZ20	UCPOL2
(0xD1)	UCSR2B	RXCIE2	TXCIE2	UDRIE2	RXEN2	TXEN2	UCSZ22	RXB82	TXB82
(0xD0)	UCSR2A	RXC2	TXC2	UDRE2	FE2	DOR2	UPE2	U2X2	MPCM2
(0xCF)	Reserved	–	–	–	–	–	–	–	–
(0xCE)	UDR1	USART1 I/O Data Register							
(0xCD)	UBRR1H	–	–	–	–	USART1 Baud Rate Register High Byte			
(0xCC)	UBRR1L	USART1 Baud Rate Register Low Byte							
(0xCB)	Reserved	–	–	–	–	–	–	–	–
(0xCA)	UCSR1C	UMSEL11	UMSEL10	UPM11	UPM10	USBS1	UCSZ11	UCSZ10	UCPOL1
(0xC9)	UCSR1B	RXCIE1	TXCIE1	UDRIE1	RXEN1	TXEN1	UCSZ12	RXB81	TXB81
(0xC8)	UCSR1A	RXC1	TXC1	UDRE1	FE1	DOR1	UPE1	U2X1	MPCM1
(0xC7)	Reserved	–	–	–	–	–	–	–	–
(0xC6)	UDR0	USART0 I/O Data Register							
(0xC5)	UBRR0H	–	–	–	–	USART0 Baud Rate Register High Byte			
(0xC4)	UBRR0L	USART0 Baud Rate Register Low Byte							
(0xC3)	Reserved	–	–	–	–	–	–	–	–
(0xC2)	UCSR0C	UMSEL01	UMSEL00	UPM01	UPM00	USBS0	UCSZ01	UCSZ00	UCPOL0
(0xC1)	UCSR0B	RXCIE0	TXCIE0	UDRIE0	RXEN0	TXEN0	UCSZ02	RXB80	TXB80
(0xC0)	UCSR0A	RXC0	TXC0	UDRE0	FE0	DOR0	UPE0	U2X0	MPCM0
(0xBF)	Reserved	–	–	–	–	–	–	–	–
(0xBE)	Reserved	–	–	–	–	–	–	–	–
(0xBD)	TWAMR	TWAM6	TWAM5	TWAM4	TWAM3	TWAM2	TWAM1	TWAM0	-
(0xBC)	TWCR	TWINT	TWEA	TWSTA	TWSTO	TWWC	TWEN	-	TWIE
(0xBB)	TWDR	2-wire Serial Interface Data Register							
(0xBA)	TWAR	TWA6	TWA5	TWA4	TWA3	TWA2	TWA1	TWA0	TWGCE
(0xB9)	TWSR	TWS7	TWS6	TWS5	TWS4	TWS3	–	TWPS1	TWPS0
(0xB8)	TWBR	2-wire Serial Interface Bit Rate Register							
(0xB7)	Reserved	–	–	–	–	–	–	–	–
(0xB6)	ASSR	–	EXCLK	AS2	TCN2UB	OCR2AUB	OCR2BUB	TCR2AUB	TCR2BUB

Address	Name	Bit 7	Bit 6	Bit 5	Bit 4	Bit 3	Bit 2	Bit 1	Bit 0
(0xB5)	Reserved	–	–	–	–	–	–	–	–
(0xB4)	OCR2B	Timer/Counter2 Output Compare Register B							
(0xB3)	OCR2A	Timer/Counter2 Output Compare Register A							
(0xB2)	TCNT2	Timer/Counter2 (8 Bit)							
(0xB1)	TCCR2B	FOC2A	FOC2B	–	–	WGM22	CS22	CS21	CS20
(0xB0)	TCCR2A	COM2A1	COM2A0	COM2B1	COM2B0	–	–	WGM21	WGM20
(0xAF)	Reserved	–	–	–	–	–	–	–	–
(0xAE)	Reserved	–	–	–	–	–	–	–	–
(0xAD)	OCR4CH	Timer/Counter4: Output Compare Register C High Byte							
(0xAC)	OCR4CL	Timer/Counter4: Output Compare Register C Low Byte							
(0xAB)	OCR4BH	Timer/Counter4: Output Compare Register B High Byte							
(0xAA)	OCR4BL	Timer/Counter4: Output Compare Register B Low Byte							
(0xA9)	OCR4AH	Timer/Counter4: Output Compare Register A High Byte							
(0xA8)	OCR4AL	Timer/Counter4: Output Compare Register A Low Byte							
(0xA7)	ICR4H	Timer/Counter4: Input Capture Register High Byte							
(0xA6)	ICR4L	Timer/Counter4: Input Capture Register Low Byte							
(0xA5)	TCNT4H	Timer/Counter4: Counter Register High Byte							
(0xA4)	TCNT4L	Timer/Counter4: Counter Register Low Byte							
(0xA3)	Reserved	–	–	–	–	–	–	–	–
(0xA2)	TCCR4C	FOC4A	FOC4B	FOC4C	–	–	–	–	–
(0xA1)	TCCR4B	ICNC4	ICES4	–	WGM43	WGM42	CS42	CS41	CS40
(0xA0)	TCCR4A	COM4A1	COM4A0	COM4B1	COM4B0	COM4C1	COM4C0	WGM41	WGM40
(0x9F)	Reserved	–	–	–	–	–	–	–	–
(0x9E)	Reserved	–	–	–	–	–	–	–	–
(0x9D)	OCR3CH	Timer/Counter3: Output Compare Register C High Byte							
(0x9C)	OCR3CL	Timer/Counter3: Output Compare Register C Low Byte							
(0x9B)	OCR3BH	Timer/Counter3: Output Compare Register B High Byte							
(0x9A)	OCR3BL	Timer/Counter3: Output Compare Register B Low Byte							
(0x99)	OCR3AH	Timer/Counter3: Output Compare Register A High Byte							
(0x98)	OCR3AL	Timer/Counter3: Output Compare Register A Low Byte							
(0x97)	ICR3H	Timer/Counter3: Input Capture Register High Byte							
(0x96)	ICR3L	Timer/Counter3: Input Capture Register Low Byte							
(0x95)	TCNT3H	Timer/Counter3: Counter Register High Byte							
(0x94)	TCNT3L	Timer/Counter3: Counter Register Low Byte							
(0x93)	Reserved	–	–	–	–	–	–	–	–
(0x92)	TCCR3C	FOC3A	FOC3B	FOC3C	–	–	–	–	–
(0x91)	TCCR3B	ICNC3	ICES3	–	WGM33	WGM32	CS32	CS31	CS30

Address	Name	Bit 7	Bit 6	Bit 5	Bit 4	Bit 3	Bit 2	Bit 1	Bit 0
(0x90)	TCCR3A	COM3A1	COM3A0	COM3B1	COM3B0	COM3C1	COM3C0	WGM31	WGM30
(0x8F)	Reserved	–	–	–	–	–	–	–	–
(0x8E)	Reserved	–	–	–	–	–	–	–	–
(0x8D)	OCR1CH	Timer/Counter1: Output Compare Register C High Byte							
(0x8C)	OCR1CL	Timer/Counter1: Output Compare Register C Low Byte							
(0x8B)	OCR1BH	Timer/Counter1: Output Compare Register B High Byte							
(0x8A)	OCR1BL	Timer/Counter1: Output Compare Register B Low Byte							
(0x89)	OCR1AH	Timer/Counter1: Output Compare Register A High Byte							
(0x88)	OCR1AL	Timer/Counter1: Output Compare Register A Low Byte							
(0x87)	ICR1H	Timer/Counter1: Input Capture Register High Byte							
(0x86)	ICR1L	Timer/Counter1: Input Capture Register Low Byte							
(0x85)	TCNT1H	Timer/Counter1: Counter Register High Byte							
(0x84)	TCNT1L	Timer/Counter1: Counter Register Low Byte							
(0x83)	Reserved	–	–	–	–	–	–	–	–
(0x82)	TCCR1C	FOC1A	FOC1B	FOC1C	–	–	–	–	–
(0x81)	TCCR1B	ICNC1	ICES1	–	WGM13	WGM12	CS12	CS11	CS10
(0x80)	TCCR1A	COM1A1	COM1A0	COM1B1	COM1B0	COM1C1	COM1C0	WGM11	WGM10
(0x7F)	DIDR1	–	–	–	–	–	–	AIN1D	AIN0D
(0x7E)	DIDR0	ADC7D	ADC6D	ADC5D	ADC4D	ADC3D	ADC2D	ADC1D	ADC0D
(0x7D)	DIDR2	ADC15D	ADC14D	ADC13D	ADC12D	ADC11D	ADC10D	ADC9D	ADC8D
(0x7C)	ADMUX	REFS1	REFS0	ADLAR	MUX4	MUX3	MUX2	MUX1	MUX0
(0x7B)	ADCSRB	-	ACME	–	–	MUX5	ADTS2	ADTS1	ADTS0
(0x7A)	ADCSRA	ADEN	ADSC	ADATE	ADIF	ADIE	ADPS2	ADPS1	ADPS0
(0x79)	ADCH	ADC Data Register High Byte							
(0x78)	ADCL	ADC Data Register Low Byte							
(0x77)	Reserved	–	–	–	–	–	–	–	–
(0x76)	Reserved	–	–	–	–	–	–	–	–
(0x75)	XMCRB	XMBK	–	–	–	–	XMM2	XMM1	XMM0
(0x74)	XMCRA	SRE	SRL2	SRL1	SRL0	SRW11	SRW10	SRW01	SRW00
(0x73)	TIMSK5	–	–	ICIE5	–	OCIE5C	OCIE5B	OCIE5A	TOIE5
(0x72)	TIMSK4	–	–	ICIE4	–	OCIE4C	OCIE4B	OCIE4A	TOIE4
(0x71)	TIMSK3	–	–	ICIE3	–	OCIE3C	OCIE3B	OCIE3A	TOIE3
(0x70)	TIMSK2	–	–	–	–	–	OCIE2B	OCIE2A	TOIE2
(0x6F)	TIMSK1	–	–	ICIE1	–	OCIE1C	OCIE1B	OCIE1A	TOIE1
(0x6E)	TIMSK0	–	–	–	–	–	OCIE0B	OCIE0A	TOIE0
(0x6D)	PCMSK2	PCINT23	PCINT22	PCINT21	PCINT20	PCINT19	PCINT18	PCINT17	PCINT16
(0x6C)	PCMSK1	PCINT15	PCINT14	PCINT13	PCINT12	PCINT11	PCINT10	PCINT9	PCINT8

Address	Name	Bit 7	Bit 6	Bit 5	Bit 4	Bit 3	Bit 2	Bit 1	Bit 0
(0x6B)	PCMSK0	PCINT7	PCINT6	PCINT5	PCINT4	PCINT3	PCINT2	PCINT1	PCINT0
(0x6A)	EICRB	ISC71	ISC70	ISC61	ISC60	ISC51	ISC50	ISC41	ISC40
(0x69)	EICRA	ISC31	ISC30	ISC21	ISC20	ISC11	ISC10	ISC01	ISC00
(0x68)	PCICR	–	–	–	–	–	PCIE2	PCIE1	PCIE0
(0x67)	Reserved	–	–	–	–	–	–	–	–
(0x66)	OSCCAL	Oscillator Calibration Register							
(0x65)	PRR1	–	–	PRTIM5	PRTIM4	PRTIM3	PRUSART3	PRUSART2	PRUSART1
(0x64)	PRR0	PRTWI	PRTIM2	PRTIM0	–	PRTIM1	PRSPI	PRUSART0	PRADC
(0x63)	Reserved	–	–	–	–	–	–	–	–
(0x62)	Reserved	–	–	–	–	–	–	–	–
(0x61)	CLKPR	CLKPCE	–	–	–	CLKPS3	CLKPS2	CLKPS1	CLKPS0
(0x60)	WDTCSR	WDIF	WDIE	WDP3	WDCE	WDE	WDP2	WDP1	WDP0
0x3F (0x5F)	SREG	I	T	H	S	V	N	Z	C
0x3E (0x5E)	SPH	SP15	SP14	SP13	SP12	SP11	SP10	SP9	SP8
0x3D (0x5D)	SPL	SP7	SP6	SP5	SP4	SP3	SP2	SP1	SP0
0x3C (0x5C)	EIND	–	–	–	–	–	–	–	EIND0
0x3B (0x5B)	RAMPZ	–	–	–	–	–	–	RAMPZ1	RAMPZ0
0x3A (0x5A)	Reserved	–	–	–	–	–	–	–	–
0x39 (0x59)	Reserved	–	–	–	–	–	–	–	–
0x38 (0x58)	Reserved	–	–	–	–	–	–	–	–
0x37 (0x57)	SPMCSR	SPMIE	RWWSB	SIGRD	RWWSRE	BLBSET	PGWRT	PGERS	SPMEN
0x36 (0x56)	Reserved	–	–	–	–	–	–	–	–
0x35 (0x55)	MCUCR	JTD	–	–	PUD	–	–	IVSEL	IVCE
0x34 (0x54)	MCUSR	–	–	–	JTRF	WDRF	BORF	EXTRF	PORF
0x33 (0x53)	SMCR	–	–	–	–	SM2	SM1	SM0	SE
0x32 (0x52)	Reserved	–	–	–	–	–	–	–	–
0x31 (0x51)	OCDR	OCDR7	OCDR6	OCDR5	OCDR4	OCDR3	OCDR2	OCDR1	OCDR0

Address	Name	Bit 7	Bit 6	Bit 5	Bit 4	Bit 3	Bit 2	Bit 1	Bit 0
0x30 (0x50)	ACSR	ACD	ACBG	ACO	ACI	ACIE	ACIC	ACIS1	ACIS0
0x2F (0x4F)	Reserved	–	–	–	–	–	–	–	–
0x2E (0x4E)	SPDR	SPI Data Register							
0x2D (0x4D)	SPSR	SPIF	WCOL	–	–	–	–	–	SPI2X
0x2C (0x4C)	SPCR	SPIE	SPE	DORD	MSTR	CPOL	CPHA	SPR1	SPR0
0x2B (0x4B)	GPIOR2	General Purpose I/O Register 2							
0x2A (0x4A)	GPIOR1	General Purpose I/O Register 1							
0x29 (0x49)	Reserved	–	–	–	–	–	–	–	–
0x28 (0x48)	OCR0B	Timer/Counter0 Output Compare Register B							
0x27 (0x47)	OCR0A	Timer/Counter0 Output Compare Register A							
0x26 (0x46)	TCNT0	Timer/Counter0 (8 Bit)							
0x25 (0x45)	TCCR0B	FOC0A	FOC0B	–	–	WGM02	CS02	CS01	CS00
0x24 (0x44)	TCCR0A	COM0A1	COM0A0	COM0B1	COM0B0	–	–	WGM01	WGM00
0x23 (0x43)	GTCCR	TSM	–	–	–	–	–	PSRASY	PSRSYNC
0x22 (0x42)	EEARH	–	–	–	–	EEPROM Address Register High Byte			
0x21 (0x41)	EEARL	EEPROM Address Register Low Byte							
0x20 (0x40)	EEDR	EEPROM Data Register							
0x1F (0x3F)	EECR	–	–	EEPM1	EEPM0	EERIE	EEMPE	EEPE	EERE
0x1E (0x3E)	GPIOR0	General Purpose I/O Register 0							
0x1D (0x3D)	EIMSK	INT7	INT6	INT5	INT4	INT3	INT2	INT1	INT0
0x1C (0x3C)	EIFR	INTF7	INTF6	INTF5	INTF4	INTF3	INTF2	INTF1	INTF0
0x1B (0x3B)	PCIFR	–	–	–	–	–	PCIF2	PCIF1	PCIF0

Address	Name	Bit 7	Bit 6	Bit 5	Bit 4	Bit 3	Bit 2	Bit 1	Bit 0
0x1A (0x3A)	TIFR5	–	–	ICF5	–	OCF5C	OCF5B	OCF5A	TOV5
0x19 (0x39)	TIFR4	–	–	ICF4	–	OCF4C	OCF4B	OCF4A	TOV4
0x18 (0x38)	TIFR3	–	–	ICF3	–	OCF3C	OCF3B	OCF3A	TOV3
0x17 (0x37)	TIFR2	–	–	–	–	–	OCF2B	OCF2A	TOV2
0x16 (0x36)	TIFR1	–	–	ICF1	–	OCF1C	OCF1B	OCF1A	TOV1
0x15 (0x35)	TIFR0	–	–	–	–	–	OCF0B	OCF0A	TOV0
0x14 (0x34)	PORTG	–	–	PORTG5	PORTG4	PORTG3	PORTG2	PORTG1	PORTG0
0x13 (0x33)	DDRG	–	–	DDG5	DDG4	DDG3	DDG2	DDG1	DDG0
0x12 (0x32)	PING	–	–	PING5	PING4	PING3	PING2	PING1	PING0
0x11 (0x31)	PORTF	PORTF7	PORTF6	PORTF5	PORTF4	PORTF3	PORTF2	PORTF1	PORTF0
0x10 (0x30)	DDRF	DDF7	DDF6	DDF5	DDF4	DDF3	DDF2	DDF1	DDF0
0x0F (0x2F)	PINF	PINF7	PINF6	PINF5	PINF4	PINF3	PINF2	PINF1	PINF0
0x0E (0x2E)	PORTE	PORTE7	PORTE6	PORTE5	PORTE4	PORTE3	PORTE2	PORTE1	PORTE0
0x0D (0x2D)	DDRE	DDE7	DDE6	DDE5	DDE4	DDE3	DDE2	DDE1	DDE0
0x0C (0x2C)	PINE	PINE7	PINE6	PINE5	PINE4	PINE3	PINE2	PINE1	PINE0
0x0B (0x2B)	PORTD	PORTD7	PORTD6	PORTD5	PORTD4	PORTD3	PORTD2	PORTD1	PORTD0
0x0A (0x2A)	DDRD	DDD7	DDD6	DDD5	DDD4	DDD3	DDD2	DDD1	DDD0
0x09 (0x29)	PIND	PIND7	PIND6	PIND5	PIND4	PIND3	PIND2	PIND1	PIND0
0x08 (0x28)	PORTC	PORTC7	PORTC6	PORTC5	PORTC4	PORTC3	PORTC2	PORTC1	PORTC0
0x07 (0x27)	DDRC	DDC7	DDC6	DDC5	DDC4	DDC3	DDC2	DDC1	DDC0
0x06 (0x26)	PINC	PINC7	PINC6	PINC5	PINC4	PINC3	PINC2	PINC1	PINC0
0x05 (0x25)	PORTB	PORTB7	PORTB6	PORTB5	PORTB4	PORTB3	PORTB2	PORTB1	PORTB0

Address	Name	Bit 7	Bit 6	Bit 5	Bit 4	Bit 3	Bit 2	Bit 1	Bit 0
0x04 (0x24)	DDRB	DDB7	DDB6	DDB5	DDB4	DDB3	DDB2	DDB1	DDB0
0x03 (0x23)	PINB	PINB7	PINB6	PINB5	PINB4	PINB3	PINB2	PINB1	PINB0
0x02 (0x22)	PORTA	PORTA7	PORTA6	PORTA5	PORTA4	PORTA3	PORTA2	PORTA1	PORTA0
0x01 (0x21)	DDRA	DDA7	DDA6	DDA5	DDA4	DDA3	DDA2	DDA1	DDA0
0x00 (0x20)	PINA	PINA7	PINA6	PINA5	PINA4	PINA3	PINA2	PINA1	PINA0

ATmega32U4

Address	Name	Bit 7	Bit 6	Bit 5	Bit 4	Bit 3	Bit 2	Bit 1	Bit 0
(0xFF)	Reserved	–	–	–	–	–	–	–	–
(0xFE)	Reserved	–	–	–	–	–	–	–	–
(0xFD)	Reserved	–	–	–	–	–	–	–	–
(0xFC)	Reserved	–	–	–	–	–	–	–	–
(0xFB)	Reserved	–	–	–	–	–	–	–	–
(0xFA)	Reserved	–	–	–	–	–	–	–	–
(0xF9)	Reserved	–	–	–	–	–	–	–	–
(0xF8)	Reserved	–	–	–	–	–	–	–	–
(0xF7)	Reserved	–	–	–	–	–	–	–	–
(0xF6)	Reserved	–	–	–	–	–	–	–	–
(0xF5)	Reserved	–	–	–	–	–	–	–	–
(0xF4)	UEINT	–	EPINT6:0						
(0xF3)	UEBCHX	–	–	–	–	–	BYCT10:8		
(0xF2)	UEBCLX	BYCT7:0							
(0xF1)	UEDATX	DAT7:0							
(0xF0)	UEIENX	FLERRE	NAKINE	–	NAKOUTE	RXSTPE	RXOUTE	STALLEDE	TXINE
(0xEF)	UESTA1X	–	–	–	–	–	CTRLDIR	CURRBK1:0	
(0xEE)	UESTA0X	CFGOK	OVERFI	UNDERFI	–	DTSEQ1:0		NBUSYBK1:0	
(0xED)	UECFG1X	–	EPSIZE2:0			EPBK1:0		ALLOC	–
(0xEC)	UECFG0X	EPTYPE1:0		–	–	–	–	–	EPDIR
(0xEB)	UECONX	–	–	STALLRQ	STALLRQC	RSTDT	–	–	EPEN
(0xEA)	UERST	–	EPRST6:0						
(0xE9)	UENUM	–	–	–	–	–	EPNUM2:0		
(0xE8)	UEINTX	FIFOCON	NAKINI	RWAL	NAKOUTI	RXSTPI	RXOUTI	STALLEDI	TXINI
(0xE7)	Reserved	–	–	–	–	–	–	–	–

Address	Name	Bit 7	Bit 6	Bit 5	Bit 4	Bit 3	Bit 2	Bit 1	Bit 0
(0xE6)	UDMFN	–	–	–	FNCERR	–	–	–	–
(0xE5)	UDFNUMH	–	–	–	–	–	FNUM10:8		
(0xE4)	UDFNUML	FNUM7:0							
(0xE3)	UDADDR	ADDEN	UADD6:0						
(0xE2)	UDIEN	–	UPRSME	EORSME	WAKEUPE	EORSTE	SOFE	MSOFE	SUSPE
(0xE1)	UDINT	–	UPRSMI	EORSMI	WAKEUPI	EORSTI	SOFI	MSOFI	SUSPI
(0xE0)	UDCON	–	–	–	–	RSTCPU	LSM	RMWKUP	DETACH
(0xDF)	Reserved	–	–	–	–	–	–	–	–
(0xDE)	Reserved	–	–	–	–	–	–	–	–
(0xDD)	Reserved	–	–	–	–	–	–	–	–
(0xDC)	Reserved	–	–	–	–	–	–	–	–
(0xDB)	Reserved	–	–	–	–	–	–	–	–
(0xDA)	USBINT	–	–	–	–	–	–	–	VBUSTI
(0xD9)	USBSTA	–	–	–	–	–	–	ID	VBUS
(0xD8)	USBCON	USBE	–	FRZCLK	OTGPADE	–	–	–	VBUSTE
(0xD7)	UHWCON	–	–	–	–	–	–	–	UVREGE
(0xD6)	Reserved	–	–	–	–	–	–	–	–
(0xD5)	Reserved	–	–	–	–	–	–	–	–
(0xD4)	DT4	DT4H3	DT4H2	DT4H1	DT4H0	DT4L3	DT4L2	DT4L1	DT4L0
(0xD3)	Reserved	–	–	–	–	–	–	–	–
(0xD2)	OCR4D	Timer/Counter4: Output Compare Register D							
(0xD1)	OCR4C	Timer/Counter4: Output Compare Register C							
(0xD0)	OCR4B	Timer/Counter4: Output Compare Register B							
(0xCF)	OCR4A	Timer/Counter4: Output Compare Register A							
(0xCE)	UDR1	USART1 I/O Data Register							
(0xCD)	UBRR1H	–	–	–	–	USART1 Baud Rate Register High Byte			
(0xCC)	UBRR1L	USART1 Baud Rate Register Low Byte							
(0xCB)	UCSR1D	–	–	–	–	–	–	CTSEN	RTSEN
(0xCA)	UCSR1C	UMSEL11	UMSEL10	UPM11	UPM10	USBS1	UCSZ11	UCSZ10	UCPOL1
(0xC9)	UCSR1B	RXCIE1	TXCIE1	UDRIE1	RXEN1	TXEN1	UCSZ12	RXB81	TXB81
(0xC8)	UCSR1A	RXC1	TXC1	UDRE1	FE1	DOR1	PE1	U2X1	MPCM1
(0xC7)	CLKSTA	–	–	–	–	–	–	RCON	EXTON
(0xC6)	CLKSEL1	RCCKSEL3	RCCKSEL2	RCCKSEL1	RCCKSEL0	EXCKSEL3	EXCKSEL2	EXCKSEL1	EXCKSEL0
(0xC5)	CLKSEL0	RCSUT1	RCSUT0	EXSUT1	EXSUT0	RCE	EXTE	–	CLKS
(0xC4)	TCCR4E	TLOCK4	ENHC4	OC40E5	OC40E4	OC40E3	OC40E2	OC40E1	OC40E0
(0xC3)	TCCR4D	FPIE4	FPEN4	FPNC4	FPES4	FPAC4	FPF4	WGM41	WGM40
(0xC2)	TCCR4C	COM4A1S	COM4A0S	COM4B1S	COM4B0S	COM4D1S	COM4D0S	FOC4D	PWM4D

Address	Name	Bit 7	Bit 6	Bit 5	Bit 4	Bit 3	Bit 2	Bit 1	Bit 0
(0xC1)	TCCR4B	PWM4X	PSR4	DTPS41	DTPS40	CS43	CS42	CS41	CS40
(0xC0)	TCCR4A	COM4A1	COM4A0	COM4B1	COM4B0	FOC4A	FOC4B	PWM4A	PWM4B
(0xBF)	TC4H	–	–	–	–	–	Timer/Counter4 High Byte		
(0xBE)	TCNT4	Timer/Counter4: Counter Register Low Byte							
(0xBD)	TWAMR	TWAM6	TWAM5	TWAM4	TWAM3	TWAM2	TWAM1	TWAM0	–
(0xBC)	TWCR	TWINT	TWEA	TWSTA	TWSTO	TWWC	TWEN	–	TWIE
(0xBB)	TWDR	2-wire Serial Interface Data Register							
(0xBA)	TWAR	TWA6	TWA5	TWA4	TWA3	TWA2	TWA1	TWA0	TWGCE
(0xB9)	TWSR	TWS7	TWS6	TWS5	TWS4	TWS3	–	TWPS1	TWPS0
(0xB8)	TWBR	2-wire Serial Interface Bit Rate Register							
(0xB6)	Reserved	–	–	–	–	–	–	–	–
(0xB5)	Reserved	–	–	–	–	–	–	–	–
(0xB4)	Reserved	–	–	–	–	–	–	–	–
(0xB3)	Reserved	–	–	–	–	–	–	–	–
(0xB2)	Reserved	–	–	–	–	–	–	–	–
(0xB1)	Reserved	–	–	–	–	–	–	–	–
(0xB0)	Reserved	–	–	–	–	–	–	–	–
(0xAF)	Reserved	–	–	–	–	–	–	–	–
(0xAE)	Reserved	–	–	–	–	–	–	–	–
(0xAD)	Reserved	–	–	–	–	–	–	–	–
(0xAC)	Reserved	–	–	–	–	–	–	–	–
(0xB7)	Reserved	–	–	–	–	–	–	–	–
(0xAB)	Reserved	–	–	–	–	–	–	–	–
(0xAA)	Reserved	–	–	–	–	–	–	–	–
(0xA9)	Reserved	–	–	–	–	–	–	–	–
(0xA8)	Reserved	–	–	–	–	–	–	–	–
(0xA7)	Reserved	–	–	–	–	–	–	–	–
(0xA6)	Reserved	–	–	–	–	–	–	–	–
(0xA5)	Reserved	–	–	–	–	–	–	–	–
(0xA4)	Reserved	–	–	–	–	–	–	–	–
(0xA3)	Reserved	–	–	–	–	–	–	–	–
(0xA2)	Reserved	–	–	–	–	–	–	–	–
(0xA1)	Reserved	–	–	–	–	–	–	–	–
(0xA0)	Reserved	–	–	–	–	–	–	–	–
(0x9F)	Reserved	–	–	–	–	–	–	–	–
(0x9E)	Reserved	–	–	–	–	–	–	–	–
(0x9D)	OCR3CH	Timer/Counter3: Output Compare Register C High Byte							

Address	Name	Bit 7	Bit 6	Bit 5	Bit 4	Bit 3	Bit 2	Bit 1	Bit 0
(0x9C)	OCR3CL	Timer/Counter3: Output Compare Register C Low Byte							
(0x9B)	OCR3BH	Timer/Counter3: Output Compare Register B High Byte							
(0x9A)	OCR3BL	Timer/Counter3: Output Compare Register B Low Byte							
(0x99)	OCR3AH	Timer/Counter3: Output Compare Register A High Byte							
(0x98)	OCR3AL	Timer/Counter3: Output Compare Register A Low Byte							
(0x97)	ICR3H	Timer/Counter3: Input Capture Register High Byte							
(0x96)	ICR3L	Timer/Counter3: Input Capture Register Low Byte							
(0x95)	TCNT3H	Timer/Counter3: Counter Register High Byte							
(0x94)	TCNT3L	Timer/Counter3: Counter Register Low Byte							
(0x93)	Reserved	–	–	–	–	–	–	–	–
(0x92)	TCCR3C	FOC3A	–	–	–	–	–	–	–
(0x91)	TCCR3B	ICNC3	ICES3	–	WGM33	WGM32	CS32	CS31	CS30
(0x90)	TCCR3A	COM3A1	COM3A0	COM3B1	COM3B0	COM3C1	COM3C0	WGM31	WGM30
(0x8F)	Reserved	–	–	–	–	–	–	–	–
(0x8E)	Reserved	–	–	–	–	–	–	–	–
(0x8D)	OCR1CH	Timer/Counter1: Output Compare Register C High Byte							
(0x8C)	OCR1CL	Timer/Counter1: Output Compare Register C Low Byte							
(0x8B)	OCR1BH	Timer/Counter1: Output Compare Register B High Byte							
(0x8A)	OCR1BL	Timer/Counter1: Output Compare Register B Low Byte							
(0x89)	OCR1AH	Timer/Counter1: Output Compare Register A High Byte							
(0x88)	OCR1AL	Timer/Counter1: Output Compare Register A Low Byte							
(0x87)	ICR1H	Timer/Counter1: Input Capture Register High Byte							
(0x86)	ICR1L	Timer/Counter1: Input Capture Register Low Byte							
(0x85)	TCNT1H	Timer/Counter1: Counter Register High Byte							
(0x84)	TCNT1L	Timer/Counter1: Counter Register Low Byte							
(0x83)	Reserved	–	–	–	–	–	–	–	–
(0x82)	TCCR1C	FOC1A	FOC1B	FOC1C	–	–	–	–	–
(0x81)	TCCR1B	ICNC1	ICES1	–	WGM13	WGM12	CS12	CS11	CS10
(0x80)	TCCR1A	COM1A1	COM1A0	COM1B1	COM1B0	COM1C1	COM1C0	WGM11	WGM10
(0x7F)	DIDR1	–	–	–	–	–	–	–	AIN0D
(0x7E)	DIDR0	ADC7D	ADC6D	ADC5D	ADC4D	–	–	ADC1D	ADC0D
(0x7D)	DIDR2	–	–	ADC13D	ADC12D	ADC11D	ADC10D	ADC9D	ADC8D
(0x7C)	ADMUX	REFS1	REFS0	ADLAR	MUX4	MUX3	MUX2	MUX1	MUX0
(0x7B)	ADCSRB	ADHSM	ACME	MUX5	–	ADTS3	ADTS2	ADTS1	ADTS0
(0x7A)	ADCSRA	ADEN	ADSC	ADATE	ADIF	ADIE	ADPS2	ADPS1	ADPS0
(0x79)	ADCH	ADC Data Register High byte							
(0x78)	ADCL	ADC Data Register Low byte							

Address	Name	Bit 7	Bit 6	Bit 5	Bit 4	Bit 3	Bit 2	Bit 1	Bit 0
(0x77)	Reserved	–	–	–	–	–	–	–	–
(0x76)	Reserved	–	–	–	–	–	–	–	–
(0x75)	Reserved	–	–	–	–	–	–	–	–
(0x74)	Reserved	–	–	–	–	–	–	–	–
(0x73)	Reserved	–	–	–	–	–	–	–	–
(0x72)	TIMSK4	OCIE4D	OCIE4A	OCIE4B	–	–	TOIE4	–	–
(0x71)	TIMSK3	–	–	ICIE3	–	OCIE3C	OCIE3B	OCIE3A	TOIE3
(0x70)	Reserved	–	–	–	–	–	–	–	–
(0x6F)	TIMSK1	–	–	ICIE1	–	OCIE1C	OCIE1B	OCIE1A	TOIE1
(0x6E)	TIMSK0	–	–	–	–	–	OCIE0B	OCIE0A	TOIE0
(0x6D)	Reserved	–	–	–	–	–	–	–	–
(0x6C)	Reserved	–	–	–	–	–	–	–	–
(0x6B)	PCMSK0	PCINT7	PCINT6	PCINT5	PCINT4	PCINT3	PCINT2	PCINT1	PCINT0
(0x6A)	EICRB	–	–	ISC61	ISC60	–	–	–	–
(0x69)	EICRA	ISC31	ISC30	ISC21	ISC20	ISC11	ISC10	ISC01	ISC00
(0x68)	PCICR	–	–	–	–	–	–	–	PCIE0
(0x67)	RCCTRL	–	–	–	–	–	–	–	RCFREQ
(0x66)	OSCCAL	RC Oscillator Calibration Register							
(0x65)	PRR1	PRUSB	–	–	PRTIM4	PRTIM3	–	–	PRUSART1
(0x64)	PRR0	PRTWI	–	PRTIM0	–	PRTIM1	PRSPI	–	PRADC
(0x63)	Reserved	–	–	–	–	–	–	–	–
(0x62)	Reserved	–	–	–	–	–	–	–	–
(0x61)	CLKPR	CLKPCE	–	–	–	CLKPS3	CLKPS2	CLKPS1	CLKPS0
(0x60)	WDTCSR	WDIF	WDIE	WDP3	WDCE	WDE	WDP2	WDP1	WDP0
0x3F (0x5F)	SREG	I	T	H	S	V	N	Z	C
0x3E (0x5E)	SPH	SP15	SP14	SP13	SP12	SP11	SP10	SP9	SP8
0x3D (0x5D)	SPL	SP7	SP6	SP5	SP4	SP3	SP2	SP1	SP0
0x3C (0x5C)	Reserved	–	–	–	–	–	–	–	–
0x3B (0x5B)	RAMPZ	–	–	–	–	–	–	RAMPZ1	RAMPZ0
0x3A (0x5A)	Reserved	–	–	–	–	–	–	–	–
0x39 (0x59)	Reserved	–	–	–	–	–	–	–	–
0x38 (0x58)	Reserved	–	–	–	–	–	–	–	–

Address	Name	Bit 7	Bit 6	Bit 5	Bit 4	Bit 3	Bit 2	Bit 1	Bit 0
0x37 (0x57)	SPMCSR	SPMIE	RWWSB	SIGRD	RWWSRE	BLBSET	PGWRT	PGERS	SPMEN
0x36 (0x56)	Reserved	–	–	–	–	–	–	–	–
0x35 (0x55)	MCUCR	JTD	–	–	PUD	–	–	IVSEL	IVCE
0x34 (0x54)	MCUSR	–	–	USBRF	JTRF	WDRF	BORF	EXTRF	PORF
0x33 (0x53)	SMCR	–	–	–	–	SM2	SM1	SM0	SE
0x32 (0x52)	PLLFRQ	PINMUX	PLLUSB	PLLTM1	PLLTM0	PDIV3	PDIV2	PDIV1	PDIV0
0x31 (0x51)	OCDR/ MONDR	OCDR7	OCDR6	OCDR5	OCDR4	OCDR3	OCDR2	OCDR1	OCDR0
		Monitor Data Register							
0x30 (0x50)	ACSR	ACD	ACBG	ACO	ACI	ACIE	ACIC	ACIS1	ACIS0
0x2F (0x4F)	Reserved	–	–	–	–	–	–	–	–
0x2E (0x4E)	SPDR	SPI Data Register							
0x2D (0x4D)	SPSR	SPIF	WCOL	–	–	–	–	–	SPI2X
0x2C (0x4C)	SPCR	SPIE	SPE	DORD	MSTR	CPOL	CPHA	SPR1	SPR0
0x2B (0x4B)	GPIOR2	General Purpose I/O Register 2							
0x2A (0x4A)	GPIOR1	General Purpose I/O Register 1							
0x29 (0x49)	PLLCSR	–	–	–	PINDIV	–	–	PLLE	PLOCK
0x28 (0x48)	OCR0B	Timer/Counter0 Output Compare Register B							
0x27 (0x47)	OCR0A	Timer/Counter0 Output Compare Register A							
0x26 (0x46)	TCNT0	Timer/Counter0 (8 Bit)							
0x25 (0x45)	TCCR0B	FOC0A	FOC0B	–	–	WGM02	CS02	CS01	CS00
0x24 (0x44)	TCCR0A	COM0A1	COM0A0	COM0B1	COM0B0	–	–	WGM01	WGM00
0x23 (0x43)	GTCCR	TSM	–	–	–	–	–	PSRASY	PSRSYNC

Address	Name	Bit 7	Bit 6	Bit 5	Bit 4	Bit 3	Bit 2	Bit 1	Bit 0
0x22 (0x42)	EEARH	–	–	–	–	EEPROM Address Register High Byte			
0x21 (0x41)	EEARL	EEPROM Address Register Low Byte							
0x20 (0x40)	EEDR	EEPROM Data Register							
0x1F (0x3F)	EECR	–	–	EEPM1	EEPM0	EERIE	EEMPE	EEPE	EERE
0x1E (0x3E)	GPIOR0	General Purpose I/O Register 0							
0x1D (0x3D)	EIMSK	–	INT6	–	–	INT3	INT2	INT1	INT0
0x1C (0x3C)	EIFR	–	INTF6	–	–	INTF3	INTF2	INTF1	INTF0
0x1B (0x3B)	PCIFR	–	–	–	–	–	–	–	PCIF0
0x1A (0x3A)	Reserved	–	–	–	–	–	–	–	–
0x19 (0x39)	TIFR4	OCF4D	OCF4A	OCF4B	–	–	TOV4	–	–
0x18 (0x38)	TIFR3	–	–	ICF3	–	OCF3C	OCF3B	OCF3A	TOV3
0x17 (0x37)	Reserved	–	–	–	–	–	–	–	–
0x16 (0x36)	TIFR1	–	–	ICF1	–	OCF1C	OCF1B	OCF1A	TOV1
0x15 (0x35)	TIFR0	–	–	–	–	–	OCF0B	OCF0A	TOV0
0x14 (0x34)	Reserved	–	–	–	–	–	–	–	–
0x13 (0x33)	Reserved	–	–	–	–	–	–	–	–
0x12 (0x32)	Reserved	–	–	–	–	–	–	–	–
0x11 (0x31)	PORTF	PORTF7	PORTF6	PORTF5	PORTF4	–	–	PORTF1	PORTF0
0x10 (0x30)	DDRF	DDF7	DDF6	DDF5	DDF4	–	–	DDF1	DDF0
0x0F (0x2F)	PINF	PINF7	PINF6	PINF5	PINF4	–	–	PINF1	PINF0
0x0E (0x2E)	PORTE	–	PORTE6	–	–	–	PORTE2	–	–
0x0D (0x2D)	DDRE	–	DDE6	–	–	–	DDE2	–	–

Address	Name	Bit 7	Bit 6	Bit 5	Bit 4	Bit 3	Bit 2	Bit 1	Bit 0
0x0C (0x2C)	PINE	–	PINE6	–	–	–	PINE2	–	–
0x0B (0x2B)	PORTD	PORTD7	PORTD6	PORTD5	PORTD4	PORTD3	PORTD2	PORTD1	PORTD0
0x0A (0x2A)	DDRD	DDD7	DDD6	DDD5	DDD4	DDD3	DDD2	DDD1	DDD0
0x09 (0x29)	PIND	PIND7	PIND6	PIND5	PIND4	PIND3	PIND2	PIND1	PIND0
0x08 (0x28)	PORTC	PORTC7	PORTC6	–	–	–	–	–	–
0x07 (0x27)	DDRC	DDC7	DDC6	–	–	–	–	–	–
0x06 (0x26)	PINC	PINC7	PINC6	–	–	–	–	–	–
0x05 (0x25)	PORTB	PORTB7	PORTB6	PORTB5	PORTB4	PORTB3	PORTB2	PORTB1	PORTB0
0x04 (0x24)	DDRB	DDB7	DDB6	DDB5	DDB4	DDB3	DDB2	DDB1	DDB0
0x03 (0x23)	PINB	PINB7	PINB6	PINB5	PINB4	PINB3	PINB2	PINB1	PINB0
0x02 (0x22)	Reserved	–	–	–	–	–	–	–	–
0x01 (0x21)	Reserved	–	–	–	–	–	–	–	–
0x00 (0x20)	Reserved	–	–	–	–	–	–	–	–

Arduino and Compatible Products Vendors

Note that the inclusion of any particular company in this appendix does not constitute an endorsement (except for the Arduino folks, of course). It is provided as a resource only.

Arduino Products

The main source of official Arduino products is, of course, Arduino. You can find out what is available at the official website, Arduino.cc. There are also distributors that carry Arduino boards, shields, and add-on accessories.

Hardware-Compatible Boards and Shields

Name	URL	Name	URL
Adafruit	www.adafruit.com	ITEAD Studio	store.iteadstudio.com
Arduino	store.arduino.cc	Macetech	www.macetech.com/store/
Arduino Lab	www.arduinolab.us	Mayhew Labs	mayhewlabs.com
Circuit@tHome	www.circuitsathome.com	Nootropic Design	nootropicdesign.com
CuteDigi	store.cutedigi.com	Numato	numato.com
DFRobot	www.dfrobot.com	RobotShop	www.robotshop.com
DealeXtreme (DX)	www.dx.com	Rugged Circuits	www.ruggedcircuits.com
Elecfreaks	www.elecfreaks.com	SainSmart	www.sainsmart.com
Elechouse	www.elechouse.com	Seeed Studio	www.seeedstudio.com
excamera	www.excamera.com	SparkFun	www.sparkfun.com
Iowa Scaled Engineering	www.iascaled.com	Tindie	www.tindie.com
iMall	imall.itead.cc	Tronixlabs	tronixlabs.com

Software-Compatible Boards

Name	URL
Adafruit	www.adafruit.com
Circuit Monkey	www.circuitmonkey.com
BitWizard	www.bitwizard.nl

Sensors, Add-on Boards, and Modules

Name	URL	Name	URL
Adafruit	www.adafruit.com	Seeed Studio	www.seeedstudio.com
CuteDigi	store.cutedigi.com	TinyCircuits	www.tiny-circuits.com
DealExtreme (DX)	www.dx.com	Trossen Robotics	www.trossenrobotics.com
KEYES	en.keyes-robot.com	Vetco	www.vetco.net

Electronics Software

Open Source Schematic Capture Tools

Name	URL
ITECAD	http://www.itecad.it/en/index.html
Oregano	https://github.com/marc-lorber/oregano
Open Schematic Capture (OSC)	http://openschcapt.sourceforge.net
TinyCAD	http://sourceforge.net/apps/mediawiki/tinycad
XCircuit	http://opencircuitdesign.com/xcircuit

CAE Software Tools

Name	Description	URL
DesignSpark	Free, not open source	http://www.rs-online.com/designspark/electronics/
Eagle	Free, not open source	http://www.cadsoftusa.com
Fritzing	Free CAE tool	http://fritzing.org/home
gEDA	Open source CAE tools	http://www.geda-project.org
KiCad	Open source CAE tool	http://www.kicad-pcb.org

PCB Layout Tools

Name	Description	URL
FreePCB	Windows (only) PCB layout	*http://www.freepcb.com*
FreeRouting	Web-based PCB autorouter	*http://www.freerouting.net*
PCB	Linux open source layout	*http://sourceforge.net/projects/pcb/*

Hardware, Components, and Tools

Electronic Component Manufacturers

Name	URL	Name	URL
Allegro	*http://www.allegromicro.com*	Micrel	*http://www.micrel.com*
Analog Devices	*http://www.analog.com*	Microchip	*http://www.microchip.com*
ASIX	*http://www.asix.com.tw*	NXP	*http://www.nxp.com*
Atmel	*http://www.atmel.com*	ON Semiconductor	*http://www.onsemi.com*
Bluegiga	*http://www.bluegiga.com*	Panasonic	*http://www.panasonic.com*
Cypress	*http://www.cypress.com*	Silicon Labs	*http://www.silabs.com*
Digi International	*http://www.digi.com*	STMicrotechnology	*http://www.st.com*
Fairchild	*http://www.fairchildsemi.com*	Texas Instruments	*http://www.ti.com*
FTDI	*http://www.ftdichip.com*	WIZnet	*http://www.wiznet.co.kr*
Linear Technology	*http://www.linear.com*	Zilog	*http://www.zilog.com*

Electronics Distributors (USA)

Name	URL
Allied	*http://www.alliedelec.com*
Digi-Key	*http://www.digikey.com*
Jameco	*http://www.jameco.com*
Mouser	*http://www.mouser.com*
Newark/Element14	*http://www.newark.com*
Parts Express	*http://www.parts-express.com*
State	*http://www.potentiometer.com*

Discount and Surplus Electronics

Name	URL
All Electronics	http://www.allelectronics.com
Alltronics	http://www.alltronics.com
American Science & Surplus	http://www.sciplus.com
BGMIcro	http://www.bgmicro.com
Electronic Surplus	http://www.electronicsurplus.com
Electronic Goldmine	http://www.goldmine-elec-products.com

Mechanical Parts and Hardware (Screws, Nuts, Bolts)

Name	URL	Name	URL
All Electronics	http://www.allelectronics.com	McMaster-Carr	http://www.mcmaster.com
Alltronics	http://www.alltronics.com	Micro Fasteners	http://www.microfasteners.com
Bolt Depot	http://www.boltdepot.com	SDP/SI	http://www.sdp-si.com
Fastenal	http://www.fastenal.com	WM Berg	http://www.wmberg.com

Electronic Enclosures and Chassis

Name	URL	Name	URL
Bud Industries	http://www.budind.com	LMB Heeger	http://www.lmbheeger.com
Context Engineering	http://contextengineering.com/index.html	METCASE/OKW Enclosures	http://www.metcaseusa.com
ELMA	http://www.elma.com	Polycase	http://www.polycase.com
Hammond Manufacturing	http://www.hammondmfg.com/index.htm	Serpac	http://www.serpac.com
iProjectBox	http://www.iprojectbox.com	TEKO Enclosures	http://www.tekoenclosures.com/en/home

Tools

Name	URL	Name	URL
Adafruit	http://www.adafruit.com	Maker Shed	http://www.makershed.com
Apex Tool Group	http://www.apexhandtools.com	MCM Electronics	http://www.mcmelectronics.com
CKB Products	http://www.ckbproducts.com	SainSmart	http://www.sainsmart.com
Circuit Specialists	http://www.circuitspecialists.com	SparkFun	http://www.sparkfun.com
Electronic Goldmine	http://www.goldmine-elec-products.com	Stanley	http://www.stanleysupplyservices.com
Harbor Freight Tools	http://www.harborfreight.com	Velleman	http://www.vellemanusa.com

Test Equipment

Name	URL	Name	URL
Adafruit	http://www.adafruit.com	SparkFun	http://www.sparkfun.com
Electronic Goldmine	http://www.goldmine-elec-products.com	Surplus Shed	http://www.surplusshed.com
MCM Electronics	http://www.mcmelectronics.com	Velleman	http://www.vellemanusa.com

Printed Circuit Board Supplies and Fabricators

Most major electronics distributors sell things like etchant and single- and double-sided copper clad PCB blanks with photoresist applied. If you aren't comfortable with the chemicals and procedures, consider using a commercial prototype PCB house.

Prototype and Fast-Turnaround Fabricators

Name	URL
Advanced Circuits	http://www.4pcb.com
ExpressPCB	http://www.expresspcb.com
Gold Phoenix PCB Co.	http://www.goldphoenixpcb.com
Sunstone Circuits	http://www.sunstone.com/
Sierra Circuits	https://www.protoexpress.com

PCB Kit Sources

Vendor name	URL	Products
Jameco Electronics	http://www.jameco.com	Conventional acid etch and supplies
Think & Tinker, Ltd.	http://www.thinktink.com	Various supplies for making PCBs
Vetco Electronics	http://www.vetco.net	Conventional acid etch kit

Other Sources

The companies in this appendix are just a sampling of what you can find with a little bit of searching. Distributors like Amazon and Mouser carry various Arduino and Arduino-compatible products. eBay is always a good place to look for bargains from Asian vendors (be sure to check the vendor ratings, but they are almost always good). And last but not least, there is Google. A search for "Arduino" on Google returned nearly 40 million results at the time of writing.

Recommended Reading

The Arduino is a popular subject for technical authors (myself included), and there are numerous books available. Some describe a specific range of applications, and others are more along the lines of a collection of projects. In addition to some titles that specifically deal with the Arduino, I have also included books on the AVR microcontroller, C and C++ programming, general electronics, interfaces, instrumentation, and printed circuit boards.

Arduino

- Massimo Banzi. *Getting Started with Arduino*. O'Reilly. 2009. ISBN 978-0596155513

- Patrick Di Justo and Emily Gertz. *Atmospheric Monitoring with Arduino*. (Maker Media). 2013. ISBN 978-1449338145

- Emily Gertz and Patrick Di Justo. *Environmental Monitoring with Arduino*. (Maker Media). 2012. ISBN 978-1449310561

- Simon Monk. *Programming Arduino: Getting Started with Sketches*. McGraw-Hill. 2011. ISBN 978-0071784221

- Jonathan Oxer and Hugh Blemings. *Practical Arduino*. Apress. 2009. ISBN 978-1430224778

AVR

- Timothy Margush. *Some Assembly Required*. CRC Press. 2011. ISBN 978-1439820643

- Elliot Williams. *Make: AVR Programming*. Maker Media. 2014. ISBN 978-1449355784

C and C++ Programming

- Brian Kernighan and Dennis Ritchie. *The C Programming Language*. Prentice Hall. 1988. ISBN 978-0131103627

- K. N. King. *C Programming: A Modern Approach*. Norton. 1996. ISBN 978-0393969450

- Stanley Lippman. *C++ Primer*. Addison-Wesley. 2012. ISBN 978-0321714114

- Stephen Prata. *C++ Primer Plus*. Addison-Wesley. 2011. ISBN 978-0321776402

General Electronics

- Analog Devices. *Data Conversion Handbook*. Newnes. 2004. ISBN 978-0750678414

- Howard Berlin. *The 555 Timer Applications Sourcebook*. Howard W. Sams. 1976. ISBN 978-0672215381

- Richard Dorf (Ed.) *The Electrical Engineering Handbook*. CRC Press LLC. 1997. ISBN 978-0849385741

- Allan Hambley. *Electronics*, 2nd Edition. Prentice Hall, 1999. ISBN 978-0136919827

- Paul Horowitz and Winfield Hill. *The Art of Electronics*, 2nd Edition. Cambridge University Press. 1989. ISBN 978-0521370950

- J. M. Hughes. *Practical Electronics: Components and Techniques*. O'Reilly. 2015. ISBN 978-1449373078

- Walter G. Jung. *IC Op-Amp Cookbook*. Howard W. Sams. 1986. ISBN 978-0672224534

- Randy Katz. *Contemporary Logic Design*, 2nd Edition. Prentice Hall. 2004. ISBN 978-0201308570

- William Kleitz. *Digital Electronics: A Practical Approach*. Regents/Prentice Hall. 1993. ISBN 978-0132102870

- Charles Platt. *Make: Electronics*. Maker Media. 2009. 978-0596153748

- Arthur Williams and Fred Taylor. *Electronic Filter Design Handbook*, 4th Edition. McGraw-Hill. 2006. ISBN 978-0071471718

Interfaces

- Jan Axelson. *Parallel Port Complete*. Lakeview Research LLC. 2000. ISBN 978-0965081917

- Jan Axelson. *Serial Port Complete*. Lakeview Research LLC. 2007. ISBN 978-1931448062

- Jan Axelson. *USB Complete*. Lakeview Research LLC. 2007. ISBN 978-1931448086

- Nick Hunn. *Essentials of Short-Range Wireless*. Cambridge University Press. 2010. ISBN 978-0521760690

- Benjamin Lunt. *USB: The Universal Serial Bus*. CreateSpace. 2012. ISBN 978-1468151985

- Charles E. Spurgeon and Joann Zimmerman. *Ethernet: The Definitive Guide*, 2nd Edition. O'Reilly Media, Inc. 2014. ISBN 978-1449361846

Instrumentation

- J. M. Hughes. *Real World Instrumentation with Python*. O'Reilly. 2010. ISBN 978-0596809560

Printed Circuit Boards

- Jan Axelson. *Making Printed Circuit Boards*. Tab Books. 1993. ISBN 978-0830639519

- Simon Monk. *Fritzing for Inventors*. McGraw-Hill. 2016. ISBN 978-0071844635

- Simon Monk. *Make Your Own PCBs with Eagle*. McGraw-Hill. 2014. ISBN 978-0071819251

- Matthew Scarpino. *Designing Circuit Boards with EAGLE*. Prentice Hall. 2014. ISBN 978-0133819991

Arduino and AVR Software Development Tools

This book has focused primarily on the Arduino IDE and the AVR-GCC toolchain, but those aren't the only tools available, by any means. There are many different tools for assembling, compiling, linking, and loading executable code into an AVR MCU. Some are open source, others are commercial, and some are more capable and polished than others.

Compilers/Assemblers

Atmel AVR Toolchain for Windows (http://bit.ly/atmel-avr)
  An open source suite of tools, including an assembler, ported to Windows. Includes both GNU-licensed software and tools developed by Atmel.

AVR-GCC (http://www.nongnu.org/avr-libc)
  A full suite of toolchain components for cross-compiling AVR executable code from C or C++ sources. See Chapter 6 for an overview.

SDCC (http://sdcc.sourceforge.net)
  An open source ANSI C compiler targeted for a variety of microcontrollers.

WinAVR (http://winavr.sourceforge.net)
  An open source ports of components from the AVR-GCC toolchain to the Windows environment. See Chapter 6 for an overview.

Integrated Development Environments (IDEs)

Arduino IDE (https://www.arduino.cc)
> The official IDE for Arduino hardware from the Arduino.cc team. Runs on Windows, Linux, and Mac OS X. Open source and free to download. See Chapter 5 for an overview.

Atmel Studio 7 (http://bit.ly/atmel-studio-7)
> Integrated C/C++ compiler and IDE. Free to download; for Windows 7 or later only. See Chapter 6 for an overview.

Eclipse Plugin (http://bit.ly/avr-eclipse)
> An open source AVR-oriented plug-in for the popular Eclipse (*https://eclipse.org*) open source IDE. Eclipse is Java-based and runs on Windows, Linux, and Mac OS X.

IAR Embedded Workbench (http://bit.ly/iar-workbench)
> Highly integrated suite of proprietary tools. License pricing by quotation, trial version available (30 days). Windows only.

MikroElektronika mikroC (http://www.mikroe.com/mikroc/avr)
> Commercial ANSI C compiler with IDE. Single-user license is $249; for Windows XP and later.

ImageCraft JumpStart (https://www.imagecraft.com/devtools_AVR.html)
> Commercial ANSI C compiler. Based partly on GPL open source software. $249 for a "standard" license for Windows. License dongle available.

Rowley CrossWorks (http://www.rowley.co.uk/avr)
> Commercial multiplatform ANSI C compiler with IDE. License cost varies from $150 to $2,250, depending on use of the product. Will run on Windows, Mac OS X, and Linux.

Programming Tools

PlatformIO (http://platformio.org)
> Command line–based AVR-GCC toolchain interface for Windows, Linux, and Mac OS X. See Chapter 6 for a brief description.

Ino (http://inotool.org)
> Command line–based AVR-GCC toolchain interface for Linux and Mac OS X. See Chapter 6 for a brief description.

Simulators

AMC VMLAB (http://www.amctools.com/vmlab.htm)
Freeware graphical AVR simulator for Windows.

GNU AVR Simulator (http://sourceforge.net/projects/avr)
Open source AVR simulator with Motif-based graphical interface. Runs on Linux/Unix.

Labcenter Proteus (http://www.labcenter.com/products/vsm/avr.cfm)
Novel schematic capture–based AVR simulator. Full graphical interface. Licenses start at $248. Windows only.

OshonSoft AVR Simulator (http://www.oshonsoft.com/avr.html)
Graphical AVR simulator with optional add-on modules. Personal license is $32.

SimulAVR (http://www.nongnu.org/simulavr)
Open source command line–based AVR simulator for Linux/Unix systems.

Index

Symbols

(hash symbol), indicating comments, 117
16-bit timer/counters, 21
315/433 MHz RF modules, 332
32-bit AVR processors, 17
3D printers
    SainSmart RAMPS 1.4 RepRap shield for, 272
555 timers, 350
8-bit RISC microcontrollers, 14
    (see also AVR family of microcontrollers)
8-bit timer/counters, 19

A

AC (alternating current)
    high-voltage AC control circuits, 478
    ratings for relay shields, 230
accelerometers, 327
acceptance testing, 364
    GreenShield custom shield, 397
    in Switchinator project, 427
access point names (APNs), 180
actions
    in makefiles, 139
    mapping to relays, 377
actuators, 2
    on Esplora board, 207
AD9850 DDS chip, 433
AD9850-based DDS module, 343
    for programmable signal generator, 439
    output frequency, 457
    passing calibration coefficient to, 456
Adafruit
    Arduino-compatible GSM shields, 174

    WiFi shield, 202
Adafruit 16-Channel 12-bit PWM/Servo
    Shield, 255
Adafruit 2.8" TFT Touch Shield, 263
Adafruit Data Logging Shield, 264
Adafruit DIY Shield Kit, 253
Adafruit LCD Shield Kit, 261
Adafruit Mega Proto Shield, 251
Adafruit Patch Shield, 232
Adafruit Proto-ScrewShield (Wingshield), 268
Adafruit Stackable R3 Proto Shield, 250
Adafruit Ultimate GPS Logger Shield, 264
adapter shields, 266
ADCs (see analog-to-digital converters)
add-on boards, vendors of, 576
air conditioning (A/C), 473
    (see also HVAC systems)
    humidity and, 475
Amazon, 579
analog comparator (AVR microcontrollers), 21
    ATmega1280/2560 inputs, 42
    ATmega168/328 inputs, 34
    ATmega32U4 inputs, 53
analog inputs
    ATmega1280/2560 microcontrollers, 43
    ATmega168/328 microcontrollers, 35
    ATmega32U4 microcontrollers, 53
    GreenShield project, 375
        reading data from, 383
        testing, 398
    on AVR MCU, using as digital I/O pins, 482
    Switchinator project, 402, 403, 405, 411
analog joysticks, 344
analog signal outputs, 342

About the Author

John M. Hughes is an embedded systems engineer with over 30 years of experience in electronics, embedded systems and software, aerospace systems, and scientific applications programming. He was responsible for the surface imaging software on the Phoenix Mars Lander and was part of the team that developed a novel synthetic heterodyne laser interferometer for calibrating the position control of the mirrors on the James Webb Space Telescope. Over the years he has worked on digital engine control systems for commercial and military aircraft, automated test systems, radio telescope data acquisition, 50+ gigapixel imaging systems, and real-time adaptive optics controls for astronomy. On his own time (when he has any) he likes to do cabinetry and furniture design; build microcontroller-based gadgets for use with greenhouses, bees, and backyard urban chickens; and write books.

Colophon

The animal on the cover of *Arduino: A Technical Reference* is a toy duck. This kind of pull-along wooden toy was one of the first ever manufactured by the LEGO Group. Released in 1935, the duck was originally constructed by the company's founder, Danish carpenter Ole Kirk Christiansen, who began producing wooden toys in the 1930s at the onset of Denmark's Great Depression.

The LEGO duck sat on a rectangular platform with wheels, and its beak opened and closed as it was pulled along on a string. The original duck was painted red, with a black head, tail, and wings, though later models varied in color.

LEGO ceased production of the wooden toy sets in the 1960s, making the iconic duck exceptionally rare. In 2011, the company released a modern version of the duck, made of the plastic bricks we know today.

Many of the animals on O'Reilly covers are endangered; all of them are important to the world. To learn more about how you can help, go to *animals.oreilly.com*.

The cover image is from *The Great American Antique Toy Bazaar* by Ronald Barlow. The cover fonts are URW Typewriter and Guardian Sans. The text font is Adobe Minion Pro; the heading font is Adobe Myriad Condensed; and the code font is Dalton Maag's Ubuntu Mono.

About the Author

John M. Hughes is an embedded systems engineer with over 30 years of experience in electronics, embedded systems and software, aerospace systems, and scientific applications programming. He was responsible for the surface imaging software on the Phoenix Mars Lander and was part of the team that developed a novel synthetic heterodyne laser interferometer for calibrating the position control of the mirrors on the James Webb Space Telescope. Over the years he has worked on digital engine control systems for commercial and military aircraft, automated test systems, radio telescope data acquisition, 50+ gigapixel imaging systems, and real-time adaptive optics controls for astronomy. On his own time (when he has any) he likes to do cabinetry and furniture design; build microcontroller-based gadgets for use with greenhouses, bees, and backyard urban chickens; and write books.

Colophon

The animal on the cover of *Arduino: A Technical Reference* is a toy duck. This kind of pull-along wooden toy was one of the first ever manufactured by the LEGO Group. Released in 1935, the duck was originally constructed by the company's founder, Danish carpenter Ole Kirk Christiansen, who began producing wooden toys in the 1930s at the onset of Denmark's Great Depression.

The LEGO duck sat on a rectangular platform with wheels, and its beak opened and closed as it was pulled along on a string. The original duck was painted red, with a black head, tail, and wings, though later models varied in color.

LEGO ceased production of the wooden toy sets in the 1960s, making the iconic duck exceptionally rare. In 2011, the company released a modern version of the duck, made of the plastic bricks we know today.

Many of the animals on O'Reilly covers are endangered; all of them are important to the world. To learn more about how you can help, go to *animals.oreilly.com*.

The cover image is from *The Great American Antique Toy Bazaar* by Ronald Barlow. The cover fonts are URW Typewriter and Guardian Sans. The text font is Adobe Minion Pro; the heading font is Adobe Myriad Condensed; and the code font is Dalton Maag's Ubuntu Mono.

Get even more for your money.

Join the O'Reilly Community, and register the O'Reilly books you own. It's free, and you'll get:

- $4.99 ebook upgrade offer
- 40% upgrade offer on O'Reilly print books
- Membership discounts on books and events
- Free lifetime updates to ebooks and videos
- Multiple ebook formats, DRM FREE
- Participation in the O'Reilly community
- Newsletters
- Account management
- 100% Satisfaction Guarantee

Signing up is easy:

1. Go to: oreilly.com/go/register
2. Create an O'Reilly login.
3. Provide your address.
4. Register your books.

Note: English-language books only

To order books online:
oreilly.com/store

For questions about products or an order:
orders@oreilly.com

To sign up to get topic-specific email announcements and/or news about upcoming books, conferences, special offers, and new technologies:
elists@oreilly.com

For technical questions about book content:
booktech@oreilly.com

To submit new book proposals to our editors:
proposals@oreilly.com

O'Reilly books are available in multiple DRM-free ebook formats. For more information:
oreilly.com/ebooks

O'REILLY®

Milton Keynes UK
Ingram Content Group UK Ltd.
UKHW010730280824
447543UK00003B/3